Rationality and Logic

Rationality and Logic

Robert Hanna

A Bradford Book
The MIT Press
Cambridge, Massachusetts
London, England

MIT Press books may be purchased at special quantity discounts for business or sales promotional use. For information, please email special_sales@mitpress.mit.edu or write to Special Sales Department, The MIT Press, 55 Hayward Street, Cambridge, MA 02142.

This book was set in Sabon by SPI Publisher Services and was printed and bound in the United States of America.

Library of Congress Cataloging-in-Publication Data

Hanna, Robert, 1957–.
Rationality and logic / Robert Hanna.
   p. cm.
"A Bradford book."
Includes bibliographical references and index.
ISBN 0-262-08349-3—ISBN 978-0-262-08349-2 (hc : alk. paper)
1. Logic. 2. Psychologism. 3. Reasoning (Psychology). I. Title.

BC53.H36 2006
128'.33—dc22   2005058402

10 9 8 7 6 5 4 3 2 1

To MTH and ETH
Le coeur a ses raisons que la raison ne connâit point

Contents

Preface and Acknowledgments   ix
Introduction   xi

1 | Psychologism Revisited   1

2 | *E pluribus unum*   29

3 | The Logocentric Predicament   53

4 | Cognition, Language, and Logic   77

5 | The Psychology of Reasoning   115

6 | Our Knowledge of Logic   155

7 | The Ethics of Logic   201

Notes   233
Bibliography   285
Index   309

Preface and Acknowledgments

This book is about human rationality, logic, and the connection between them. On my view, this connection is both constitutive and mutual. More precisely, I defend the broadly Kantian thesis that logic is the result of the constructive operations of an innate protological cognitive capacity that is necessarily shared by all rational human animals, and governed by categorically normative principles. Working out and writing up this idea has involved many extended visits to the domains of logical theory and cognitive psychology. But although I am a philosopher who by virtue of a deep interest in human rationality is also deeply interested in logic and cognition, I am neither a professional logician nor a professional cognitive psychologist. So I want to make it very clear in advance that I am drawing and relying even more heavily than is usual for philosophers on the theoretical expertise of others. I hope to make my contribution at the synoptic level of the Big Picture, and then turn this project back over to the specialists as a new and important joint research program.

I am very grateful to the following people for conversations or correspondence on and around my topic: Sean Anderson, Luc Bovens, Nicholas Denyer, Christopher Green, Neil Manson, Arlo Murphy, Graham Oddie, Alex Oliver, Eric Olson, Onora O'Neill, James Russell, Peter Strawson, Evan Thompson, Dana Vanzanten, John Vejsada, and Jessica Wilson. Shards of the material were presented to appropriately and helpfully skeptical audiences in talks at Cambridge University; King's College London; Trinity College Dublin; and York University, Canada. Several of the central arguments were first sketched or talked out during a visiting fellowship at Clare Hall, Cambridge, in Michaelmas term 1998.

Institutionally speaking, the Social Sciences and Humanities Research Council of Canada and York University generously gave me research grants for a sabbatical leave from York in 1998–1999; Fitzwilliam College, Cambridge, generously gave me visiting fellowships for Lent term 2000, Lent term 2001, and the academic year 2003–2004; and the Faculty of Philosophy at the University of Cambridge generously gave me the unique opportunity to lecture on Wittgenstein's *Tractatus* and *Investigations*, back to back, during Michaelmas term 2003 and Lent term 2004.

Personally speaking, Graham Oddie gave me encouragement at a crucial moment. Thanks mate.

I also owe special debts of gratitude to Ruth Barcan Marcus and Michael Potter. Had I not had the good luck to study logic with the former and to be her logic teaching assistant when I was a graduate student at Yale in the 1980s, this book would never have been started. And had I not had the good luck in more recent years to be pushed hard by the latter to clarify, reformulate, and rethink my fuzzy thoughts on the logical and the psychological, this book would never have been finished. Needless to say, neither can be held in any way responsible for the views I defend here.

One last pair of philosophical acknowledgments—oddly enough, to Francisco Goya and Blaise Pascal. The caption of the most famous of Goya's drawings in the *Los caprichos* reads: *el sueño de la razon produce monstruos*. The sleep of reason produces monsters. In other words, without rational guidance we inevitably commit atrocities. Goya's stark and uncompromising pronouncement on human folly and how to prevent it, however, should always be juxtaposed with the quotation on the dedication page, which is taken from Pascal's *Pensées* (section 4, no. 277). The heart has reasons of its own that reason knows nothing about. In other words, the rational guidance of human conduct is inevitably embedded in and inevitably constrained by our equally fundamental pursuits of happiness, personal integrity, and empathic connectedness with others. These are pursuits that may run, most perversely, contrary to our rationality, yet at the same time they drive rationality itself. Taken separately these two remarks capture, for me, the moral depth and the moral limits of human rationality. And taken together they state far better than I ever could my motivations for writing this book.

Introduction

A syllogism is language [*logos*] in which, certain things being asserted, something else follows of necessity from their being so.
—Aristotle[1]

Logic is a science of reason . . . a science a priori of the necessary laws of thought, not in regard to particular objects, however, but to all objects in general;—hence a science of the correct use of the understanding and of reason in general, not subjectively, however, i.e., not according to empirical (psychological) principles for how the understanding does think, but objectively, i.e., according to principles a priori for how it ought to think.
—Immanuel Kant[2]

[T]he word Logic in its primal sense means the Science of the Laws of Thought as expressed. Considered in this sense, Logic is conversant about all thought which admits of expression; whether that expression be effected by the signs of common language or by the symbolic language of the mathematician.
—George Boole[3]

If sheer logic is not conclusive, what is?
—W. V. O. Quine[4]

The logical notions are embedded in our deepest nature, in the very form of our language and thought, which is presumably why we can understand some kinds of logical systems quite readily, whereas others are inaccessible to us without considerable effort . . . if at all.
—Noam Chomsky[5]

For all we now know, cognition is saturated with rationality through and through.
—Jerry Fodor[6]

This book is a philosophical study of the relation between human rationality and logic. Its two central claims are (i) that logic is cognitively constructed by rational animals and (ii) that rational human animals are essentially logical animals. The dual idea expressed by these claims—that logic is intrinsically psychological, and that human psychology is intrinsically logical—has a long and troubled history in the philosophical, logical, and psychological traditions alike. From Pierre Arnauld and Jean Nicole's *Art of Thinking* (1662), through Immanuel Kant's *Jäsche Logic* (1800), J. S. Mill's *System of Logic* (1843), and George Boole's *Investigation of the Laws of Thought* (1854), right up to the appearance of Gottlob Frege's revolutionary *Begriffsschrift* (1879), logic and psychology seemed to be, if not precisely the same subject, then at least theoretically married to one another. But the much-celebrated attack on "logical psychologism"—the explanatory reduction of logic to empirical psychology—at the end of the nineteenth century brought about a nasty divorce. According to the leaders of the attack, Frege and Edmund Husserl, this parting of the ways was a simple matter of irreconcilable differences: the principles or laws of logic are absolutely necessary, whereas the laws of empirical psychology are only contingent generalizations; logic is true, whereas empirical psychology deals only with human belief; logic is a fully formal or "topic-neutral" science, whereas empirical psychology focuses only on the species-specific or individual contents of mental states; logical knowledge is a priori or independent of all sense experience, whereas empirical psychological knowledge is a posteriori or dependent on experience;[7] and so on. Thereafter "pure logic," pursued in armchairs by philosophers and philosophically minded mathematicians, went one way, and "experimental psychology," pursued in laboratories by men in white coats, went diametrically another. To make things worse, as Elliott Sober aptly observes, "while the psychologists were leaving, the philosophers were slamming the door behind them."[8]

Of course philosophy, logic, and psychology have changed a lot since those days. Most philosophers gave up classical analysis and replaced it with scientific naturalism: the doctrine that all metaphysical, epistemic, and methodological questions can ultimately be answered by the natural sciences alone, without appeal to supernatural facts.[9] Most logicians went from thinking that all logic is classical or elementary[10] to thinking that logic can be conservatively "extended"[11] or radically "deviant,"[12] or even mind-blowingly "paraconsistent"[13] or "dialetheic."[14] And most psychologists dropped behaviorism and adopted cognitivism: the thesis that the rational human mind is essentially an active innately specified information-processor.[15] In other words, the philosophers, logicians, and psychologists loosened up significantly and moved on. But the old myths die hard. Even now, it remains an almost unchallenged axiom of conventional philosophical wisdom that the logical and the psychological are intrinsically incompatible.

In my opinion, the view that logic and psychology are fundamentally at odds with one another could not be more mistaken. On the contrary, if I am correct there is an essential link between logic and psychology, despite the fact that logical psychologism is self-refuting and hence false. This brings me back to the first central claim of this book: *logic is cognitively constructed by rational animals*, in the sense that all and only rational animals—including, of course, all rational humans—possess a cognitive faculty that is innately set up for representing logic, because it contains a single universal "protologic," distinct in structure from all classical and nonclassical logical systems, that is used for the construction of all logical systems. I call this claim *the logic faculty thesis*. The logic faculty thesis draws explicitly but not uncritically on some ideas of Kant, Boole, Quine, Chomsky, and Fodor.

But what is logic? This question can mean two very different things. The first is: what is the science of logic? And the second is: what is the nature of logic? The first is a question internal to the logical enterprise itself, whereas the second is a specifically philosophical question.

The internal question can be answered fairly easily, at least in a preliminary way. Aristotle discovered the science of logic by discovering the science of syllogisms. A syllogism, in turn, is "language in which, certain things being asserted, something else follows of necessity from their being so." Here, for example, are three syllogisms:

All politicians are crooks.
Dubya is a politician.
Therefore, Dubya is a crook.

All politicians are crooks.
All crooks are liars.
Therefore, all politicians are liars.

If all politicians are crooks and all crooks are liars,
then all politicians are liars.
All politicians are crooks and all crooks are liars.
Therefore, all politicians are liars.

Actually, Aristotle focuses not on concrete or complete syllogisms like these three, but instead on abstracted or *schematic* syllogisms; and for special metaphysical and epistemological reasons, he was interested fundamentally in general propositions and the logical import of general referring terms. But, for expository convenience, we can also add to the Aristotelian notion of the syllogism the later Stoic interests in the logical behavior of truth-functional connectives (such as "not," "and," and "if . . . then") and the logical import of names, and derive schematic versions of the three syllogisms listed above:

All As are Bs.
a is an A.
Therefore, a is a B.

All As are Bs.
All Bs are Cs.
Therefore, all As are Cs.

If P, then Q.
P.
Therefore, Q.

Here the schematization of a concrete or complete syllogism is obtained by uniformly substituting distinct capital letters (say, near the middle of the alphabet) for distinct sentences;[16] distinct capital letters (say, near the beginning of the alphabet) for distinct predicates; and distinct lowercase letters (ditto) for distinct individual names. The alphabetic letters are *nonlogical constants*. The words left over after the uniform substitution of nonlogical constants for predicates, sentences, or individual names are *logical constants*. Within the domain of logical constants we could further distinguish between "object language" logical constants like "all" (the universal quantifier) and "if . . . then" (the conditional) on the one hand, and "metalinguistic" logical constants like "therefore" (provability or consequence) on the other; but this subtlety can be left dormant for the time being.

What is of leading importance, in any case, is that each and every concrete or complete syllogism fitting into one of these schemata is such that some sentence follows with necessity from the assertion of some other sentences, together with the assumption of their being so or (what is the same) their being true. To say that a sentence follows with necessity from the assertion of some other sentences, together with the assumption of their being true, is to say that it is impossible that the sentences assumed to be true will carry over to a false sentence. In other words, truth is necessarily preserved.

If we abstract away altogether now from the special syllogistic framework of Aristotle's logic, we can say two things. First, the science of logic is about "schematizable" language, that is, orderly sequences of sentences linked together by fixed interpretations of the logical and nonlogical constants occurring in them. Second, and more precisely, the science of logic is about schematizable language in which some sentences are asserted and another sentence is asserted that in fact follows with necessity from the assumed truth of the asserted sentences. Schematizable language in which some sentences are asserted and another sentence is asserted that is held to follow from the others is an *argument*. The asserted sentences are the *premises* of the argument. The asserted sentence that is held to follow from the others is the *conclusion* of the argument. The fact (whenever it is a fact) that truth is necessarily preserved from the premises to the conclusion is the *validity* of the argument. And the necessary connection between the premises and conclusion of a valid argument is the relation of *consequence*.[17] Thus logic is the science of the necessary relation of consequence.

So much for a preliminary internal characterization of the science of logic. But what about the specifically philosophical question about the nature of logic? My answer is that the nature of logic is explained by the logic faculty thesis: logic is cognitively constructed by rational animals.

Obviously the fundamental notion lying behind this thesis is that of a rational animal. For my purposes animals are *sentient living organisms* and for simplicity's sake I shall assume unless otherwise specified that all animals are *sound*, that is, intact and mature. Even so, only some animals in this sense are rational. On my view, rational animals are conscious, rule-following,[18] intentional (that is, possessing capacities for object-directed cognition and purposive action), volitional (possessing a capacity for willing),[19] self-evaluating, self-justifying, self-legislating, reasons-giving, reasons-sensitive, and reflectively self-conscious—or, for short, "normative-reflective"[20]—animals, whose inner and outer lives alike are sharply constrained by their possession of concepts expressing strict modality. Modality in the philosophical sense comprises the concepts of necessity, possibility, and contingency. Strict modality, in turn, includes the concepts of logical necessity (truth in all logically possible worlds),[21] epistemic necessity (certainty or indubitability), and deontic necessity (unconditional obligation or "the ought"). So, to put my first central claim yet another way, logic is cognitively constructed by all and only those normative-reflective animals who are also in possession of concepts expressing strict modality.

This approach to rational animals substantively invokes the concept of rationality. An unfortunate but pervasive feature of the *philosophy* of rationality, however, is that it does not operate with either a univocal or generally accepted sense of the term 'rationality'.[22] Reasonable people, including specialists on rationality, are both muddled and also in sharp disagreement about the very concept of rationality. So in order to avoid troublesome ambiguity and state my commitments explicitly I need to make some basic distinctions, and orient my view in relation to them.

The first basic distinction is between (a) the *mentalistic* sense of rationality and (b) the *procedural* sense of rationality. In the mentalistic sense, rationality is a complex psychological capacity for logical inference and insight, and also for practical deliberation and decision making. By contrast, in the procedural sense, rationality is a complex formal property of a certain class of mechanical, mathematical, computational, or logical processes, namely the property of being (i) well formed and (ii) either provable and recursive (Turing-computable), valid (truth-preserving), or sound (valid with true premises).[23] The crucial difference here is that rationality in the mentalistic sense is such that all of its manifestations are conscious, whereas some process can quite easily be rational in the procedural sense without being in any way conscious.

(For later purposes, it is also quite useful to distinguish, within the mentalistic sense of rationality, between (a_1) the rationality of *animals*, (a_2), the rationality of mental *episodes* or *acts*, and (a_3) the rationality of mental *states*. The important contrast here is that it is possible for something to be a rational animal by having an overall mental capacity for rationality, yet fail to be occurrently rational with respect to some of its mental episodes or mental states, as in the case of someone who completely loses his temper temporarily. Conversely, it is possible for an animal to be occurrently rational with respect to some of its mental episodes or states, but lack an overall mental capacity for rationality, as in the case of certain sorts of mental illness. This point in turn implies another useful distinction, again within the mentalistic sense of rationality, between (a_5) an animal's mental *capacity* for rationality, and (a_6) *occurrent* rationality with respect to the mental episodes or mental states of an animal. And finally, for completeness, we can also distinguish, within occurrent mentalistic rationality, between (a_7) the occurrent rationality of mental episode or state *types*, and (a_8) the occurrent rationality of mental episode or state *tokens*. Here the contrast is that it is possible for a certain mental episode or state type—say, righteous anger—to be rational when tokened in some contexts, but fail to be rational when tokened in others.)

The second basic distinction is beween (c) the *meeting-the-minimal-standards* sense of rationality, and (d) the *meeting-the-maximal-or-ideal-standards* sense of rationality. In the meeting-the-minimal-standards sense, rationality means either possessing a psychological capacity for rationality or meeting the well-formedness conditions for being a rational procedure of the relevant sort. By contrast, in the meeting-the-maximal-or-ideal-standards sense, rationality means either perfectly using a psychological capacity or else perfectly satisfying the provability/computability conditions, validity conditions, or soundness conditions of the relevant sort of rational procedure. The crucial difference here is that in the meeting-the-minimal-standards sense, irrationality means lacking the basic conditions necessary for rationality, and hence means *nonrationality*; whereas in the meeting-the-maximal-or-ideal-standards sense, irrationality merely means falling short of perfect rationality.

The third and last basic distinction is between (e) the *principled* sense of rationality, (f) the *holistic* sense of rationality, and (g) the *instrumental* sense of rationality. In the principled sense, rationality means the possession of a capacity for generating or recognizing necessary truths, a priori beliefs, strictly universal normative rules, nonconsequentialist moral obligations, and categorical "ought"-claims.[24] Put in historical terms, this is the *Kantian* conception of rationality, acccording to which "reason is the faculty of a priori principles." By contrast, in the holistic sense, rationality means the possession of a capacity for systematically seeking coherence (or, to use a contemporary term of art, "reflective equilibrium") across a network or web of beliefs, desires, emotions, intentions, and volitions.[25] In historical terms, this is the *Hegelian* conception of rationality, according to which "the truth is the whole." And finally, in the instrumental sense, rationality means the possession of a capacity for generating or recognizing contingent truths, a posteriori beliefs, contextually normative rules, consequentialist obligations, and hypothetical "ought"-claims.[26] Put historically, this is the *Humean* conception of rationality, according to which "reason is the slave of the passions."

The crucial three-way difference here is that whereas in the principled sense, rationality means generating or recognizing rules that are absolute or *unconditional*, in the holistic sense, by contrast, rationality means generating or recognizing rules or laws that are merely thoroughly interdependent or *mutually conditioned* (hence none of those rules or laws can have a greater degree of necessity or certainty, or be more binding, than the modally or epistemically weakest proposition in the total holistic network of rules or laws), and, by another contrast, rationality in the instrumental sense means generating or recognizing rules that are merely empirically regular or *conditional* (hence none of those rules or laws can be fully necessary or certain or binding).

Unless otherwise noted, in what follows I will focus primarily on the mentalistic, meeting-the-minimal-standards, and principled senses of rationality. This is not to say that I reject or wish to depreciate in any way the procedural, meeting-the-maximal-or-ideal-standards, holistic, or instrumental senses of rationality. On the contrary, I am saying only that rationality in the senses I am primarily interested in *should not be confused with* other fundamentally different senses of rationality.

The class of normative-reflective animals in possession of concepts expressing strict modality would appear to be at least extensionally equivalent with the class of rational humans; and even if (as seems very likely) it is not intensionally equivalent for the simple reason that the cognitive capacities required for the possession of concepts expressing strict modality are multiply embodiable,[27] nevertheless those humans who are rational constitute a central case or paradigm. I am assuming that it is a primitive fact, yielded directly by the reader's capacity for introspection, that there *are* some rational humans. So I am proposing to explain the nature of logic by taking human rationality seriously. More precisely, I am proposing to explain the nature of logic by taking rationality seriously, and to take rationality seriously by taking human rationality seriously. And what we reach at the end of this explanation is the thesis that something protological is built innately into human rationality itself. This leads me back to my second central claim: *rational human animals are essentially logical animals*, in the sense that a rational human animal is defined by its being an animal with an innate constructive modular capacity for cognizing logic, a competent cognizer of natural language, a real-world logical reasoner, a competent follower of logical rules, a knower of necessary logical truths by means of logical intuition, and a logical moralist. This is what I call *the logic-oriented conception of human rationality*.

Is it possible to be a skeptic or an eliminativist about human rationality? Yes. But there is clearly something reflexively odd and even cognitively self-stultifying, if not outright self-contradictory, about having *reasons* for doubting or getting rid of rationality. I will address that point in chapter 7. But it must also be frankly admitted that if someone upon serious reflection simply does not believe that there is any such thing as rationality per se or human rationality in particular, or that these are pseudo-concepts that ought to be eliminated, then there is probably little I can do to convince him. I am assuming that the existence of rationality and human rationality are primitive and irreducible facts, and that as a consequence the prima facie case for their reality and conceptual integrity is far more compelling than any attempt to reject or eliminate them. Nevertheless, even allowing that it is cognitively coherent to try to challenge rationality, a rationality-skeptic or rationality-eliminativist might still find it interesting to inquire into the extent that the nature of logic could be explained, *if* one were to take human rationality seriously.

Whether or not there are rational animals other than humans, rational human animals as a matter of fact constitute the basic class of cognizers or thinkers studied by cognitive psychology. So if I am correct about the connection between rationality and logic, it follows that the nature of logic is significantly revealed to us by cognitive psychology. Correspondingly, I call the overall view expressed by the conjunction of my two central claims *logical cognitivism*.

Logical cognitivism has two important and rather controversial consequences. First, the philosophers must reopen their door and civilly invite the psychologists back in. As some people have been saying for two or three decades now, we are all colleagues working in the very same metadiscipline: cognitive science. On this picture, analytic philosophy is at bottom the same as the philosophy of rational human cognition. Second, however, and perhaps even more controversially, a reconciliation between philosophy and psychology by way of logical cognitivism must also be expected to change cognitive science itself quite radically. Wittgenstein pregnantly remarks in the *Tractatus* that "logic *precedes* every experience—that something is *so*."[28] My way of glossing this is to say that logic is *not* strictly determined by the contingent or natural facts: by which I mean that it is *not* the case that logic is nothing over and above all physical facts plus all sensory experiential facts. In other words, logic is not basically physical and a posteriori. That, I believe, is the correct way to understand the fundamental lesson taught us by Frege's and Husserl's critique of logical psychologism.

Thirty years later, Wittgenstein equally pregnantly remarks in the *Investigations* that

[I]f language is to be a means of communication there must be agreement not only in definitions but also (queer as this may sound) in judgments. This seems to abolish logic, but does not do so.[29]

My way of glossing this is to say that logic is necessarily bound up with the human activity of linguistic communication and in particular with the human activities of making theoretical and practical judgments. Or in other words, logic is intrinsically normative. On this extended picture, analytic philosophy is at bottom the same as *rational anthropology*.[30] So it follows from antipsychologism, together with the necessary connection between logic and language, together with logical cognitivism, that cognitive science is not at bottom a natural science. Instead it is both an objective or truth-oriented science and *also* what some nineteenth-century philosophers rather quaintly called a "moral science"—that is, a normative human science or *Geisteswissenschaft*—just like logic itself.

This does not mean that the natural sciences are not highly relevant to cognitive science. Of course they are highly relevant! It means simply that the natural sciences cannot in and of themselves provide the foundations of cognitive science. It is significantly odd that contemporary conventional philosophical wisdom should include, simultaneously, strong commitments to scientific naturalism and to the assumption that the logical and the psychological are incompatible.[31] I am interested in trying to formulate and defend the broadly Kantian theory[32] of human rationality and logic that results if we firmly reject both of these assumptions.

It may be useful to the reader, before pressing on, to have a sketch of the overall argument in front of her.

In the first three chapters I explore three different philosophical approaches to the nature of logic, each in the form of a basic problem. Chapter 1 deals with the problem of logical psychologism: what is the relation between the logical and the psychogical? Here I argue that logical psychologism is a species of scientific naturalism; that scientific naturalism about logic is false; and that logical cognitivism can effectively avoid both logical scientific naturalism and the equally but oppositely flawed doctrine of logical platonism. Chapter 2 addresses what I call the *e pluribus unum* problem: how can we reconcile the unity of logic with the plurality of logical systems distinct from classical or elementary logic? I argue that, despite their deep differences, all logical systems—whether classical, extended, or deviant—must presuppose a single universal protologic, distinct in structure from all classical or nonclassical systems, that is used to construct those systems. I then propose that this protologic is contained in the logic faculty. If correct, my proposal implies that the precise structural description of this protologic can be turned over to logicians and cognitive psychologists as a new and important joint research program. Chapter 3 deals with a deep problem called *the logocentric predicament*, which arises from the very unsettling fact that in order to explain any logical theory, or justify any deduction, logic is presupposed and used—so logic appears to be both inexplicable and unjustified. On the assumption that logic must have a nonlogical and nonpsychological foundation, the logocentric predicament is insoluble and devastating. But on the alternative assumption that logic has a *logico-psychological* foundation in the fact that a single universal protologic is innately contained in the logic faculty, the logocentric predicament loses its sting: it is merely another way of expressing the first half of logical cognitivism.

In the next three chapters I turn from philosophical logic to human rationality. In chapter 4, I argue that human thinking conforms to what I call *the standard cognitivist model of the mind*, a model which has its remote origins in Kant's transcendental psychology and its proximal sources in Chomsky's psycholinguistics. This model includes representationalism or intentionalism, innatism or nativism, constructivism, modularity, and a mental language or language of thought. I then critically refine the model, and also extend it to include the thesis that the language of thought presupposes a *mental logic* or *logic of thought*. This chapter also offers a defense of the logic-oriented conception of human rationality, and of the thesis that rational human animals are defined by their possession of logical abilities and are necessarily also linguistic animals, but that not all linguistic humans are rational, nor are all linguistic animals rational. In chapter 5 I develop an empirical argument in favor of logical cognitivism by first critically sifting through classical and recent work on the psychology of human reasoning, then second defending a doctrine I call *the protological competence theory*, then third and finally applying this doctrine to the heated debate about human rationality in recent cognitive science and philosophy. In chapter 6 I sketch the outlines of a theory of logical knowledge, based on logical intuition, in response to natural extensions of Wittgenstein's famous worry about "following a rule," and Paul Benacerraf's almost equally famous worry about reconciling our face-value or standard semantics of mathematical truth with our best epistemology of intuitive knowledge.

The seventh and last chapter subsumes the themes of the earlier chapters under a discussion of the normativity of logic. The central claim is that logic is a moral or "prescriptive" science and not merely a factual or "descriptive" one, because the principles and concepts of the single universal protologic, whatever they turn out to be, must be intrinsically categorically normative—unconditionally obligatory—for human reasoning. The proper construal of this claim leads to three further claims. First, the obvious fact that humans persistently make logical gaffes does not count in any way against their being logical animals but, on the contrary, counts all the more strongly in favor of it: only a logical animal would ever care about committing fallacies, just as only a moral animal would ever care about committing sins. Second, the obvious gap between abstract logical systems and concrete human reasoning does not entail, as Gilbert Harman has argued, that logic has little or nothing to do with reasoning. Third and finally, attempts by neo-Nietzscheans (and also by some contemporary cognitive scientists) to defend the skeptical thesis that humans are irrational and could at least in principle become *logic-liberated animals*, because their logical reasoning abilities are nothing but expressions of "the will-to-power" (or: "mechanisms of natural selection"), and because logic itself is nothing but a social construct (or: the result of using "social contract schemas"), while surprisingly resistant to philososophical refutation, ultimately fail because they are cognitively self-defeating.

Rationality and Logic

1 | Psychologism Revisited

Although at one time it was quite usual to suppose that the principles of logic are "the laws of thought" . . . , Frege's vigorous critique was so influential that there has been rather little support, of late, for "psychologism" in any shape or form. However, Frege's arguments against psychologism are, I suspect, less conclusive, and at least some form of psychologism more plausible, than it is nowadays fashionable to suppose.
—Susan Haack[1]

1.0 Introduction

In this chapter I revisit the late nineteenth- and early twentieth-century debate about logical psychologism. It is clear that this debate significantly determined the subsequent development of philosophy and psychology alike. Neither the emergence of analytic philosophy from Kant's idealism[2] nor the emergence of experimental or scientific psychology from Brentano's phenomenology[3] could have occurred without it. It is also clear that Frege and Husserl routed the "psychologicists." What is much less clear, and what I want critically to rethink and reformulate, is the philosophical upshot of this seminal controversy.

In section 1.1, I look at what Frege and Husserl say about and against logical psychologism. Logical psychologism boils down to the thesis that logic is explanatorily reducible to empirical psychology. Identifying a cogent Fregean or Husserlian argument against psychologism proves to be difficult, however, because their antipsychologistic arguments are question-begging. In section 1.2, I propose that logical psychologism can be most accurately construed as a species of scientific naturalism, and more particularly as a form of scientific naturalism about logic. If logical psychologism is a form of scientific naturalism about logic, then Frege's and Husserl's antipsychologism is also a species of antinaturalism.

This leads me in section 1.3 to go in search of a cogent argument against scientific naturalism, by looking at G. E. Moore's near-contemporary attack on ethical naturalism in *Principia Ethica*. But again our high hopes are dashed to the ground: for Moore's celebrated critique of the "naturalistic fallacy" fails in two ways. First, in arguing against the identification of any natural property with the property Good, Moore assumes an absurdly high standard of property-individuation; and second, although somewhat more ironically, he incoherently combines his antinaturalism with the thesis that intrinsic-value properties are logically strongly supervenient on (or explanatorily reducible to) natural facts. Yet all is not lost—we can go to school on Moore's mistakes. This leads me to the formulation of a new general argument against scientific naturalism. In section 1.4, I apply this general argument specifically to scientific naturalism about logic, and thereby also to logical psychologism.

It is not implausible to take Frege to be the most thoroughgoing opponent of logical psychologism. And Frege has often been taken to be a platonist. So one might easily assume that any rejection of logical psychologism entails logical platonism. According to logical platonism, the "standard" (or Tarskian, referential) semantics of natural language, together with the plausible idea that the semantics of logic should be "homogeneous" or uniform with the rest of natural language, requires (i) the existence of objectively real (intersubjectively knowable, nonmental), abstract (nonspatiotemporal) logical objects, and (ii) the human knowability of these objects. I argue in section 1.5 that logical platonism is false. The fundamental problem with logical platonism is not, however, as Paul Benacerraf has argued in connection with the same problem about the semantics of mathematics, that the causal inertness of abstract objects contradicts the further assumption of a "reasonable epistemology," to the effect that knowledge requires causal contact with the object known. Benacerraf's argument has three questionable steps in it. Instead the fundamental problem is that logical platonism yields the metaphysical alienation of the human mind from logic, which is inconsistent with two very plausible commonsense beliefs: that we humans actually have some logical knowledge, and that logic is intrinsically normative and perhaps even unconditionally obligatory for actual human reasoning processes.

Nevertheless, as I argue in section 1.6, it is possible consistently to hold (i) that logical psychologism is false, (ii) that logical platonism is false, and (iii) that logic is cognitively constructed by rational animals, in the sense that every rational animal—including every rational human animal—possesses a cognitive faculty that is innately configured for the representation of logic. In other words, logic is explanatorily and ontologically dependent on rational animals, but logical facts are not reducible to the natural facts. The view expressed by (iii) is what I call *the logic faculty thesis*, which in turn is the first of two basic parts of the doctrine of *logical cognitivism*. Given logical cognitivism, we can consistently reject logical psychologism on the one hand while also rejecting logical platonism on the other, and yet in a certain qualified sense still endorse a psychological theory of the nature of logic.

1.1 Frege, Husserl, and Logical Psychologism

According to Michael Dummett's crisp and compelling formulation, recent and contemporary philosophy is "post-Fregean philosophy," in the sense that Frege is arguably the most important figure in the early development of the mainstream Euro-American twentieth-century tradition in analytic philosophy.[4] It seems equally true that contemporary logic is "post-Fregean logic," in the sense that Frege is arguably the most important figure in the early development of pure—that is, mathematical and symbolic—logic.[5] These two historical facts are not of course unconnected. As Jean Van Heijenoort observes: "Frege's philosophy is analytic in the sense that logic has a constant control over his philosophical investigations."[6] So pure logic constantly controls Frege's philosophy, and in turn Frege's logically oriented philosophy constantly controls the analytic tradition. The chain of command is clear. What we need to understand better is the nature of pure logic.

In this section I focus on a fundamental element in Frege's conception of pure logic: his critique of logical psychologism. This critique was later codified and deepened by Husserl. Here are some characteristic samples of Frege's arguments against the psychologicists:

Never let us take . . . an account of the mental and physical conditions on which we become conscious of a proposition for a proof of it. A proposition may be thought, and again it may be true; let us never confuse these two things.[7]

We suppose . . . that concepts sprout in the mind like leaves on a tree, and we think to discover their nature by studying their birth: we seek to define them psychologically, in terms of the nature of the human mind. But this account makes everything subjective, and if we follow it through to the end, does away with truth.[8]

[T]he expression 'law of thought' seduces us into supposing that these laws govern thinking in the same way as laws of nature govern events in the external world. In that case they can be nothing but laws of psychology: for thinking is a mental process. And if logic were concerned with these laws it would be a part of psychology. . . . Then one can only say: men's taking something to be true conforms on the average to these laws . . . ; thus if one wishes to correspond with the average one will conform to these. . . . Of course—if logic has to do with something's being taken to be true, rather than its being true! And these are what the psychological logicians confuse.[9]

Psychological treatments of logic arise from the mistaken belief that a thought (a judgement as it is usually called) is something psychological like an idea. . . . Now since every act of cognition is realized in judgements, this means the breakdown of every bridge leading to what is objective.[10]

With the psychological conception of logic we lose the distinction between the grounds that justify a conviction and the causes that actually produce it. This means that a justification in the proper sense is not possible. . . . If we think of the laws of logic as psychological, we shall be inclined to raise the question whether they are somehow subject to change. . . . The laws of truth, like all thoughts, are always true if they are true at all. . . . Since thoughts are not mental in nature, every psychological treatment of logic can only do harm. It is rather the task of this science to purify logic of all that is alien and hence of all that is psychological. . . . Logic is concerned with the laws of truth, not with the laws of holding something to be true, not with the question of how men think.[11]

Not everything is an idea. Otherwise psychology would contain all the sciences within it, or at least would be the supreme judge over the sciences. Otherwise psychology would rule even over logic and mathematics. But nothing would be a greater misunderstanding of [logic or] mathematics than making it subordinate to psychology.[12]

Even just a quick skim through these texts reveals that philosophically there is quite a lot going on in them. It is evident that in different places Frege employs somewhat different characterizations of logical psychologism, and somewhat different criticisms of it too.[13] Given this complexity, along with the reasonable hunch that we might find the same or at least a similar complexity in Husserl's critique of psychologism, I will refrain from glossing the Fregean texts until I have also sketched Husserl's critique.

In 1894 Frege published a devastating review of the first volume of Husserl's *Philosophie der Arithmetik*, an investigation into the basic concepts of arithmetic that was heavily influenced by Brentano's *Psychology from an Empirical Standpoint*. Among other things, Frege accused Husserl of committing the cardinal sin of logical psychologism. Husserl obviously received the message loud and clear, because he never wrote the second volume. By the turn of the century, however, Husserl had gotten his revenge: he not only converted whole-heartedly to antipsychologism in the late 1890s, thus joining his erstwhile accuser, but he also effectively "out-Frege-ed" Frege by publishing the *Prolegomena to Pure Logic*. As Martin Kusch has shown, the *Prolegomena* had the highly significant double effect of simultaneously (1) establishing the pure logic tradition in early twentieth-century European philosophy, and (2) creating the discipline of experimental or scientific psychology by providing a reason (or more accurately, an excuse) to banish the nonconforming psychologicists from the leading German philosophy departments.[14] It also introduced several original points into the debate about logical psychologism. So for both sociological and purely philosophical reasons, the *Prolegomena* rapidly became the bible on antipsychologism. Ironically— and tragically, given Russell's shattering contemporaneous discovery of the paradox of classes in his own and Frege's logical systems[15]—Frege's logical and logico-philosophical writings were almost entirely ignored by his contemporaries.[16]

The *Prolegomena* is massively documented and carefully argued. Yet in one respect it develops a rather simple story line by dividing philosophers of logic neatly into three groups:

(i) what we might call the "eternally damned" psychologicists (Richard Avenarius, Benno Erdmann, Theodor Lipps, Ernst Mach, J. S. Mill, Christian Sigwart, and Herbert Spencer);

(ii) the "eternally saved" antipsychologicists (Leibniz and Bernard Bolzano— note Frege's conspicuous absence!); and

(iii) those precariously balanced between the hell of psychologism and the heaven of antipsychologism (Kant, Johann Herbart, Hermann Lotze, Paul Natorp, and Wilhelm Wundt).

It also contains an interesting and original critique of normative conceptions of logic[17] and ingeniously connects logical psychologism directly with cognitive relativism[18]—indeed, Husserl appears to have coined the term 'relativism'. Of course the main task of the *Prolegomena* is to identify and refute psychologism:

No natural laws can be known *a priori*, nor established by sheer insight. The only way a natural law can be established and justified, is by an induction from the singular facts of experience. . . . [If psychologism is correct, then] logical laws must accordingly, without exception rank as mere probabilities. Nothing, however, seems plainer than that the laws of 'pure logic' all have a priori validity. They are established and justified, not by induction, but by apodeictic self-evidence.[19]

How plausible the ready suggestions of psychologistic reflection sound! Logical laws are laws for validations, proofs. What are validations but peculiar human trains of thought, in which, in normal circumstances, the finally emergent judgments seem endowed with a necessarily consequential character. This character is itself a mental one, a peculiar mode of mindedness and no more. . . . How could anything beyond empirical generalities result in such circumstances? Where has psychology yielded more? We reply: Psychology certainly does not yield more, and cannot for this reason yield the apodeictically self-evident, and so non-empirical and absolutely exact laws which form the core of all logic.[20]

The psychologistic logicians ignore the fundamental, essential, never-to-be-bridged gulf between ideal and real laws, between normative and causal regulation, between logical and real necessity, between logical and real grounds. No conceivable gradation could mediate between the ideal and the real.[21]

These points are, manifestly, very similar in content to Frege's and reveal a similar multifariousness. The *Prolegomena* has two advantages over Frege's critique of logical psychologism, however. First, Husserl deftly compresses the different versions of psychologism into a single formula:

Let us place ourselves for the moment on the ground of the psychologistic logic, and let us assume that the essential theoretical foundations of logic lie in psychology. However the latter discipline may be defined . . . it is universally agreed that psychology is a factual and therefore an empirical science.[22]

Second, he also deftly compresses the different worries about psychologism into a single objection:

The basic error of Psychologism consists, according to my view, in its obliteration of the fundamental distinction between pure and empirical generality, and in its misinterpretation of the *pure* laws of logic as *empirical* laws of psychology.[23]

Here we can see that what Frege and Husserl both reject by rejecting logical psychologism is the claim that empirical psychology provides "the essential theoretical foundations of logic." I take it that a science X contains the essential theoretical foundation of a science Y if and only if Y can be explanatorily reduced to X. Explanatory reduction is the strongest sort of reduction. As standardly construed, reduction can be either (i) explanatory or (ii) ontological.[24] Explanatory reduction involves expressing the "higher-level"—or less basic—concepts of one science in terms of the "lower-level"—or more basic—concepts of another, without any appreciable loss of meaning or cognitive significance. Assuming that concepts pick out corresponding properties,[25] and that facts are instantiations of properties, then an explanatory reduction entails either the identity of higher-level properties/facts with lower-level properties/facts, or else the logical strong supervenience (which at this point we can construe as asymmetric or one-way logically necessary dependence—I will spell out the notion of logical strong supervenience more explicitly in section 1.2) of higher-level properties/facts on lower-level properties/facts. Logical strong supervenience is consistent with the identity of higher-level and lower-level properties or facts and is also consistent with their nonidentity. But in either case, an explanatory reduction of Y to X shows that the concepts and corresponding properties/facts of Y are "nothing over and above" those of X. Ontological reduction, by contrast, involves showing only that higher-level properties/facts are identical with lower-level properties/facts. So given an ontological reduction of Y to X, there can still be an "explanatory gap" between Y and X, in the sense that concepts and corresponding properties/facts of Y are not analytically definable in terms of the concepts and corresponding properties/facts of X. For example, it is possible to claim that mental properties are identical with physical properties (say, of the brain), while also asserting that there is an explanatory gap between mentalistic concepts and physicalistic concepts.[26] Thus every explanatory reduction is also an ontological reduction, but a reduction can be ontological without also being explanatory.

Empirical psychology is the same as experimental or scientific psychology. At the end of the 19th century, of course, scientific psychology was only in its infancy. And even today it remains an open question whether (and if so, in what sense) special sciences like cognitive psychology are reducible to the fundamental sciences: biology, chemistry, and especially physics.[27] For my purposes, however, empirical psychology can be indifferently construed as an introspective science of the mental ("introspectionist psychology"), as a social science of the mental ("folk psychology" in Wundt's original sense of that term), as a behavioral-ethological science of the mental ("behavioral psychology"), as a computer-driven science of the mental ("computational psychology"), as psychobiology, as psychochemistry, or as psychophysics. The bottom line for Frege and Husserl, and the bottom line for me, is the psychologicist's assertion that logic has its essential foundations contained in, and is therefore explanatorily reducible to, empirical psychology.

In direct opposition to logical psychologism, Frege and Husserl both explicitly insist that logic is *pure*, by which they mean that logic is necessary, objectively true, fully formal or topic-neutral, and a priori. This is nicely captured in Frege's assertion in the *Foundations of Arithmetic* that

[w]hat is of concern to logic is not the special content of any particular relation, but only the logical form. And whatever can be asserted of this, is true analytically and known a priori.[28]

Thus Frege's and Husserl's Ur-objection to logical psychologism is that it obliterates the fundamental distinction between the necessary, objectively true, fully formal or topic-neutral, and a priori character of pure logic on the one hand, and the contingent, belief-based, topic-biased, and a posteriori character of empirical psychology on the other, thereby wrongly reducing the former to the latter.

That Ur-objection in turn breaks down into these four sub-objections:[29]

(1) *Modal downsizing* Psychologism wrongly reduces the necessity and strict universality of logical laws to the contingent generality of empirical laws.

(2) *Cognitive relativism* Psychologism wrongly reduces objective logical truth to mere (individual, socially constituted, or species-specific) belief.

(3) *Topic bias* Psychologism wrongly reduces the full formality or topic-neutrality of logic to the topic bias of (individualistic, socially constituted, or species-specific) mental content.

(4) *Radical empiricism* Psychologism wrongly reduces the apriority of logical knowledge to the aposteriority of empirical methods of belief-acquisition and belief-justification.

Of course it is one thing to have some serious worries about logical psychologism, and quite another to have compelling arguments against it. Suppose that psychologism entails modal downsizing, cognitive relativism, topic bias, and radical empiricism. Does it follow automatically that psychologism is false? No. Notice that the formulation of each sub-objection includes the crucial word 'wrongly'. This begs the question. Pointing out that logical psychologism entails modal downsizing, and so on, does not amount to a refutation unless one has independent arguments to show that logic *really is* necessary, objectively true, topic-neutral, and a priori; or unless one has independent arguments to show that one or more of the four reductions leads directly to falsity or absurdity. But as far as I can determine, Frege and Husserl only ever *assert* that logic is absolutely necessary, and so on, and never try to prove those claims independently; nor do they ever make any serious attempts to reduce the psychologistic reductions to falsity or absurdity. Therefore, even if they are entirely correct about the nature of logical psychologism and its consequences, ultimately they provide no *noncircular* arguments against psychologism, which is to say that ultimately they provide no *cogent* arguments against psychologism.

1.2   Antipsychologism as Antinaturalism

Historically considered, logical psychologism is the product of mid- to late-nineteenth-century European philosophy, especially including three overlapping subtraditions: (i) the German neo-Kantian tradition; (ii) the positivist tradition in England, France, Germany, and Austria; and (iii) J. S. Mill's empiricism, as expressed in his *System of Logic*. By the middle of the twentieth century, moreover, these three subtraditions had achieved a stable fusion or synthesis with the pragmatic tradition in the United States. This stable synthesis of neo-Kantianism, positivism, empiricism, and pragmatism is epitomized by the writings of Quine.[30] In turn, the underlying theme and theoretical engine of the three-headed tradition that originally gave rise to logical psychologism, hence equally the underlying theme and theoretical engine of Quine's synthesis, is scientific naturalism.[31]

Scientific naturalism includes four basic elements: (1) anti-supernaturalism, (2) scientism, (3) physicalist metaphysics, and (4) radical empiricist epistemology. I will look briefly at each of them in turn.

(1) *Anti-supernaturalism*   is the rejection of any theoretical appeal to nonphysical, nonmaterial, or nonspatiotemporal entities, properties, and causes (e.g., platonic universals or God). The motivating thought here is that only what is either specifically material, or more generally part of the spatiotemporal and causal order of things, can be truly real.

(2) *Scientism*   says that the *exact sciences*—mathematics and the fundamental natural sciences, especially physics—are the leading sources of knowledge about the world, the leading models of rational method, and collectively the basic constraint on all other sciences and on the acquisition and justification of all genuine knowledge. In other words, nothing in the world falls outside the theoretical purview of the exact sciences.

(3) *Physicalist metaphysics*   says that the physical facts strictly determine all the facts. Let the term 'the physical facts' stand for every fact in the world about the instantiation of physical properties. There are two types of physical facts, and two corresponding types of physical properties. First, there are *basic* physical facts, or facts about the instantiation of the first-order physical

properties of fundamental physical entities, processes, and forces, which in turn are the proper objects of the fundamental natural sciences.[32] And second, there are *nonbasic* physical facts, or facts about the instantiation of second-order physical properties that specify how first-order physical facts are causally configured or patterned in relation to one another: more precisely, these nonbasic physical facts are all *functional organizations* of one sort or another. The nonbasic physical facts are logically strongly supervenient on the basic physical facts. So, otherwise put, according to the scientific naturalist thesis of physicalist metaphysics, all facts are either identical to or logically strongly supervenient on the basic physical facts.

(4) *Radical empiricist epistemology* says that all knowledge whatsoever originates in individual sensory experience, derives its significant content from sensory experiential sources, and is ultimately verified and justified by empirical means and methods alone. In other words, all epistemic facts are strictly determined by—are logically strongly supervenient on—the sensory experiential facts.

To summarize, then, scientific naturalism says (a) that reality is ultimately whatever the exact sciences tell us it is, (b) that all properties and facts in the world are ultimately nothing over and above first-order physical properties and basic physical facts, and (c) that all knowledge is ultimately empirical.

As will already be evident, the technical notion of logical strong supervenience[33] is important to my treatment of logical psychologism, so I had better pause to spell it out a little more carefully. The "very idea" of supervenience is that it captures a modal dependency relation between types of properties that is somewhat weaker than identity, hence consistent with the denial of identity between properties of the relevant types, and thereby consistent with "property dualism" of some sort. So we can separate properties into two distinct classes: the *lower-level* or more basic properties, and the *higher-level* or less basic properties. Call the lower-level properties "A-properties" and the higher-level properties "B-properties." Then we can say that B-properties supervene on A-properties if and only if:

(1) necessarily, anything that has some property G among the B-properties also has some property F among the A-properties (or equivalently: no two things can share all their A-properties unless they also share all their B-properties; or again equivalently: no two things can differ in any of their B-properties without also having a corresponding difference in their A-properties); and

(2) necessarily, anything's having F is sufficient for its also having G.

This two-part supervenience relation is what Jaegwon Kim aptly calls "strong supervenience."[34] The label is apt because we can characterize at least two modally weaker supervenience relations by slightly modifying the concept of strong supervenience. On the one hand, we can characterize a *weak* supervenience[35] by dropping the second occurrence of 'necessarily,' thus making the supervenience an intraworld or merely coextensive relation instead of an interworld or cross-possible-world relation. And on the other hand, retaining the cross-possible-world character of supervenience, we can instead characterize what I will call a *moderate* supervenience by asserting feature (1) alone without feature (2). According to moderate as opposed to strong supervenience, it is merely the case that there can be no B-property difference without an A-property difference.[36] The crucial difference between moderate supervenience and strong supervenience is that strong implies the *existential modal dependence* of B-properties on A-properties, whereas moderate does not. So the relation of moderate supervenience is consistent with the existence of possible worlds in which the A-properties exist but the B-properties do not.

Now back to strong supervenience itself. In this context, feature (1) of strong supervenience is known as the *necessary covariation* of the A-properties with the B-properties, and feature (2) is known as the *upward dependence* of the B-properties on the A-properties. If we further assume that the A-properties are first-order physical properties and that the B-properties are, at least when taken at face value, nonphysical properties of some sort (say, mental properties, normative properties, or modal properties), then this yields a materialist or physicalist strong supervenience.[37] It is also sometimes held—for example, by Kim—that a properly reductive physicalist strong supervenience must incorporate the proviso that feature (1) and (2) are further constrained by nomological connections running between the A-properties and the B-properties.[38] When this extra constraint is added, materialist strong supervenience is called *superdupervenience*,[39] because it captures the idea that the lower-level or basic physical properties necessarily determine the higher-level properties in a thoroughly lawlike and adequately systematic fashion. Given superdupervenience, the higher-level properties are really "nothing but" or "nothing over and above" the lower-level physical properties. Or in other words, the higher-level properties are *fully reducible* to the lower-level physical properties, without being identical to them.

The notion of full reduction brings me to the notion of *logical* strong supervenience. Logical strong supervenience means that the two occurrences of "necessarily" in the formulation of strong supervenience are to be read as "logically or analytically necessarily," as opposed, for example, to either "nonlogically or synthetically necessarily" or "physically, nomologically, or naturally necessarily," which pick out more restricted modalities.[40] As David Chalmers has pointed out, the philosophical importance of the notion of logical strong supervenience is precisely its entailment of (indeed, necessary equivalence with) the notion of explanatory reduction.[41] If B-properties logically strongly supervene on A-properties, then B-properties follow logically or analytically from A-properties and thereby provide a reductive explanation of those properties, because an ideally rational thinker could, from her (possibly a posteriori) knowledge of the A-properties together with her (possibly a posteriori) knowledge of any nomological connections between the A-properties and the B-properties, *logically infer or deduce* all the B-properties.[42] Otherwise put, the explanatory reduction is the result of *conceptual analysis* (possibly assisted by empirical investigation).

I will call the total conjunction of all the basic physical facts and all the sensory experiential facts *the natural facts*. Then scientific naturalism can be most compactly expressed as the thesis that *all facts logically strongly supervene on the natural facts*. This formulation captures the anti-supernaturalism, scientism, physicalist metaphysics, and radical empiricist epistemology of scientific naturalism all in one go. Three further things should be noted about scientific naturalism, however.

First, it needs to be reemphasized that although scientific naturalism is consistent with the identity of higher-level properties with lower-level properties, it does not absolutely require the identity but rather only the logical strong supervenience of the former on the latter. So scientific naturalism is consistent with various nonidentity theses such as, for example, that functionally defined mental properties are not identical with first-order physical properties, or that evolutionarily grounded normative properties are not identical with first-order physical properties.

Second, although scientific naturalism generally requires that the lower-level or A-properties on which the higher-level or B-facts logically strongly supervene must be contingent facts, those A-facts can be *either* first-order physical facts *or* sensory experiential facts. So although the scientific naturalist by virtue of her physicalist metaphysics is committed to the thesis that all facts ultimately logically strongly supervene on the first-order physical facts, she need not hold that the sensory experiential facts are themselves identical to the first-order physical facts: the sensory experiential facts can be nonidentical with but still logically strongly supervenient on the first-order physical facts.

Third and finally, it is crucially important to recognize that not everything that goes by the name of "naturalism" is scientific naturalism. So I want especially to emphasize that what I am calling "scientific naturalism" does not capture *every* form of philosophical naturalism, but only those views that are in the exact-science-oriented tradition of the neo-Kantians, the positivists, Mill, and Quine, and those that are explicitly or implicitly committed to anti-supernaturalism, scientism, physicalist metaphysics, and radical empiricist epistemology, as well as the logical strong supervenience of all facts on the natural facts. Many weaker forms of philosophical naturalism also exist,[43] and some of these are perfectly consistent with the view I will eventually spell out and defend: logical cognitivism. Indeed, as we will see, logical cognitivism explicitly accepts anti-supernaturalism, and also asserts a nonreductive explanatory and ontological dependence of logic on the innate cognitive capacities of rational animals. It is obvious that necessarily, all rational animals—whether human or nonhuman—are animals. Then, since animals, as sentient living organisms, are surely *natural* beings if anything is, we can quite accurately say that logical cognitivism implies what I will call an *embodied rationalistic naturalism* about logic, although it rejects scientific naturalism as defined above.

In any case, the concept of scientific naturalism allows us to achieve a deeper reading of the psychologistic thesis. As we have seen, logical psychologism is the thesis that logic is explanatorily reducible to empirical psychology. And we have also seen that the explanatory reduction of logic to empirical psychology entails scientific naturalism about logic. Thus logical psychologism is nothing more and nothing less than a species of "naturalized logic," or a form of scientific naturalism about logic.[44] Scientific naturalism, in turn, is the thesis that all facts are logically strongly supervenient on the natural facts.

Now, in my opinion, the most philosophically illuminating formulation of logical psychologism is the thesis that logic is logically strongly supervenient on the natural facts. This is because although there are in fact more recent versions of scientific naturalism about logic that do not appeal specifically to empirical psychology[45]—thereby showing indirectly the overwhelming historical success of the Frege–Husserl critique of logical psychologism, even in the face of the rise of scientific naturalism in the latter half of the twentieth century—these do not differ at all from logical psychologism in respect of their basic explanatory, ontological, epistemological, or methodological commitments. Correspondingly, then, the antipsychologism proposed by Frege and Husserl is for all intents and purposes equivalent to the following direct denial of scientific naturalism about logic: logic is *not* logically strongly supervenient on the natural facts.

This formulation may seem to have an air of paradox. Suppose that one assumes, along with Frege and Husserl, that logic is necessary, objectively true, topic-neutral, and a priori. Then logic is logically derivable from anything and everything, and even logically derivable from nothing at all, and thus it is trivially true that logic is logically strongly supervenient on the natural facts. Then Frege and Husserl are denying a trivial truth! But as we have seen, one cannot simply *assume* that logic is necessary, objectively true, topic-neutral, and a priori without begging the question; and of course this is just what the defender of logical psychologism or of any other version of scientific naturalism about logic *denies*: the psychologicist or other logical scientific naturalist is claiming that logic is *neither* necessary, *nor* objectively true, *nor* topic-neutral, *nor* a priori, precisely because logic is explanatorily reducible to the natural facts. So in asserting antipsychologism, Frege and Husserl are denying a substantive and controversial thesis.

1.3   Moore, Antipsychologism, and Antinaturalism

We are currently in search of a cogent argument against logical psychologism, because Frege's and Husserl's famous antipsychologistic arguments, sadly, beg the question. I have proposed that logical psychologism is a species of scientific naturalism. It makes good sense, then, to look at leading arguments against scientific naturalism. But where to look?

All things considered we probably cannot do better than to go back to G. E. Moore's writings, since Moore was a near-contemporary of both Frege and Husserl, since he explicitly argued against both psychologism and naturalism, and since those arguments later became part of the conventional wisdom of the analytic tradition. Given his unfamiliarity with the works of Frege at that time, Moore appears to have more or less independently invented antipsychologism, although in a nonlogical context. In his amazing essays "The Nature of Judgment" (1898) and "The Refutation of Idealism" (1903), and in the even more amazing *Principia Ethica* (1903), he went after psychologism in two ways: from the standpoint of epistemology, and from the standpoint of ethics.

Moore's first concern is with psychologistic epistemology in the neo-Kantian, neo-Hegelian, and Millian traditions. His objection is that their epistemology involves a fundamental confusion between two senses of the "content" of a cognition: (i) content as that which literally belongs to the conscious mental act of cognizing (the psychologically immanent content, or act-content); and (ii) content as that at which the mental act is directed, or which it is "about" (the psychologically transcendent content, or objective content). The communicable meaning and truth value of the judgment belong strictly to objective content. But psychologism assimilates the objective content to the act-content. This is what Moore glosses as

the fundamental contradiction of modern Epistemology—the contradiction involved in both distinguishing and identifying the *object* and the *act* of Thought, 'truth' itself and its supposed *criterion*.[46]

Given this "contradiction," the communicable meaning and truth value of the content of cognition are both reduced to the point of view of a single subject. The unpalatable consequences are that meaning becomes unshareably private (which is a form of topic bias) and that truth turns into mere personal belief (which is a form of cognitive relativism).

Moore's *Principia* contains another and much more famous objection to psychologism. His general target is what he explicitly calls "naturalism" in ethics:

[Naturalism] consists in substituting for 'good' some one property of a natural object or of a collection of natural objects; and in thus replacing Ethics by some one of the natural sciences. In general, the science thus substituted is one of the sciences specially concerned with man. . . . In general, Psychology has been the science substituted, as by J. S. Mill.[47]

And his objection centers on the famous naturalistic fallacy:

[T]he naturalistic fallacy . . . [is] the fallacy which consists in identifying the simple notion which we mean by 'good' with some other notion.[48]

[The naturalistic] fallacy, I explained, consists in the contention that good *means* nothing but some simple or complex notion, that can be defined in terms of natural qualities.[49]

In other words, according to Moore ethical naturalism is the claim that the property[50] of being good is identical with some simple or complex natural property (which for our purposes we can construe as either a first-order physical property, a second-order physical property, or a sensory experiential property); and the naturalistic fallacy consists precisely in accepting such an identification of properties. So far, so good—awful pun intended. But now for the sad part of the story.

Most post-Moorean analytic philosophers have accepted Moore's characterization of ethical naturalism as well as his antinaturalistic conclusions; yet his main argument in support of its putative fallaciousness—the "open question argument"—is generally held to be a notorious failure. Here is the argument:

The hypothesis that disagreement about the meaning of good is disagreement with regard to the correct analysis of a given whole, may be most plainly seen to be incorrect by consideration of the fact that, whatever definition be offered, it may always be asked, with significance, of the complex so defined, whether it is itself good.[51]

We must not, therefore, be frightened by the assertion that a thing is natural into the admission that it is good: good does not, by definition, mean anything that is natural; and it is always an open question whether anything that is natural is good.[52]

For convenience I will call the fundamental ethical property of being good *the Good*. The open question argument says that any attempt to explain the Good solely in terms of some corresponding natural property N (say, the property of being a pleasurable state of mind) falls prey to the decisive objection that even if X is an instance of N it can still be significantly asked whether X is good: that is, it can be significantly postulated that X is an instance of N but is not good. Moore's rationale for this is that the only case in which it would be altogether nonsensical to postulate that X is an instance of N but is not good is the case in which it is strictly impossible or contradictory to hold that X is not good, that is, when X is, precisely, good. So if it is significant to ask whether X is N but not good, then N is not identical to the Good. And Moore finds it to be invariably the case that it is significant to ask whether X is N but not good, hence invariably it is the case that N is not identical to the Good. He concludes that the Good is an indefinable or unanalyzable nonnatural property, and that it is a fallacy to try to identify the Good with any natural property.

The open question argument is doomed, I think, because of a mistake Moore has made about the individuation of properties. The problem as I see it is that the argument implies a criterion of property-identity that is absurdly strict.[53] Familiar criteria for the identity of two properties include (i) equivalence of their analytic definitions, (ii) synonymy of their corresponding predicates, and (iii) identity of their cross-possible-worlds extensions. But Moore's criterion is importantly different:

> [W]hoever will attentively consider with himself what is actually before his mind when he asks the question 'Is pleasure (or whatever it may be) after all good?' can easily satisfy himself that he is not merely wondering whether pleasure is pleasant. And if he will try this experiment with each suggested definition in succession, he may become expert enough to recognise that in every case he has before his mind a unique object, with regard to the connection of which with any other object, a distinct question can be asked. Everyone does in fact understand the question 'Is this good?' When he thinks of it, his state of mind is different from what it would be, were he asked 'Is this pleasant, or desired, or approved?' It has a distinct meaning for him, even though he may not recognize in what respect it is distinct. Whenever he thinks of 'intrinsic value', or 'intrinsic worth', or says that a thing 'ought to exist', he has before his mind the unique object—the unique property of things—which I mean by 'good'. . . . 'Good', then, is indefinable.[54]

Moore's criterion is that two properties are identical if and only if the intentional contents of the states of mind in which the properties are recognized are phenomenally indistinguishable.[55] Consequently, even two properties that are by hypothesis definitionally equivalent—for example, the property of being a bachelor and the property of being an adult unmarried male—will come out nonidentical according to this test. The intentional content of the state of mind of someone who says or thinks that X is a bachelor is clearly phenomenally distinguishable from that of the same person when she says or thinks that X is an unmarried adult male. I might not wonder even for a split second whether a bachelor is a bachelor, yet find myself mentally double-clutching as to whether a bachelor is an unmarried adult male. But then according to that test it is not nonsensical to ask whether X is an unmarried adult male but not a bachelor: from which we must conclude by Moorean reasoning that the property of being a bachelor is indefinable, and that it is a fallacy to try to identify any property with any other property, *including the property that expresses its definition*. Obviously this cannot be correct. It is patently absurd to constrain property identity so very, very tightly.[56]

Moore's antinaturalism also contains another less noticed but equally serious difficulty. This difficulty stems from his explicit commitment to a certain strict modal connection between intrinsic-value properties and natural facts:

> I have tried to shew, and I think it is too evident to be disputed, that such appreciation [of intrinsically valuable, or good, qualities] is an organic unity, a complex whole;

and that, in its most undoubted instances, part of what is included in this whole is *a cognition of material qualities*, and particularly of a vast variety of what are called *secondary* qualities. If, then, it is *this* whole, which we know to be good, and not another thing, then we know that material qualities, even though they be perfectly worthless in themselves, are yet essential constituents of what is far from worthless . . . . [A] world, from which material qualities were wholly banished, would be a world which lacked many, if not all, of those things, which we know most certainly to be great goods.[57]

[I]f a given thing possesses any kind of intrinsic value in a certain degree, then not only must that same thing possess it, under all circumstances, in the same degree, but also anything *exactly like it*, must, under all circumstances, possess it in exactly the same degree. Or, to put it in the corresponding negative form: it is not *possible* that of two exactly similar things one should possess it and the other not, or that one should possess it in one degree, and the other in a different one.[58]

According to Moore (1) every intrinsic-value property has some complex set of natural qualities as its "essential constituents," and (2) for any natural thing that "possesses any kind of intrinsic value in a certain degree, then not only must that same thing possess it, under all [logically possible] circumstances, in the same degree, but also anything *exactly like it*, must, under all [logically possible] circumstances, possess it in exactly the same degree." In other words, intrinsic-value properties are both *constituted by* and *logically strongly supervenient on* natural properties. It follows that the Good is, incoherently, both natural and nonnatural. I say "incoherently" rather than "inconsistently" because, as we have seen, strictly speaking it is possible to say that two sets of properties are nonidentical even though one of those sets of properties is logically strongly supervenient on the other set of properties. But since logical strong supervenience implies explanatory reduction, and since the philosophical upshot of Moore's ethical antinaturalism is surely not the mere *nonidentity* of the Good with any other property, but rather the *explanatory irreducibility* of the Good to any other property, his overall view is in conflict with itself.

We have just seen that Moore's antinaturalism is a double failure. But all is not lost, for this double failure teaches us two important lessons. First lesson: do not make your argument against scientific naturalism rest on questionable assumptions about property-individuation or property-identity! Second lesson: you must attack the logical strong supervenience thesis of scientific naturalism directly! Taking these two post-Moorean dicta to heart, here is a new general argument against scientific naturalism.

Prove: That scientific naturalism is false

1. Scientific naturalism asserts that facts about strict modality (e.g., facts about logical necessity, certainty, and obligation) logically strongly supervene on the natural facts. (Premise.)

2. So, if scientific naturalism is true, then in every logically possible world in which the natural facts exist, facts about strict modality exist (From (1).)

3. But natural facts are logically contingent facts, that is, they logically could have been otherwise. (Premise.)

4. If the natural facts logically could have been otherwise, then it is logically possible that facts about strict modality do not exist. (From (3).)

(Elucidation of step 4: This does not mean that necessarily, if there is a change in the lower-level properties/facts, then there is also a corresponding change in the higher-level properties/facts. Rather, it means that necessarily, if there is a change in the lower-level properties/facts, then *possibly* there is also a corresponding change in the higher-level properties/facts. If, as logical strong supervenience implies, there is a logically necessary covariation relation between the higher-level properties/facts and the lower-level properties/facts, then it must be the case that changes in the lower-level properties/facts are *logically consistent with* corresponding changes in the higher level properties/facts. For example, if the higher-level properties/facts happen to be identical with the lower-level properties/facts, then obviously changes in the lower-level properties/facts will also yield changes in the higher-level properties/facts.)

5. Logical strong supervenience is a strict modal relational fact. (Premise.)

6. So, if the natural facts logically could have been otherwise, then it is logically possible that logical strong supervenience does not exist. (From (4) and (5).)

7. If it is logically possible that logical strong supervenience does not exist, then it is logically possible that it is logically possible that all the natural facts remain the same and strict modality does not exist. (From (6).)

8. If it is logically possible that it is logically possible that all the natural facts remain the same and strict modality does not exist, then it is logically possible that all the natural facts remain the same and strict modality does not exist. (From (7).)

9. So, if scientific naturalism is true, then strict modality does not logically strongly supervene on the natural facts. (From (1) and (8).)

10. So, if scientific naturalism is true, then scientific naturalism is false. (From (1), (2), and (9).)

11. Therefore, scientific naturalism is false, by reductio. (From (10).)

The key elements in this argument are the concepts of logical strong supervenience, strict modality (whether logical, epistemic, or deontic), logical contingency, and the plausible modal principle deployed in step (8)—directly derivable from one of the axioms of C. I. Lewis's modal system S4—that if it is logically possible that it is logically possible that S, then it is logically possible that S (or in the Hughes and Cresswell symbolism, $MMp \supset Mp$).[59]

Correspondingly, the key move in the argument is to display the absurd consequences of making facts about strict modality logically strongly supervenient on the natural facts. The very idea of strict modality implies *logical independence from any particular logically possible world, including the actual world*, while contrariwise the very idea of logical strong supervenience on the natural facts implies *logical dependence on the actual world*.

1.4 Antinaturalism as Antipsychologism

We should probably remind ourselves where we are in the overall argument of the chapter. In section 1.1, we saw that Frege and Husserl desperately wanted to reject logical psychologism—the thesis that logic is explanatorily reducible to scientific psychology—but that in fact they presented no cogent arguments against it. In section 1.2 we saw that logical psychologism is a species of scientific naturalism, and consequently that antipsychologism is antinaturalism. In section 1.3 we saw that Moore's open question argument against naturalism fails. And in section 1.4 I offered a new general argument against naturalism that is designed to avoid Moore's mistakes. Now it is time to apply that same general argument specifically to scientific naturalism about logic.

Prove: That scientific naturalism about logic is false

(1) Scientific naturalism about logic asserts that logic logically strongly supervenes on the natural facts. (Premise.)

(2) So, if scientific naturalism about logic is true, then in every logically possible world in which the natural facts exist, logic exists. (From (1).)

(3) But natural facts are logically contingent facts, that is, they logically could have been otherwise. (Premise.)

(4) If the natural facts logically could have been otherwise, then it is logically possible that logic does not exist. (From (3).)

(5) Logical strong supervenience is a logical relational fact. (Premise.)

(6) So, if the natural facts logically could have been otherwise, then it is logically possible that logical strong supervenience does not exist. (From (4) and (5).)

(7) If it is logically possible that logical strong supervenience does not exist, then it is logically possible that it is logically possible that all the natural facts remain the same and logic does not exist. (From (6).)

(8) If it is logically possible that it is logically possible that all the natural facts remain the same and logic does not exist, then it is logically possible that all the natural facts remain the same and logic does not exist. (From (7).)

(9) So, if scientific naturalism about logic is true, then logic does not logically strongly supervene on the natural facts. (From (1) and (8).)

(10) So, if scientific naturalism about logic is true, then scientific naturalism about logic is false. (From (1), (2), and (9).)

(11) Therefore, scientific naturalism about logic is false, by reductio. (From (10).)

I conclude that logic is not scientifically naturalizable. And since logical psychologism is a form of scientific naturalism about logic, it follows that logical psychologism is false.

1.5 The Perils of Platonism

Frege is not implausibly taken by many philosophers to be the "compleat" antipsychologicist, the most thoroughgoing opponent of logical psychologism. This belief is well supported by the Frege quotations we surveyed in section 1.2. Furthermore, Frege is often taken to be logical platonist. According to logical platonism, the "standard" (or Tarskian, referential) semantics of natural language, together with the plausible idea that the semantics of logic should be "homogeneous" or uniform with the rest of natural language, requires (i) the existence of objectively real (i.e., intersubjectively knowable and non-mind-dependent), abstract (i.e., nonspatiotemporal) logical objects, and (ii) the human knowability of these objects. Thus, logical platonism is a version of logical supernaturalism. On the face of it, Frege certainly *seems* to be a logical platonist, and thereby a logical supernaturalist, because he explicitly says in "Thoughts" that logical entities must exist in an ontologically distinct domain he calls the "third realm,"[60] distinct from the mental and physical realms. So one might easily assume that any rejection of logical psychologism entails logical platonism. Frege scholars—

and here I am thinking specifically of Oxford-trained Frege scholars influenced by Michael Dummett—may demur. But for our purposes it does not matter whether the historical Frege was a logical platonist or not. What matters is the thesis that antipsychologism entails logical platonism. I want to reject this thesis on the grounds that although antipsychologism is true (by the argument I just sketched in section 1.4), nevertheless logical platonism is false. In other words, the thesis asserts a non sequitur.

What is wrong with logical platonism? Philosophers have tended to approach this problem indirectly, by way of mathematical platonism. Mathematical platonism, by the same argument that applied to logical platonism above, says that the semantics of mathematical truth requires the existence of humanly knowable, real, abstract mathematical objects. In response, Paul Benacerraf has put forward a highly influential argument against mathematical platonism, which I have rationally reconstructed as follows:[61]

(1) If mathematical platonism is true, then mathematical objects are causally inert because (i) they are abstract, hence not in spacetime, and (ii) all causally relevant[62] (not to mention causally efficacious) entities are in spacetime.
(2) Our best overall theory of knowledge, as applied to mathematics, requires a sense-perception-like capacity to account for our cognitive access to mathematical objects.
(3) Sense perception requires an efficacious causal link, involving direct physical contact, between the object perceived and the perceiver.
(4) So, if mathematical platonism is true, then mathematical objects cannot be known by any sort of sense perception.
(5) Therefore, if mathematical platonism is true, mathematical knowledge is impossible.

There are, however, three apparent problems with Benacerraf's argument.

First, Benacerraf assumes that an entity can be causally relevant only if it is "in" spacetime. This could mean different things, but for the purposes of argument I will take it to mean that the entity has a *unique location* in spacetime. So he is saying that an entity can be causally relevant only if it has a unique location in spacetime. But that seems false. Causal laws and functional organizations, for example, have causal relevance—indeed, fundamental causal relevance—because the existence and application of causal laws is a necessary and sufficient condition of all causal relations, and because functional organizations, which specify patterns or configurations of causation in the material world, are necessarily instantiated whenever and wherever causal processes occur: yet causal laws and functional organizations are not uniquely located in spacetime. Causal laws obtain without spatial or temporal bias throughout spacetime. And functional organizations are multiply realizable across spacetime. Indeed, causal laws and functional organizations alike are plausibly held to be *abstract* in the sense that they are not uniquely located in spacetime. Yet they are fundamentally causally relevant. So step (1) is questionable.

Second, it is not at all obvious that our best overall theory of knowledge, as applied to mathematics, requires a sense-perception-like capacity to account for our cognitive access to mathematical entities. Let's call this cognitive access "mathematical intuition." To be sure, philosophers have often assumed that mathematical intuition is sense-perception-like. But unless they have some further independent argument, I see no good reason why mathematical intuition could not operate *non*perceptually: say, like memory, imagination,[63] or conceptual understanding. So step (2) is questionable.

Third and finally, even granting momentarily for the purposes of argument that entities can have causal relevance only if they are in spacetime, Benacerraf further assumes that an entity can have an efficacious causal influence on another only by direct physical contact. But that seems false too if we adopt either a counterfactual analysis[64] or a probabilistic analysis[65] of causation, since these do not require direct physical contact between cause and effect. So step (3) is questionable.

Therefore, at least on the face of it (but see section 6.5 and section 6.6 for a more in depth analysis of Benacerraf's argument), we need another argument, distinct from Benacerraf's, against mathematical platonism and a fortiori against logical platonism. Such a non-Benacerrafian argument can, I think, be found in the fairly simple idea that logical platonism *metaphysically alienates the human mind from logic*. What I mean is this. The human mind is an animal mind—more specifically, the human mind is the mind of a sentient living organism, a finite mortal creature that is uniquely located in spacetime. But if on the one hand logical entities must exist in a nonmental and abstract or nonspatiotemporal world in the platonic sense of *transcending* spacetime, and on the other hand we humans are animals fully *in* spacetime, then the nature of logic apparently neither presupposes, requires, implies, nor in any other way saliently connects with actual or possible human thinkers. This difficulty will still hold even if the real nature of knowledge does not require any causal relation whatsoever between the knower and the object known. The predicament that the human mind apparently has no salient connection with logic is what I mean by its metaphysical alienation from logic. I do not mean to imply that the metaphysical alienation of the human mind from mathematics or logic is somehow radically or even basically different from what Benacerraf is driving at: on the contrary, what I mean is that when we peel away some questionable aspects of Benacerraf's argument, the metaphysical alienation of the human mind from mathematics or logic is its simple bottom line. So, in that simple, bottom-line sense, I am fully in agreement with Benacerraf. Nor do I mean to imply that logical platonism definitely *does* metaphysically alienate the human mind from logic, but rather only that *on the face of it* there is an intelligible and important worry that the platonist must respond to.

Here is the worry. The supposition that the human mind is alienated from logic has two very implausible consequences. If the human mind is not in any way saliently connected with logic, then how could humans ever have knowledge of logic? And if the human mind is not in any way saliently connected with logic, then how could logic ever be normative and perhaps even unconditionally obligatory for human reasoning processes? In other words, if the human mind is metaphysically alienated from logic, then human logical knowledge and human logical reasoning both appear to be impossible. But this is directly inconsistent with two plausible commonsense beliefs: that we human animals do have some logical knowledge, and that logic is normative and perhaps even unconditionally obligatory for our human reasoning processes. These beliefs are, it seems to me, confirmed each time someone teaches an introductory logic class and marks her students' work accordingly. I conclude that until logical platonists have shown us that they have some acceptable way of avoiding the metaphysical alienation of the human mind from logic, we should reject logical platonism.

1.6   Logical Cognitivism Briefly Introduced

Up to this point, my account may seem distressingly negative and critical. Frege and Husserl were basically right about antipsychologism, but their argument against it is wrong; Moore was basically right about antinaturalism, but his argument against it is wrong; scientific naturalism is wrong; scientific naturalism about logic is wrong, so logical psychologism is wrong;

Benacerraf was basically right about antiplatonism, but his argument against it is wrong; and logical platonism is wrong, or at least it is currently unacceptable.

So am I nothing but a nattering nabob of negativity about psychologism and platonism? Fortunately the upshot of this chapter is positive. By way of conclusion and as a segue to later chapters I want to state my own view about the nature of rationality and logic, namely logical cognitivism. This brief introduction is by no means a proper argument for logical cognitivism; that will come later. All I want to do right now is indicate that logical cognitivism is well positioned to build on the results of this chapter; that it has traditional, recent, and contemporary theoretical motivations; and that it is prima facie supported by a considerable body of empirical work in cognitive psychology.

Logical cognitivism says (i) that logic is cognitively constructed by rational animals, and (ii) that rational human animals are essentially logical animals. For the moment I will concentrate on the first claim. To say that logic is cognitively constructed by rational animals is to say that rational animals—including all rational human animals—possess a cognitive faculty that is innately configured for representing logic and is the means by which all actual and possible logical systems are constructed. This claim is what I call *the logic faculty thesis*. If the logic faculty thesis is correct, then logic is both explanatorily and ontologically dependent on rational animals. It should be particularly noted that the logic faculty is a mental *faculty* and not a mere mental *capacity*, because it is a modular[66] capacity for producing mental representations; and it is innate in the dual sense that it is an intrinsic part of the mind of a rational animal and also universally embodied in mature, healthy, fully equipped humans. But the logic faculty is not necessarily restricted to humans. On the contrary, the logic faculty is multiply embodiable, or instantiable across many different biological species, since it seems quite conceivable and thus logically possible that there could be Martian logicians and perhaps even logical animals belonging to other earthly species.

As regards its provenance, the logic faculty thesis is the fusion of three fairly familiar philosophical ideas: (1) the traditional idea, drawn from Kant[67] and Boole,[68] that logic is the a priori science of the "laws of thought"; (2) the mid-twentieth-century idea, drawn from Quine, that logic has a universal, indispensable, and unrevisable basis, namely, "sheer logic";[69] and (3) the contemporary idea, drawn from Chomsky's psycholinguistics and Fodor's rational psychology, that the human animal carries out all its specifically rational cognitive activities in a fully meaningful inner language or "language of thought," which in turn is sufficient to account for our cognition of natural language. These three ideas, in turn, seem to be supported by a significant body of empirical work in psycholinguistics and the cognitive psychology of reasoning.[70]

I need to emphasize that I am *not* saying that just because I have dropped some important names and theory-labels it is in any way proven that either the logic faculty thesis in particular or logical cognitivism more generally is true. My job in the rest of this book is to argue from independent grounds that logical cognitivism is true. Right now I want to stress just two points: (1) the prima facie intelligibility of logical cognitivism; and (2) the fact that my indebtedness to the laws-of-thought tradition, to the sheer-logic tradition, to the language-of-thought tradition, and to the psychology-of-reasoning tradition, is certainly explicit but not in any way uncritical.

As regards point (1): I will get to that very shortly.

As regards point (2): I do not accept Kant's idealism or Boole's theism. Nor do I hold that their very limited conceptions of logical theory are defensible without serious qualification. Also, I am fully aware that the claim that something counts as a "sheer logic" needs to be reconciled on the one hand with the insistent claims of those—paradigmatically, for example, Quine—who take the One True Logic to be classical or elementary logic (or some relatively minor variation on it such as monadic first-order logic),[71] and on the other hand with the equally insistent claims of those who point to the patent existence of a plurality of nonclassical (whether extended or deviant) logics.[72] I do not accept the biologically based scientific naturalism that is sometimes added to Chomsky's psycholinguistics.[73] I do not accept Fodor's computational or machine functionalism,[74] his view that the language of thought must be written in a single code,[75] or his view that *every* mental module is "informationally encapsulated."[76] And I am fully aware that the empirical psychology of reasoning is fraught with controversy, and needs to be critically unpacked and interpreted.[77]

What I am most concerned with right now, in any case, is point (1). I mean that it is perfectly consistent to hold (i) that logical psychologism is false, (ii) that logical platonism is false, and (iii) that logical cognitivism is true. This is because according to my conception of rationality, rational animals are normative-reflective animals in possession of scientifically–naturalistically intractable notions expressing strict modality, among which are concepts expressing logical necessity, epistemic certainty, and unconditional obligation. The concept of logical necessity in turn belongs to sheer logic via the notion of *consequence* and is contained innately in the logic faculty. So, human beings, precisely insofar as they are rational, not only possess concepts expressing logical necessity but are also capable of making a priori knowledge claims about logic and of taking logic to be normative and perhaps even unconditionally obligatory for their reasoning processes. This, finally, implies that logical cognitivism smoothly conforms to my arguments for the claims that logic is not scientifically naturalizable and that logical platonism is false.

At this point you are no doubt asking yourself this highly relevant critical question: Is logical cognitivism ultimately a form of psychologism? My answer is that it depends on what one means by the word 'psychologism'. If we are being historically precise and take logical psychologism to be the view that logic is explanatorily reducible to empirical psychology, then logical cognitivism is most definitely *not* a form of psychologism, since psychologism entails scientific naturalism whereas logical cognitivism assumes the denial of scientific naturalism and is nonreductive. Nevertheless, if we allow ourselves a temporary historical imprecision, and for the moment take psychologism to be any theory that asserts an essential connection between the logical and the psychological, then we can say that logical cognitivism is indeed a form of psychologism.

Furthermore, there is an important intellectual benefit to be gained by temporarily loosening our historical scruples about the use of the term 'psychologism'. We are as a consequence able to recognize that the destruction of psychologism carried out by Frege, Husserl, and (to a lesser extent) Moore was the legitimate rejection of every form of scientific naturalism about logic, including logical psychologism—but not the legitimate rejection of *every* psychological theory of logic. Not every psychological theory of logic is a form of scientific naturalism.[78] In my opinion, their collective problem was that they did not take human rationality seriously enough. On the contrary, by seriously underestimating the nature, scope, and limits of human rationality they strongly encouraged a misguided tendency to jump straight from the rock of logical psychologism over to the hard place of logical platonism, thereby metaphysically alienating the human mind from logic. But I believe that human rationality and logic are essentially related: I believe that logic is cognitively constructed by rational animals (the logic faculty thesis), and also that rational human animals are essentially logical animals (the logic-oriented conception of human rationality). So I believe that by taking human rationality seriously we can vindicate Haack's highly prescient suspicion that "at least some form of psychologism [is] more plausible, than it is nowadays fashionable to suppose."

Which logic do we use to assess the consequences of different logics? . . . Regress threatens. Is the super-logic . . . a priori, or incorrigible?
—Stewart Shapiro[1]

2.0 Introduction

The upshot of chapter 1 was that logical cognitivism offers us a theory of the nature of logic that avoids logical psychologism and logical platonism alike, thereby steering a safe course between the perils of psychologism on the one hand (modal downsizing, cognitive relativism, topic bias, radical empiricism, and self-refuting scientific naturalism), and the perils of platonism on the other (the metaphysical alienation of the human mind from logic). That constitutes a fairly strong case for logical cognitivism. This chapter develops another even stronger case for logical cognitivism and also, you will be happy to know, uses fewer words ending in 'ism'.

Logic, as we have seen in a preliminary way, is the science of the necessary relation of consequence. But there are many different and seemingly incommensurable logical systems. So one outstanding philosophical problem about the nature of logic is how to preserve the unity of logic while accepting the manifest multiplicity of logical systems distinct from classical or elementary logic. This is what I call the *e pluribus unum* ("out of many, one") problem.

My solution to this problem consists in a development of Stewart Shapiro's two deep thoughts: (i) that assessing the consequences of different logics requires a superlogic, on pain of a vicious regress of increasingly synoptic logics, and (ii) that the superlogic is a priori and incorrigible. My way of developing these two thoughts is this: (i*) that the internal analysis and cross-logic evaluation of the many logics presuppose a superlogic—or more precisely, a "protologic"—that is structurally distinct from all classical and nonclassical systems, but *unifies* the many logics through its role in the construction of all logical systems, thus resolving the *e pluribus unum* problem; and (ii*) that the superlogic or protologic is incorrigible or unrevisable and a priori, precisely because it is both constructively and epistemically presupposed by every logical system.

I will also argue that the protologic is contained innately in a cognitive faculty for logical representation, the logic faculty. This of course is the logic faculty thesis that I briefly described in chapter 1, and it is the first of the two basic parts of logical cognitivism. And because logical cognitivism, via the logic faculty thesis, coherently resolves the problem of psychologism, we already have in hand a good reason to accept logical cognitivism. So obviously we also already have in hand the very same good reason to accept the logic faculty thesis. Conversely, the fact that the logic faculty thesis also coherently resolves the *e pluribus unum* problem yields another good reason to accept logical cognitivism. Finally, then, the fact that the logic faculty thesis coherently resolves *both* the psychologism and *e pluribus unum* problems constitutes an even stronger case for logical cognitivism.

In sections 2.1 to 2.4 I motivate the *e pluribus unum* problem by surveying some of the main varieties of logic. In section 2.5 I argue for the existence of a superlogic, or protologic, as the best solution of the *e pluribus unum* problem. Then in section 2.6 I put forward the logic faculty thesis as an extension of Chomsky's psycholinguistics and argue that the logic faculty thesis provides a coherent double resolution of both the psychologism and *e pluribus unum* problems.

2.1 Formal Logic and Nonformal Logic

Logic, as the science of the necessary relation of consequence, is one and the same as *formal logic*. This allows us to distinguish between logic in the proper sense—or formal logic—and nonformal logic.

Nonformal logic is the science of arguments not strictly governed by consequence, or, put slightly differently, nonformal logic is the science of arguments that need not be formally valid in order to be acceptable. A useful distinction can then be drawn between (a) nonformal logic that has a significant mathematical component (i.e., inductive logic: the science of probable or statistical inference), and (b) nonformal logic that is largely nonmathematical (i.e., informal logic: the science of speech-context-dependent inference and practical inference). The essential thing about nonformal logic of both types, however, is that they permit nonvalid arguments to count as good arguments. For example, a perfectly good induction can be falsified by some as yet unsurveyed contingent fact; and a perfectly good informal argument can lead from truth to falsity in some contexts. On the other hand, it is also the case that some formally valid arguments are not acceptable in nonformal logic. For example, a circular argument—an argument whose conclusion is to be found among its premises—is inadmissible in most informal contexts, even though it is formally valid.[2]

2.2 Aristotle's Logic and Frege's Logic

Back now to formal logic, or logic in the proper sense. Aristotle discovered formal logic, but he developed it in a special way. His way was based on the following ideas: (1) the triadic (major premise–minor premise–conclusion) argument form of the syllogism; (2) the fundamental grammatical distinction between subjects (individuals or kinds) and monadic or one-place predicates (concepts or properties); (3) the assumption that all true universal claims have existential commitment; and (4) the universality of the (strong) laws of bivalence, excluded middle, and noncontradiction.[3] Aside from some anticipations of truth-functional logic (logic in which the consequence relation is based on the meanings of the logical constants 'or', 'and', 'not', 'if . . . , then . . .', and 'if and only if') by Philo and the Stoics, and of intensional logic (logic in which the consequence relation is based on the form or content of the concepts or properties expressed by predicates) by Leibniz and Kant, the edifice of Aristotelian formal logic—effectively maintained with only minor renovations by the Scholastic logicians—remained firmly in place until the mid-nineteenth century. In the 1780s, Kant quite correctly claimed that logic had not advanced in any of its essentials since Aristotle.

From a modern standpoint, however, Kant's remark was to say the least an unhappy one: all of the special features of Aristotle's logic I just mentioned have been in one way or another rejected by nineteenth- and twentieth-century logicians, for whom Frege's logic is the historical paradigm.[4] Frege's logic recognizes many different kinds of argument form beyond that of the Aristotelian syllogism, especially those found in Boolean truth-functional logic. Frege's logic also subsumes the simple subject–monadic predicate distinction under the more comprehensive distinction between objects and functions (which include (i) Fregean concepts, i.e., functions from objects to truth-values; (ii) the concepts designated by relational or polyadic predicates, i.e., functions from ordered *n*-tuples of objects to truth-values; and (iii) second-order functions) and adds to it a theory of multiple quantification (iterated and embedded uses of 'all' and 'some') based on the crucial notion of a variable. Finally, Frege's logic construes the "A" or *Affirmo* sentences in the Aristotelian–Scholastic Square of Opposition (sentences of the form 'All As are Bs') as universal quantifications over the material conditional and therefore as lacking existential commitment.

On the other hand, however, Aristotle and Frege share a deep[5] adherence to the (strong) laws of bivalence, excluded middle, and noncontradiction. This sharply separates them from most contemporary logicians. As we will see shortly, one of the fundamental ways in which post-1950s logical theory has moved beyond the Fregean paradigm is by countenancing logics that incorporate rejections of one or more of these laws.

2.3 Mathematical Logic and Symbolic Logic

Traditional logic is Aristotelian–Scholastic logic. But modern logic is, first and foremost, mathematical logic. It was pioneered by Boole's *Laws of Thought* (1854), but above all by Frege's *Begriffsschrift* (1879) and *Basic Laws of Arithmetic* (1893–1903), and it reached full maturity with Whitehead and Russell's *Principia Mathematica* (1910–1918). There are three main ideas behind it: (1) the idea of a formal logic, (2) the idea of an algebra or calculus of logic, and (3) the idea of a logic of mathematics.

(1) *Mathematical logic as a formal logic*   A mathematical logic, like every other sort of formal logic, is a science of the necessary relation of consequence. (2) *Mathematical logic as an algebra or calculus of logic*   Mathematical notions can be expressed in the medium of an algebra or calculus. This involves two factors. First, an algebra or calculus is an *artificial* language: its syntax or grammar, which determines what counts as a well-formed formula or "wff," is created by explicit stipulative definition or fiat; and every sign in the language is arbitrarily assigned an unambiguous meaning (whether a semantic value, or a semantic function mapping values to values). Second, an

algebra or calculus is an *operational* language: the symbols of the language are systematically transformable or interreplaceable according to strict rules, while still preserving their original interpretations. To say that mathematical logic expresses logical notions in the medium of an algebra or calculus is to say, more specifically, that mathematical logicians abstract, construct, operate, and evaluate. They abstract away from features of natural language not directly relevant to logical consequence, logical truth, and logical proof. They then construct artificial languages that encode consequence, truth, and proof. They then operate their languages by doing proofs. Then, finally, they evaluate their languages by metalogically proving various results—completeness, soundness, consistency, compactness, and so on—for them. In other words, a mathematical logic is a *logical system*.

(3) *Mathematical logic as a logic of mathematics*   A mathematical logic is a logic that is *about* whatever it is that mathematics is about. This can mean one of two things: either some of its terms refer directly to prima facie mathematical objects (e.g., numbers or geometric figures), concepts or properties (e.g., equality or congruence), or operations (e.g., addition or geometric transformation); or else those prima facie mathematical objects, concepts, and operations are systematically reduced to purely logical correlates (e.g., sets, open sentences, quantified sentences, and functions). In this way a mathematical logic is necessarily committed to a mathematical *ontology*, whether that commitment be reductive or nonreductive.

The third idea I just described makes it possible to draw a distinction between a "symbolic logic" and a mathematical logic, as C. I. Lewis very usefully does in his *Survey of Symbolic Logic:*

Symbolic logic is the development of the most general principles of rational procedure, in ideographic symbols, and in a form which exhibits the connection of these principles with one another. Principles which belong exclusively to one type of rational procedure—e.g., to dealing with number and quantity—are hereby excluded, and generality is designated as one of the marks of symbolic logic.[6]

Lewis's point is that symbolic logic is even more formal, topic-neutral, or ontologically uncommitted than mathematical logic. Both kinds of logic are sciences of the necessary relation of consequence, and both involve an algebra or calculus of logic. Hence both symbolic logic and mathematical logic deal essentially with logical systems. But whereas a symbolic logic is about consequence, truth, and proof, without any *further* restrictions as to subject matter, a mathematical logic is also specifically committed to the existence of a world filled with interesting bits and pieces of prima facie mathematical, or reductively logico-mathematical, furniture.

A symbolic logic, as we have just seen, is an algebra or calculus that is essentially about the topic-neutral or ontologically uncommitted notions of consequence, truth, and proof. But there are also two further ways in which it is sharply different from other algebras and calculi.

First and foremost, the very idea of a symbolic logic invokes a fundamental distinction between *signs* and *symbols*. A symbolic logic is not merely a rule-governed collection of "dead" signs (whether types or tokens) or uninterpreted linguistic counters.[7] On the contrary, it is a system of *"living" signs* or symbols, by which I mean interpreted or meaningful signs, that is, signs that are either used or usable by rational animals. As Wittgenstein crisply puts it in the *Tractatus*:

In order to recognize the symbol [*Symbol*] in the sign [*Zeichen*] we must consider the significant use.[8]

Second, as opposed to other sorts of algebras or calculi, a symbolic logic involves a unique kind of "sign design," or syntactic architecture. More precisely, as Frege noted, a symbolic logic is a *Begriffsschrift* or "conceptual notation." A symbolic logic uses ideographic or pictorial signs to encode within its syntax the greatest possible amount of information concerning the fundamentally different sorts of semantic values of its nonlogical or logical constants. Furthermore, its signs are designed in such a way as to encode within its syntax the most lucid or cognitively effective methods for displaying logical truth and logical proof. These essentially iconic features of symbolic logic are often overlooked and sometimes even explicitly rejected.[9] But they are nicely highlighted by Lewis:

[T]he really distinguishing mark of symbolic logic is the approximation to a certain *form*, regarded as ideal. . . . The important characteristics of this form are: (1) the use of ideograms instead of the phonograms of ordinary language; (2) the deductive method—which may be here taken to mean simply that the greater portion of the subject matter is derived from a relatively few principles by operations which are "exact"; and (3) the use of variables having a definite range of significance. Ideograms have two important advantages over phonograms. In the first place, they are more compact, + than 'plus', 3 than 'three', etc. This is no inconsiderable gain, since it makes possible the presentation of a formula in small enough compass, so that the eye may apprehend it at a glance and the image of it (in visual or other terms) may be retained for reference with minimum effort. None but a very thoughtless person, or one without experience of the sciences, can fail to understand the enormous advantage of such brevity. In the second place, an ideographic notation is superior to any other in precision. Many ideas which are quite simply expressed in mathematical symbols can only with the greatest difficulty be rendered in ordinary language. Without ideograms, even arithmetic would be difficult, and higher branches impossible.[10]

And unsurprisingly, given his crucial distinction between signs and symbols, these basic points were also adopted as constitutive features of symbolic logic by the Tractarian Wittgenstein:

5.4731 Self-evidence, of which Russell has said so much, can only be discarded in logic by language itself preventing every logical mistake. That logic is *a priori* consists in the fact that we *cannot* think illogically.

6.113 It is the characteristic mark of logical propositions that one can cognize [*erkennen*] in the symbol alone that they are true; and this fact contains in itself the whole philosophy of logic.[11]

I shall return in chapter 6 to Lewis's deep idea that the ideographic compactness and precision of a symbolic logic is closely connected with our cognitive capacity for "apprehending" and "retaining" mental images, and also to Wittgenstein's even deeper idea that a properly sign-designed logical symbolism is itself the very medium of our a priori knowledge of logical truths and logical proofs. Right now, the crucial point is that a symbolic logic is not only a "living" or fully interpreted symbolic algebra or a calculus of consequence, truth, and proof, as opposed to a mere game played with dead or uninterpreted signs; it is also a lucid ideograph of its unique subject matter and an effective cognitive vehicle of a priori knowledge. These essentially iconic features of sign architecture above all make a symbolic logic an "ideal language."

2.4 Tarski's Reconstruction: From Logicism to Elementary Logic

Following C. I. Lewis, I have teased apart the notions of mathematical logic and symbolic logic. But historically speaking, those notions were almost seamlessly amalgamated in the system of *classical* or *elementary logic*: that is, bivalent truth-functional first-order quantified polyadic logic plus identity. This highly familiar and apparently eternal logic is, in fact, nothing but the carefully reconstructed version of the logic of *Principia Mathematica* that is to be found in Tarski's semantic writings on logic:

The terms 'logic' and 'logical' are used [by most logicians] in a broad sense, which has become almost traditional in the last decades; logic is here assumed to comprehend the whole theory of classes and relations (i.e., the mathematical theory of sets). *For many different reasons* I am personally inclined to use the term 'logic' in a much narrower sense, so as to apply it only to what is sometimes called 'elementary logic', i.e., to the sentential calculus and the (restricted) predicate calculus.[12]

Why is Tarski being so cautious here? And what are his "many different reasons" for adopting a much narrower conception of logic than most logicians of his day? The answers lie in the rise and fall of the project of logicism.

Logicism says that all of mathematics is explanatorily reducible to logic. And mathematical logic was in fact created expressly to meet the stringently reductive demands of logicism. Frege, Russell, Wittgenstein, and Carnap all defended one or another version of logicism. But since Frege believed that arithmetic is reducible to logic but that geometry is *not* so reducible, we can take the crux of logicism to be whether or not arithmetic can be explanatorily reduced to logic. Focusing, then, on the arithmetic-oriented conception of logicism, we find five basic constraints or conditions of adequacy on the successful achievement of its program: (1) that the (e.g., Dedekind or Peano) axioms of arithmetic be expressible as "logical definitions," or purely logical principles supplementary to the laws of symbolic logic; (2) that the concept *Number* be expressible in strictly logical terms without appreciable loss of meaning or cognitive significance; (3) that numbers, as entities, be systematically constructible as logical entities—paradigmatically, as sets or classes of equinumerous (one-to-one correlated), classes; (4) that the symbolic system or ideal language representing mathematics be wholly free of contradictions; and (5) that all the logically true or valid sentences (tautologies) of the ideal language be logically provable sentences (theorems) of that language and conversely.

The first three conditions are individually necessary and jointly sufficient for the explanatory reduction of arithmetic to logic. The last two conditions are also individually necessary, but lie implicit in the first two conditions. The fourth condition guarantees the consistency of the logical system that expresses arithmetic. And the fifth condition guarantees its completeness and soundness. Unfortunately for the logicists, however, four problems got directly in the way of meeting these conditions of adequacy: (A) the problem of analyticity, (B) the paradox of classes; (C) the Liar paradox, and (D) Gödel's incompleteness theorems. Before getting back to Tarski, we should take a very quick look at these troublemakers.

(A)   **The problem of analyticity**   If the concept *Number* is expressible in strictly logical terms, then all the truths of arithmetic will come out as logical truths. In turn, Frege held that all the logical truths, including of course those that express arithmetical truths, are "analytic" in a sense that is supposedly continuous with Kant's classical doctrine of analyticity.[13] But neither Frege, nor Russell, nor Wittgenstein, nor Carnap was able to show that all logical truths are analytic. Frege's own definition of an analytic truth as a truth that is deducible solely from the laws of logic plus logical definitions rests on an essentially unclear notion of a "logical definition."[14] Early Russell, by contrast, held the view (from 1900 to 1918, except for a brief period in 1905) that both logical and mathematical truths are synthetic, *not* analytic.[15] Wittgenstein's Tractarian conception of an analytic truth as a tautology, or a truth guaranteed vacuously by the use of correct logical symbolism alone, holds only for the truth-functional part of logic.[16] And finally, Carnap's conception of an analytic truth as a sentence guaranteed by either conventional syntactical rules alone,[17] or conventional semantical rules alone,[18] was shown by Quine to be viciously circular in two ways. On the one hand, conventional syntactical rules for defining logical truth cannot be implemented without using logic;[19] and on the other hand, conventional semantic rules—which in particular are supposed to explain nonlogical or synonymy-based analyticity—covertly presuppose synonymy.[20] More generally, all versions of the thesis that arithmetic is explanatorily reducible to logic remain questionable as long as no good theory of analyticity is available.

(B)   **The paradox of classes**   In 1901 Russell discovered the paradox of classes or sets: the class of all classes not members of themselves (call it "the Russell Class") is a member of itself if and only if it is not a member of itself.[21] This seriously hampers satisfaction of the second condition of adequacy, since it casts doubt on the very idea of a class or set by showing the antinomous character of an unrestricted principle of class formation. In 1903 and 1908 Russell proposed *the theory of types* as the solution to the paradox.[22] In 1905 and 1906, however, he briefly opted for the *no-class* or *substitution* theory instead. And from time to time he also brought in *the theory of descriptions* and *the vicious-circle principle* for extra help. But it was all to no avail. The theory of types organizes predicates into a hierarchy of levels such that an otherwise paradox-forming predicate occurs, harmlessly, one level higher than whatever it properly applies to. But the theory of types is not only inordinately complicated but also committed to several unproven assumptions, including *the axiom of infinity* (that the world contains some infinite collections) and *the axiom of reducibility* (that "there is a type (*r* say) of *a*-functions such that, given any *a*-function, it is formally equivalent to some function of the type in question"[23]). According to the no-class theory, there really are no classes, and hence no Russell Class, since classes can be systematically reduced to other things, more specifically, to propositional functions. But obviously there is no airtight guarantee that there cannot be an analogue of the paradox of classes for propositional functions too.[24] The theory of descriptions systematically translates all singular terms, including those that stand for paradoxical items, into syncategorematic parts of general propositions;[25] but the thesis that all ordinary proper names are disguised definite descriptions has been effectively undermined by *the direct reference theory*.[26] And finally, the vicious-circle principle bans the construction of collections by means of self-reference. But this principle implausibly rules out *all* self-reference, even seemingly benign forms of it. Later theorists have avoided Russell's intellectual travails and indefinitely put off worries about the idea of a class or set by simply opting for some or another suitably restricted principle of class- or set-formation.[27]

(C)   **The Liar paradox**   Even if one or more of Russell's solutions for the paradox of classes could be made workable, and even if a suitable restriction on principles of class-formation is the cure-all for worries about the foundations of set theory, there is still the Liar paradox. Liar sentences say of themselves that they are false; so they are true if and only if they are false. This paradox turns out not to be limited to truth-predicates, however, but can also be regenerated in slightly different versions for a seemingly indefinitely large number of syntactic and semantic predicates.[28] Hence it appears that any logic that permits syntactic or semantic predicates to occur within its symbolism—for example, the system of *Principia Mathematica*—is inconsistent.

(D)   **Gödel's incompleteness theorems**   Worst of all for logicism, Kurt Gödel demonstrated in 1931 that completeness is impossible for logical systems of the *Principia*-type, and many other systems too—in fact for any logical system rich enough to include Peano's axioms of arithmetic. Any such system is consistent only if it is incomplete (the first incompleteness theorem); and in any such system, there are true but unprovable—also undecidable—

sentences (the second incompleteness theorem).[29] On the assumption of completeness, the unprovability of a sentence yields its falsity. But by hypothesis the unprovable and undecidable sentence is true in the system. So it follows that all such systems are consistent if and only if they are incomplete.

By the mid-1930s, these results taken together as a package seemingly provided a sufficient reason for abandoning logicism: the Great Expectations of logicism were nothing but a Grand Illusion. To be sure, important late twentieth-century work by "neo-Fregeans"—especially Bob Hale and Crispin Wright[30]—has shown that in fact much of the original logicist project can be carried out without resorting to the set-theoretic reduction of numbers. So logicism is still a serious going concern, even at the beginning of the twenty-first century. Nevertheless, the gradual breakup of the first wave of logicism produced some remarkable logical results. More precisely, Tarski managed to solve the inconsistency problem and to accommodate Gödel's result in one fell swoop.[31] In turn, that one fell swoop contains two subswoops.

First, since the Liar requires that the predicates 'true' and 'false' (and more generally, all syntactic and semantic predicates) in a given language L be directly applied to the sentences in L that contain those very predicates, Tarski argued that truth cannot be defined in a language that contains its own truth-predicate. Instead, truth must always be defined for a given language $L1$ (the "object language"), which contains no truth-predicate of its own, in a distinct language $L2$ (the "metalanguage") that is used to talk about $L1$ and thereby contains the truth-predicate for $L1$. This leads to the general idea of a well-ordered hierarchy of metalanguages and object languages, bottoming out in some object language that is not itself a metalanguage, such that each metalanguage contains the syntactic or semantic predicates for the next-lowest language in the hierarchy. Enriched by the Tarskian hierarchy of languages, *Principia*-style logic is consistent. By Gödel's theorems, however, it follows that it is also incomplete. But that result can now be swallowed without choking.

This is because, second, according to Tarski, the concept of truth itself is a semantic notion, not a syntactic notion: truth consists essentially in the fit or lack of fit between what a sentence says and what actually or indeed is. The informal notion of "what actually or indeed is," in turn, is formally analyzable as a special configuration, or structure, of the objects in the domain of discourse such that this structure satisfies, or makes true, the relevant sentence.

In short, this structured set of objects is a *model* of that sentence, and the same general idea of a model can be extended to arbitrarily constructed sets of sentences and to whole theories. So the concept of truth is, in a precise and systematic way, logically independent of the concept of proof.

This brings us back to Tarski and elementary logic. He finesses the problem of the paradox of classes by restricting logic to quantification over individuals. He solves the inconsistency problem by means of his strict hierarchy of object languages and metalanguages. He accommodates Gödel's incompleteness theorems by adopting a semantic conception of truth. And finally, in response to some deep worries about how to isolate and define the logical constants, he also rejects the analytic–synthetic distinction.[32] Logicism went dormant for fifty years. But thanks to Tarski's brilliant reconstruction, elementary logic just kept on going and going and going, like a classic automobile, essentially intact and in mint condition after millions of miles on the road. Indeed it is the logic currently taught in every introductory- and intermediate-level logic course, in every university's or college's philosophy department or mathematics department, in every country in the world.

2.5   From Nonclassical Logic to the Protologic

Since at least the time of Lewis's *Survey*, it has been known that the "second-order" or "higher-order" logic of *Principia* (so called because it permits quantification into predicates and over properties, functions, and sets) can be modified in various ways.[33] This remains true, and true in spades, for elementary logic. In her pathbreaking 1974 book *Deviant Logic*, Susan Haack helpfully collected these modifications together under the comprehensive heading of "alternative" or "nonclassical" logics.[34] And she also very helpfully distinguished, within nonclassical logics, between (1) extensions of classical or elementary logic and (2) deviant logics.

Briefly put, extensions of elementary logic introduce nontrivial changes (changes other than mere notational variation) that preserve all the logical constants, valid sentences, theorems, valid inferences, and laws of elementary logic. By contrast, deviants of elementary logic introduce nontrivial changes that do *not* preserve all the classical or elementary logical constants, valid sentences, theorems, valid inferences, and laws.[35]

Less briefly put, an extension of classical or elementary logic involves the addition, deletion, or redefinition of classical logical constants, interpretation rules, axioms, or inference rules such that all the tautologies, theorems, valid inferences, and laws of elementary logic still hold, along with some additional ones. For instance, the "calculus of strict implication" devised by C. I. Lewis,[36] and the various axiomatic modal logics developed by Lewis,[37] Kripke,[38] and others, are extensions in this sense.

By contrast, a deviant logic involves the addition, deletion, or redefinition of classical logical operators, interpretation rules, axioms, or inference rules such that *not* all the tautologies, theorems, valid inferences, and laws of classical or elementary logic still hold. Deviant logics include intuitionist logics, which reject the classical or strong law of excluded middle;[39] relevance logics, which reject the classical conditional or classical validity;[40] three-valued, many-valued, truth-value-gapped, and fuzzy logics, which all reject the classical or strong law of bivalence;[41] "free" and "Meinongian" logics, which reject unrestricted existential generalization;[42] paraconsistent logics, which allow contradictions to occur as theorems without entailing every sentence; and dialetheic logics, which are both paraconsistent *and* reject the classical or strong laws of noncontradiction and bivalence by allowing some truth-value "gluts" or true contradictions.[43] Extensions of elementary logic, while promoting some genuine changes to classical logic, are conservative. But a deviant logic is radical and *flouts* classical or elementary logic in one way or another.

This plethora of logics vividly raises the question: given the various non-classical logics and especially the deviant logics, what, then, should we say about the nature of logic?

Most philosophers of logic hold that there are three and only three mutually exclusive options for answering this question. The first option is to be a *diehard classicist* and insist that classical or elementary logic is the One True Logic and that the nonclassical logics are not "really and truly" logic. The second option is to be a *diehard nonclassicist* and insist that one or another of the nonclassical logics is the One True Logic and that all other logics including classical or elementary logic are not "really and truly" logic. And the third option is to be an *unconstrained pluralist* by (a) denying that there is any One True Logic, whether classical or nonclassical, and (b) asserting that all the logics are equally acceptable.[44]

Perhaps the strongest argument in favor of the diehard classicist option is Quine's thesis that all linguistically competent human speakers find the logical constants, logical truths, and proof procedures of classical or elementary logic to be "obvious, actually or potentially,"[45] because they are intrinsic to the very practice of translation:

Take the . . . case of trying to construe some unknown language on the strength of observable behavior. If a native is prepared to assent to some compound sentence but not to a constituent, this is a reason not to construe the construction as conjunction. If a native is prepared to assent to a constituent but not the compound, this is a reason not to construe the construction as alternation. We impute our orthodox logic to him, or impose it on him, by translating his language to suit. We build the logic into our manual of translation. Nor is there cause here for apology. We have to base translation on some kind of evidence, and what better?[46]

It follows directly from this that if someone attempts to deny some fundamental law of elementary logic (say, the classical or strong law of noncontradiction), then she cannot be intelligibly regarded as offering a genuine challenge to classical logic. If classical or elementary logic is built into our manual of translation, then the challenger must instead be regarded by us as meaning something different by the logical words she uses than we do by the very same words:

What if someone were to reject the law of non-contradiction and so accept an occasional sentence and its negation both as true? An answer one hears is that this would vitiate all science. Any conjunction of the form '$p . \sim p$' logically implies every sentence whatever; therefore acceptance of one sentence and its negation as true would commit us to accepting every sentence as true, and thus forfeiting all distinction between true and false. In answer to this answer, one hears that such a full-width trivialization could perhaps be staved off by making compensatory adjustments to block this indiscriminate deducibility of all sentences from an inconsistency. Perhaps, it is suggested, we can so rig our new logic that it will isolate its contradictions and contain them. My view of this dialogue is that neither party knows what he is talking about. They think they are talking about negation, '$\sim$', 'not'; but surely the notation ceased to be recognizable as negation when they took to regarding some conjunctions of the form '$p . \sim p$' as true, and stopped regarding such sentences as implying all others. Here, evidently, is the deviant logician's predicament: when he tries to deny the doctrine he only changes the subject.[47]

Against Quine's diehard classicism, however, it seems to me that it cannot be seriously denied that both extended and deviant logic (even such highly deviant systems as dialetheic logic) are really and truly logic. This is because we must accept, I think, that any nonclassical logic (NCL, for short) is really and truly logic just in case the following three conditions are satisfied:

(I) *The formal logic condition*   The NCL is a science of the necessary relation of consequence.

(II) *The representational adequacy condition*   The NCL's proposed extension of, or deviation from, classical logic is based on its being able accurately to represent in the format of symbolic logic some apparent linguistic facts that are not represented within classical logic: for example, strict implication or modality, constructibility of proofs, relevance, vagueness, future contingency, nonexistent objects, paradoxes, and so forth.

(III) *The localization of application condition*   The NCL's scope of application is restricted to all and only those language domains containing the apparent nonclassical linguistic facts that it represents.

Conditions (I) to (III) say, truistically, that nonclassical logics are logics that apply to the nonclassical linguistic domains they adequately represent. So nonclassical logic is really and truly logic, and diehard classicism is wrong. But diehard nonclassicism has the same problem in the reverse direction. It cannot be denied that classical or elementary logic is really and truly logic, since the three conditions of formal logic, representational adequacy, and localization of application are all trivially satisfied by elementary logic for apparent classical logical linguistic facts.

Are we then forced into the all-embracing arms of the unconstrained pluralist option? No. It seems to me that Quine overstates his case in favor of diehard classicism, and that a slightly weaker version of Quine's argument in fact strongly supports a distinct fourth option that is neither diehard classicism, nor diehard nonclassicism, nor unconstrained pluralism. This fourth option asserts that *neither* is it true that there is One True Logic, whether elementary logic or nonclassical logic, *nor* is it true that the several different logics are all equally acceptable.

Suppose, then, that we accept that *some one single logic* is "obvious, actually or potentially" because it is (among other things) built into the very practice of translation, but also that this logic is itself neither strictly speaking classical or elementary logic nor strictly speaking nonclassical logic (whether extended or deviant), because it is structurally distinct from any classical or nonclassical logical system. Indeed, it is not a logical *system* as such, but rather a single set of *schematic logical structures*, in the form of a coherent repertoire of metalogical principles and logical concepts. Furthermore, it is presupposed by every logical system whatsoever.[48] This is because it is a *protologic*, in the sense that it is used for the construction of every actual or possible logical system. Such a universally presupposed constructive logic would be *somewhat* like Quine's "sheer logic," but with his diehard classicism subtracted out. More precisely, it would be *quite* like Shapiro's "super-logic," the logic that we use for the internal analysis and cross-logic evaluation of the plurality of logics, which is also unrevisable and a priori. The protologic, as I am conceiving it, is unrevisable and a priori precisely because its total set of schematic logical structures determines what will *count* as a possible logical system, and because some knowledge of this set of structures must also be consciously *available* to thinkers if they are to be able to justify assertions or claims made about any classical or nonclassical logic. So the protologic is both constructively and epistemically presupposed by every logical system.

This fundamental point requires a bit more elaboration. To say that some single universal set of schematic structures is constructively presupposed by every member of a plurality of formal systems of some definite sort (say, mathematical systems or linguistic systems) does not in and of itself guarantee that this set of schematic structures is unrevisable and a priori. Take, for example, the crucially analogous case of natural language. The mere fact that, say, Chomsky's universal grammar is constructively presupposed by every natural language does not in and of itself guarantee that the universal grammar is consciously available to competent speakers of a natural language, whether in the form of noninferential beliefs (also called "intuitions" if they are either prima facie compelling or intrinsically compelling) or inferential beliefs. But the case of logic is crucially different. This is because competent thinkers must be able to (try to) *justify* the assertions they make, and logical principles and concepts are always more or less consciously explicit in the connections between reasons for believing and the beliefs based on those reasons. What I mean is that the 'because' in 'I believe that Q, because of the fact that P (or: because it seems to me very likely that P, etc.)' is necessarily a *logical* 'because': it says that something (i.e., that Q) *logically follows* from something else (i.e., that P). So all justification is, to some extent, conscious logical justification. And this obviously also covers the specific case in which justification is concerned with conscious beliefs *about* logic. To justify any assertion is to invoke conscious logical beliefs; so to justify any assertion about logic is to invoke conscious logical beliefs about logic; but nothing will count as a logic unless it presupposes the protologic; therefore in order to justify any assertions about any logic, we must invoke conscious logical beliefs about the protologic. Unrevisability is guaranteed by the special nature of the dual presupposition relation between the protologic and the many logics. (For more on this idea, see chapter 3.) In this way the protologic is unrevisable and a priori because it is not only constructively but also epistemically presupposed by every logic.

If this proposal is correct, then the unconstrained pluralist option is also ruled out, because the single universally (constructively and epistemically) presupposed protologic is *not* on a par with any of the many logical systems. On the contrary, the single universally presupposed protologic is—to use Kantian language for a moment—*the condition of the possibility of there being informative natural language discourse (including of course translation) and reasoning (including of course theories) in the first place.* (This idea will also be further developed in chapters 3 to 5.)

Now, of course, the $64,000 question becomes: what is the precise structural description of the protologic? This, however, is a question that cannot be answered within the scope of this book. I will allow myself to admit here that it *seems* to me that the following four metalogical principles, together with the logical concepts implicit in them, are at least good *candidates* for belonging to the protologic:

(i) *The weak principle of validity*   An argument is valid if it is impossible for all of its premises to be true and its conclusion false.
(ii) *The weak principle of noncontradiction*   Not every sentence is both true and false.[49]
(iii) *The weak principle of logical truth*   A sentence is logically true if it comes out true under every possible uniform reinterpretation of its nonlogical constants.
(iv) *The weak principle of the transfer from logical truth to valid proof*   A proof from a set of premises to a conclusion is valid if the corresponding classical conditional of its underlying argument is logically true.

Each of these principles is a weak version of a basic principle of classical logic. My rationale for tentatively proposing the inclusion of these weak classical metalogical principles in the protologic is the (to me) very plausible thought that although every extended or deviant variant on classical logic adds something to or subtracts something from classical logic, *no* logical system can reject absolutely *all* of classical logic and still remain a logic. This in turn would suggest that the protologic, among other things, captures a minimal classical "core" that is preserved in every classical or nonclassical system. So the protologic would also to that extent capture the *core* of classical logic. But that is as far as I am willing to go.

This is not an evasion: it is simply the scientific division of labor, and a little theoretical modesty too. My intention in this chapter has been to argue that a protologic must exist, and to spell out the basic features that the protologic will intrinsically possess: it is a single universal set of schematic logical structures, in the form of a repertoire of metalogical principles and logical concepts, that possesses an overall structure distinct from every classical or nonclassical logical system, and is used for the construction of every classical or nonclassical system; and it is unrevisable and a priori precisely because it is both constructively and epistemically presupposed by every logical system. But this ambitious claim having been (let us assume for a moment!) accepted, then there must then be a scientific division of labor. This is because the question of the precise structural description of the protologic is an intrinsically *formal-logical* question, not a *philosophical* question, and this book—to the extent that it is about logic, as one of its two basic topics—deals only with philosophical logic and the philosophy of logic, and is *not* a treatise in formal logical theory. Furthermore, I am not a professional logician, so I am not competent to undertake this further investigation, even if I wanted to. It is also the case that if what I argue in the next section is sound, then the question of the precise structural description of the protologic is in part an irreducibly *empirical* question, in that it will depend on some factual results in the psycholinguistic part of cognitive psychology. These somewhat negative points should be regarded as carrying a definitely positive scientific upshot, however: if I am correct, the $64,000 question of the precise structural description of the protologic constitutes a new and important joint research program for logicians and cognitive psychologists.

2.6 Chomsky and Me: How to Prove the Logic Faculty Thesis

Let me suppose for the purposes of argument that the existence of the protologic has been successfully established and the *e pluribus unum* problem solved. I now want to propose that the protologic is innately contained in a cognitive faculty for logical representation, namely, the logic faculty.

As I have already mentioned several times, the logic faculty thesis is the first of two basic parts of logical cognitivism. So, to demonstrate the logic faculty thesis is to demonstrate half of logical cognitivism. That's the good news. But at the same time, the logic faculty thesis is an ambitious and controversial doctrine that is not likely to be demonstrated decisively by any single line of argument. That's the bad news. Nevertheless, even the bad news is consistent with the possibility that the logical faculty thesis be demonstrable by the combined force of several different but interlocking lines of argument. And that's *not* bad news. So my general strategy will be to develop a *five-pronged cumulative argument* for the logic faculty thesis. In this section I am going to argue for the logic faculty thesis, first, by an extension of Chomsky's psycholinguistics (first prong), and second, by an inference to the best explanation, where this best explanation is that the logic faculty thesis, and apparently the logic faculty thesis alone, provides a coherent double resolution of the psychologism and the *e pluribus unum* problems (second prong). In chapter 3, I will extend this inference-to-the-best-explanation argument by arguing that the logic faculty thesis, and apparently the logic faculty thesis alone, provides a coherent *triple* resolution of the psychologism problem, the *e pluribus unum* problem, and the logocentric predicament (third prong). In chapter 4, I will supplement the argument for the logic faculty thesis from Chomsky's psycholinguistics, with an argument from the language-of-thought (LOT) theory of human cognition (fourth prong). And in chapter 5, I will further supplement the arguments from Chomsky's psycholinguistics and the LOT theory by adding an empirical argument from contemporary cognitive psychology (fifth prong).

Now for the first prong. According to Chomsky's psycholinguistics, whether in its early "transformational" version or in its later and canonical "principles and parameters" version,[50] what is called a "phrase-structure" grammar (a grammar that decomposes all sentences into immediate constituents such as noun phrases, verb phrases, adjectival phrases, prepositional phrases, etc., according to rules) applies to each natural language.[51] The phrase-structure grammar of each natural language differs significantly from every other. If, by analogy with the divisions of logic that I sketched in section 2.5, we think of the phrase-structure grammar of our own language as "classical" or "elementary," then necessarily every other natural-language grammar is either an "extension" of it (if it preserves the same set of phrase-structure rules as our language but adds more—as in idiolects and dialects) or else a "deviant" of it (if it lacks or modifies some of the phrase-structure rules of our language—as in foreign languages). Yet Chomsky thinks that all of these as it were classical or elementary, extended, and deviant phrase-structure grammars presuppose a single set of underlying principles and concepts. This set of principles and concepts constitutes a generative[52] universal grammar (UG)[53] that is realized in different ways in all particular natural language grammars, by setting certain formal parameters differently for each natural language.

Let me now apply this Chomskyan idea to the protologic. I have argued that the protologic consists in a single set of schematic logical structures, in the form of a coherent repertoire of metalogical principles and logical concepts, that is constructively presupposed by every logical system whatsoever, and is unrevisable and a priori (or epistemically presupposed). My proposal is that *the protologic stands to the many classical or elementary, extended, and deviant logics, precisely as Chomsky's UG stands to the many native, idiolectic or dialectic, and foreign natural languages.* It needs to be particularly reemphasized that the protologic is *not* to be identified with classical or elementary logic, or indeed with any nonclassical logic, just as the UG is *not* to be identified with the grammar of one's native language, or indeed with the grammar of any other language. Therefore just as the UG is a supergrammar or sheer grammar that is nonempirically implicit in the construction, internal analysis, and cross-language comparison (which may be either consciously or nonconsciously available to speakers) of all natural languages, so too the protologic is a super-logic or sheer logic that is not merely nonempirically but also unrevisably implicit in the construction, internal analysis, and cross-logic comparison (which must be consciously as well as nonconsciously available to thinkers) of all classical or nonclassical logical systems.

According to Chomsky's view, cognitive mastery or knowledge of a natural language is a person's "linguistic competence," which expresses her possession of an innate, universally shared "language faculty" within her total cognitive capacity:

[O]ne of the faculties of the mind, common to the species, is a faculty of language that serves the two basic functions of rationalist theory: it provides a sensory system for the preliminary analysis of linguistic data, and a schematism that determines, quite narrowly, a certain class of grammars. Each grammar is a theory of a particular language, specifying formal and semantic properties of an infinite array of sentences. These sentences, each with its particular structure, constitute the language generated by the grammar. The languages so generated are those that can be "learned" in the normal way. The language faculty, given appropriate stimulation, will construct a grammar; the person knows the language generated by the constructed grammar. This knowledge can then be used to understand what is heard and to produce discourse as an expression thought within the constraints of the internalized principles, in a manner appropriate to situations as these are conceived by other mental faculties, free of stimulus control.[54]

On Chomsky's view, linguistic competence in this sense is to be sharply distinguished from *performance,* or the actual use of language in concrete contexts.[55] Performance is notoriously variable and partially determined by external, contingent factors. Competence, by contrast, consists in the construction and possession of a comprehensive mental representation of a natural language, which Chomsky also calls an "internalized language" or "I-language." The I-language is constructed by means of the innate "schematism" that is the generative UG:

> Let us define "universal grammar" (UG) as the system of principles, conditions, and rules that are elements or properties of all human languages not merely by accident but by necessity. . . . Thus UG can be taken as expressing "the essence of human language." UG will be invariant among humans. UG will specify what language learning must achieve, if it takes place successfully. . . . What is learned, the cognitive structure attained, must have the properties of UG, though it will have other properties as well, accidental properties. Each human language will conform to UG. If we were to construct a language violating UG, we would find that . . . it would not be learnable under normal conditions of access and exposure to data.[56]

For the purposes of my argument in this section I am quite prepared to assume without further ado that Chomsky's psycholinguistics is on the whole well supported both philosophically and empirically—except in one respect. And that is the scientific naturalism that is sometimes added to Chomsky's theory.[57] Chomsky explicitly holds the thesis that both competence and the UG are in principle explicable in terms of the biology of the human species:

> Linguistic theory, the theory of UG, construed in the manner just outlined, is an innate property of the mind. In principle, we should be able to account for it in terms of human biology.[58]

This *might* mean that the UG in particular, and the language faculty more generally, are (1) intrinsic to the mind, and (2) neurobiologically strictly determined, in the sense of being logically strongly supervenient on human biology. And if that is what it *does* mean, then the second sentence in the quote clearly does not follow from the first sentence.[59] It is conceivable (as a commonplace of science fiction, for example) and therefore possible that some species of creatures made of a very different material or compositional stuff from ours (say, Martians) could instantiate our language faculty. So our language faculty is multiply embodiable. It is also conceivable and therefore possible that some species of creatures possessing the very same biological constitution as ours, in a world with very different laws of nature, could fail to instantiate our language faculty. Such creatures would be "language zombies."

No doubt there are natural laws linking the language faculty with human biology (and possibly this is all that Chomsky means: if so, then please think of the cognitivist philosopher I am criticizing here as Chomsky*). But natural laws, since they are at best physically necessary (true in every logically possible world sharing the same underlying physical-causal architecture found in this world), and not logically or conceptually or metaphysically necessary, cannot determine what is possible or not possible with respect to language mastery per se. Given the conceivable and thus logically possible situations described in the previous paragraph, human biology is therefore neither strictly necessary nor strictly sufficient for linguistic competence, and hence cannot reductively explain linguistic competence. From this it also follows immediately that neither can the Darwinian or evolutionary theory of human biology reductively explain linguistic competence.[60] So I am prepared to accept Chomsky's psycholinguistics with the special proviso that the language faculty is intrinsically variable with respect to the type of animal embodiment: otherwise put, the language faculty is not only multiply embodiable (so linguistic properties are not identical to first-order biophysical properties), but also not logically strongly supervenient on human biology.

In any case, what is most important for my present argument is the fact that Chomsky holds (in effect, if not precisely in name) that what I am calling the protologic is built right into the UG, hence built innately into our innate language faculty. As he puts it:

The logical notions are embedded in our deepest nature, in the very form of our language and thought, which is presumably why we can understand some kinds of logical systems quite readily, whereas others are inaccessible to us without considerable effort . . . if at all.[61]

To avoid confusion, we must distinguish between (i) the most deeply cognitively embedded logical notions being described by Chomsky here, which I am identifying with the innate protologic, and (ii) what Chomsky specifically calls "Logical Form" or LF[62] in the context of the current version of his psycholinguistic theory, the minimalist program.[63] LF is a level of natural-language representation in Chomsky's current model of linguistic competence that (i) combines both the underlying logical form of the sentences of a given natural language and their underlying semantic interpretations and (ii) is related to phonological structure by means of syntactic structure. Still, LF *presupposes* the most deeply cognitively embedded logical notions, or what I am calling the innate protologic. Incidentally, LF is also not the same as what philosophical logicians call "logical form,"[64] although back in the 1970s some philosophers attempted to identify the logician's logical form with what Chomsky then called the "deep structure" of grammatical transformations.[65]

In any case, the Chomskyan idea of "logical notions [that] are embedded in our deepest nature, in the very form of our language and thought," or the idea (to use my twist on it) of the innate protologic, leads directly to the further idea of a "logical competence"[66] that is presupposed by linguistic competence. On this extended Chomskyan picture, every linguistically competent being constructs an internalized logic or "I-logic"[67] for the representation of the "natural logic"[68] of her own natural language, just insofar as she constructs an I-language for the representation of the grammar and semantics of her own natural language. In other words, every linguistically competent being constructs a *logic of thought* just insofar as she constructs a *language of thought*. For later reference, I will call this *the logic of thought thesis*. The crucial point for now is that the innately grounded logic of thought, as well as every other logical system whatsoever, both constructively and epistemically presupposes the protologic, just as any creature's innately grounded I-language, as well as every other language whatsoever, constructively presupposes the UG.

This concludes my initial argument for the logic faculty thesis from Chomsky's psycholinguistics. Again, the nub of the argument is that on the (empirical) assumption that Chomsky's psycholinguistics is true, it entails the truth of the logic faculty thesis. This also means, however, that there is an irreducibly empirical component in the logic faculty thesis, and that the precise structural description of the protologic will depend in part on empirical results in cognitive psychology. So, if I am correct, there is a new and important joint research program for logicians and cognitive psychologists, built on top of Chomskyan psycholinguistics, that consists in working out the precise structural description of the protologic.

Now suppose that every possible rational and linguistically competent animal possesses a logic faculty that is presupposed by her language faculty, and that this logic faculty innately contains the protologic. This provides a coherent double resolution of the psychologism and the *e pluribus unum* problems, as follows. The existence of a single universal protologic which is both constructively and epistemically presupposed by every logical system (hence is both unrevisable and a priori) solves the latter problem, since it accounts both for the theoretical unity of logic and for the multiplicity of logics. The inherence of the protologic in the logic faculty in turn entails that every logicical system whatsoever is cognitively constructed by the very creatures—namely, rational animals, including all rational human animals—that possess this faculty. So logic is *not* mind independent and nonspatiotemporal, that is, it is not platonic or metaphysically alienated from human cognition. But conjoining the multiple embodiability of the logic faculty across different kinds of animals together with the unrevisability and apriority of the protologic, guarantees that logic also is *not* logically strongly supervenient on the natural facts. So logic is neither scientifically naturalizable nor, more specifically, explanatorily reducible to empirical psychology. Therefore, the psychologism and *e pluribus unum* problems alike are both coherently resolved by the logic faculty thesis.

The truth of the logic faculty thesis yields the truth of the first part of logical cognitivism. But the crucial point right now is that apparently no theory of the nature of logic currently on offer, apart from the logic faculty thesis, can effectively handle both the *e pluribus unum* and psychologism problems. On the one hand, in view of the argument I sketched in section 2.5, neither diehard classicism nor diehard nonclassicism is acceptable. And on the other hand, most of the available forms of unconstrained pluralism are scientifically naturalistic or explanatorily reductive:[69] but scientific naturalism about logic is self-refuting by the argument I presented in section 1.4. And even if there are some nonreductive forms of unconstrained pluralism, then obviously, as *pluralistic*, they could not account for the theoretical *unity* of logic. Logic could be, at best, a loose-knit and fully egalitarian family of symbolic-system-constructing practices. This may sound really nice for a group of mutually respectful persons in an economically stable society, but it does not necessarily hold for science. I mean that, other things being equal, surely a theory that comprehensively explains the underlying unity of its subject matter is much better than a theory that does not. I conclude that the logic faculty thesis (and therefore the first half of logical cognitivism) is to this extent well supported by an inference to the best explanation.

"And *would* you mind, as a personal favour, considering what a lot of instruction this colloquy of ours will provide for the Logicians of the Nineteenth Century—*would* you mind adopting a pun that my cousin the Mock-Turtle will then make, and allowing yourself to be re-named *Taught-Us?*"

"As you please!" replied the weary warrior, in the hollow tones of despair, as he buried his face in his hands. "Provided that *you*, for *your* part, will adopt a pun the Mock-Turtle never made, and allow yourself to be re-named *A Kill-Ease!*"

—Lewis Carroll[1]

3.0   Introduction

In the last two chapters we encountered two basic problems about the nature of logic: (1) how to relate the logical and the psychological to one another; and (2) how to reconcile the theoretical unity of logic with the multiplicity of logical systems. I argued that the logic faculty thesis—which says that a single universal unrevisable a priori protologic is innately contained in a multiply embodiable modular constructive cognitive capacity for logical representation—coherently resolves both difficulties. The logic faculty thesis in turn is the first part of logical cognitivism, which says (i) that logic is cognitively constructed by rational animals, and (ii) that rational human animals are essentially logical animals. In this chapter we encounter a third basic problem about the nature of logic. I argue that the logic faculty thesis, and apparently the logic faculty thesis alone, solves this third problem, thereby resolving at once all *three* basic problems about the nature of logic. If my argument is cogent, it constitutes *a very strong case* for the logic faculty thesis and thus a very strong case for the first half of logical cognitivism as well.

Like the other two basic problems about logic, the third problem has a history. In 1895, Lewis Carroll (a.k.a. Charles Dodgson) asserted that the attempt to generate the total list of premises required to validly deduce the conclusion of an argument leads to a vicious regress. Carroll's strategy was resuscitated in 1936 by Quine, who claimed that the attempt to define logical (or analytic) truth on the basis of syntactic metalogical conventions alone is viciously circular in a Carrollian manner, because logic is required to generate the truths from the conventions. Quine himself was responding directly to Carnap's 1934 *Logical Syntax of Language*, which offered conventionalism as a solution to a paradoxical doctrine that had been asserted by Wittgenstein in 1921 in the *Tractatus*. Wittgenstein held that the nature of logic is not expressible via language, which is our sole means of representing the world, precisely because logic is presupposed by language and the world alike: nevertheless, paradoxically, the fact of this inexpressibility is itself linguistically expressible.

But conventionalism survived Quine's attack, in the form of Gerhard Gentzen's 1934 thesis (not explicitly considered by Quine in 1936) that the meanings of logical constants are strictly determined by the arbitrary metalogical adoption of rules of inference for sentences in which those constants occur as constituents. In 1960, A. N. Prior challenged Gentzen's thesis by arguing that the attempt to define the logical constants in terms of their inferential roles by means of metalogical conventions leads to absurdity. This nevertheless leaves open the possibility of explaining or justifying logic by some nonconventional means. In 1973 and again in 1991, Dummett argued that the obvious and indeed sole candidate for the job of justifying deduction is an appeal to metalogical soundness and completeness proofs. But in 1976 and again in 1982, Susan Haack raised a synoptic worry about the very idea of a justification of deduction by arguing (i) that all justification is either nondeductive (e.g., inductive) or deductive, and (ii) that on the one hand a nondeductive justification of deduction is too weak and on the other hand a deductive justification of deduction is circular; therefore, (iii) deduction cannot be justified.

These four worries—about valid deduction, conventionalism for logical truth, defining logical constants in terms of their inferential roles by means of metalogical conventions, and justifying deduction—share more than their worrisomeness, however. Indeed, each is but a different slant on the same underlying problem. That underlying problem is *the logocentric predicament*. In a 1926 review of the second edition of *Principia Mathematica*, Harry Sheffer (who, not altogether coincidentally, was one of Quine's teachers) aphoristically observed that the attempt to formulate the foundations of logic is rendered arduous by a . . . "logo-centric" predicament. In order to give an account of logic, we must presuppose and employ logic.[2]

As I read him, Sheffer is saying that logic is epistemically circular, in the sense that any attempt to explain or justify logic must presuppose and use some or all of the very logical principles and concepts that it aims to explain or justify. It is assumed, I think, by Sheffer and all the other participants in this debate, that epistemic noncircularity is a necessary condition of all legitimate explanations and justifications. If so, then the epistemic circularity of logic entails that logic is both inexplicable and unjustified: the circularity of logic is a *vicious* circularity. Or more starkly put: *logic is groundless.* This radically skeptical result is Carroll's "A Kill-Ease."

In sections 3.1 to 3.4 I look at the four versions of the logocentric predicament and consider some strategies that have been proposed for avoiding it. I argue that all of these avoidance-strategies fail and that there appear to be no others. If so, then it follows that we must take the logocentric predicament to be an intrinsic feature of logic. This thesis is what I call *acknowledging the predicament.* So apparently we must also accept the groundlessness of logic. In section 3.5 I consider six ways of acknowledging the predicament. The first five ways accept the groundlessness of logic, and for various reasons, I reject them. By sharp contrast, the sixth way of acknowledging the predicament, which I endorse, consists in challenging an assumption that is a necessary condition of logic's groundlessness: the assumption that epistemic noncircularity is a necessary condition of all legitimate explanations or justifications. This leads to what I call *the cognitivist solution* to the logocentric predicament. According to the cognitivist solution, logic is not groundless after all, because despite its epistemic circularity it nevertheless has a legitimate explanation and justification in the logic faculty thesis, and thereby also in logical cognitivism. To acknowledge the predicament while rejecting the groundlessness of logic is to affirm logical cognitivism. The logocentric predicament is therefore something we should learn to love.

3.1 Carroll's Tortoise

Carroll's attack on the theory of valid deduction appeared in a highly whimsical three-page essay called "What the Tortoise Said to Achilles." Philosophers, who are *deadly* serious people, have often found Carroll's beyond-the-looking-glass rhetoric somewhat of a barrier to understanding his exact intentions.[3] But in a dewhimsified gloss on the essay, Carroll observed that

> my paradox . . . turns on the fact that, in a Hypothetical, the *truth* of the Protasis, the *truth* of the Apodosis, and the *validity of the sequence*, are 3 distinct propositions. . . . Now suppose I *deny* this . . . sequence to be a valid one? . . . Surely my granting [the conclusion of the argument] must *wait* until I have been made to see the validity of this sequence . . . And so on. I think you will find that it goes on like "the house that Jack built."[4]

In other words, Carroll is saying that if I am to validly deduce the conclusion

(Z) Dubya is a crook

from the premise (= the protasis)

(A) All politicians are crooks

and the premise (= the apodosis)

(B) Dubya is a politician

I must assume the truth of

(C) The sequence 'if all politicians are crooks and if Dubya is a politician, then Dubya is a crook' is valid.

But then in order to move deductively all the way to (Z) from (A) and (B) I must also assume the truth of

(D) The sequence 'if (A) and (B) and (C), then (Z)' is valid.

Furthermore, if I were challenged (say, by the sort of talking tortoise one occasionally meets in logic classes) as to the legitimacy of my assumption of (D), then by the previous reasoning I must reply that I am also assuming the truth of

(E) The sequence 'if (A) and (B) and (C) and (D), then (Z)' is valid.

And so on ad infinitum, adding new intermediate premises without end—hence, paradoxically, *never* being able to validly deduce (Z).

Otherwise and more briefly put, here is Carroll's argument:

(1) Every valid deductive advance from the premises of an argument to its conclusion can be explained only by appeal to a principle of valid inference. (2) That principle of valid inference must therefore itself be included as a true premise in the very same argument.

(3) But now the valid deduction of the original conclusion can be explained only by an appeal to *another* principle of valid inference that includes the original premises plus the original principle of valid inference.

(4) The pattern of reasoning exemplified in (1) through (3) leads directly to a vicious regress according to which no conclusion is ever validly deduced in an argument because the list of premises needed to yield that conclusion is never complete.

It has been pointed out by many philosophers that Carroll's reasoning contains a fatal mistake in step (2).[5] Carroll confuses the appeal to principles of valid deductive inference with asserting a true (and in the best case scenario, a logically true) hypothetical or conditional premise that contains the conjoined premises of the argument in its antecedent and the conclusion in its consequent. But principles of valid deductive inference *for* a proof are not the same as true or logically true conditional premises *in* a proof. On the contrary, principles of valid deduction for a proof are nothing but rules of inference—or what Ryle aptly called "inference-tickets"—for a natural deduction calculus. That is, rules of inference are nothing but generalized permissions to draw conclusions of a certain sort, from sets of premises of a certain sort; hence they are principles for correctly operating or running the machinery of the calculus, thereby systematically transforming appropriately configured certain well-formed formulas (wffs) in it into other wffs. As generalized permissions, inference rules are not even truth-bearers.[6] Moreover, they occur in the metalanguage, not in the object language. So inference rules for a calculus are clearly not the same as true or logically true sentences in a calculus.

When principles of valid deduction are construed as rules of inference, those rules can be shown to be sound and complete by proving metalogically (1) that if a sentence S is provable from a set of sentences Γ by means of the rules, then S is a consequence of Γ, and (2) that if a sentence S is a consequence of a set of sentences Γ, then S is provable from Γ by means of the rules. So, in particular, valid deductions can be legitimately carried out within the system merely by following the (sound) rules. And in this way, assuming that the inference rules of a calculus are sound, the Carrollian vicious regress of new intermediate premises never has a chance to get underway. That is because the sound inference rules for a given calculus are properly expressible only *outside* the calculus and its internal deductive structure. As Timothy Smiley puts it:

It seems to me that the rule strategy has several advantages over its rival [i.e., the suppressed premise strategy employed by Carroll]. [They] spring from the fact that invoking a supporting rule is external to the deductive structure of an argument, while adding a premise is internal to it.[7]

Sound rules of inference stand to the calculus in which the rules are applied as a metalanguage stands to its object language. By virtue of this external logical support—the sound rules being as it were vertically imposed upon a horizontally ordered integral deductive architecture like foundational pillars—there is no risk of confusing what grounds valid deduction with any of the premises internal to a given valid deductive argument.

That seems to me a plausible critique and diagnosis of Carroll's Tortoise paradox. Nevertheless, even if Carroll has failed to see the distinction between sound rules of deductive inference on the one hand and true or logically true conditional premises on the other, it is not at all clear that the philosophical problem he was trying desperately to get "Logicians of the Nineteenth Century" to notice is solved or dissolved *merely* by invoking the post-Hilbertian notion of a metalogical proof. That is, even granting the plausible suggestion that we must avoid confusing externally given and metalogically proven sound rules of inference with internally given true or logically true conditional premises in valid deductions, it is perfectly legitimate to ask a different but nevertheless recognizably Tortoise-style question: by virtue of what logical resources are valid *metalogical* deductions to be explained or justified? If the reply is that an appeal must then be made to sound metalogical rules of inference, which are in turn shown to be sound by *meta*-metalogical soundness proofs, then it looks very much as though Carroll's vicious regress of intermediate premises can be smoothly modeled by a vicious ladder of higher and higher soundness-conferring metalogics.

This in turn strongly suggests that Carroll's paradox is only superficially the expression of a Zeno-like worry about how a logician could ever validly deduce the conclusion of an argument from its original explicit premises without getting bogged down in an infinite number of intermediate implicit premises. At a much deeper level, it is a worry about where the logical buck could *ever* stop in an adequate explanation or justification of valid deduction, at *any* level in the hierarchy of languages. That is, it is a worry about what grounds valid deduction *itself*. If, as the deeper interpretation of Carroll's paradox suggests, the logical buck can never find a stopping-place—if absolutely every attempt to explain or justify valid deduction must presuppose and employ logic—then surely the logocentric predicament must be taken to be an intrinsic feature of logic. I will return to this point in section 3.4.

3.2 Quine's New Tortoise

It is, I think, a truism that history often repeats itself with interesting minor variations due to context: so it would be equally truistic that the history of logic often does the same thing. Hence it is not surprising that Carroll's paradox was later interestingly resuscitated by Quine in a slightly different context. Carroll's large and slow-moving target was the theory of valid deduction as it stood at the end of the nineteenth century. Quine, however, used Carroll's paradox against the smaller and faster target of logical conventionalism, or more specifically, against Carnap's conventionalism in *The Logical Syntax of Language*. Carnap, in turn, was responding to a central problem in Wittgenstein's *Tractatus*.

There, Wittgenstein asserts the strange and troubling thesis that the nature of logic, quintessentially captured in what he calls "logical form," can never be explicitly stated or "said," but only ever implicitly indicated or "shown," because logic is presupposed by both language (which is our basic means of representing the world) and also the world itself:

4.12 Propositions can represent the whole reality, but they cannot represent what they must have in common with reality in order to be able to represent it—the logical form.

To be able to represent the logical form, we should have to be able to put ourselves with the propositions outside logic, that is outside the world.

4.121 Propositions cannot represent the logical form: this mirrors itself in the propositions.

That which mirrors itself in language, language cannot represent.

That which expresses *itself* in language, *we* cannot express by language.

The propositions *show* the logical form of reality.

They exhibit it.

4.1212 What *can* be shown *cannot* be said. . . .

5.6 *The limits of my language* mean the limits of my world.

5.61 Logic fills the world: the limits of the world are also its limits. . . .

6.13 Logic is not a theory but a reflexion of the world.

Logic is transcendental.[8]

Here, what we might call Wittgenstein's "logical transcendentalism" (because he holds that logic is *the condition of the possibility* of both language and the world) leads to the result that the nature of logic cannot be linguistically expressed: yet it remains possible to state the fact of this inexpressibility. But this seems self-contradictory. Or as Russell crisply puts it in his introduction to the first English translation of the *Tractatus*, Wittgenstein is fully committed to the thesis that

everything . . . which is involved in the very idea of the expressiveness of language must remain incapable of being expressed in language, and is, therefore, inexpressible in a perfectly precise sense;

yet

Mr Wittgenstein manages to say a good deal about what cannot be said, thus suggesting to the sceptical reader that possibly there may be a loophole through a hierarchy of languages, or by some other exit.[9]

This leads Russell to a positive proposal:

These difficulties [about the inexpressibility of logic] suggest to my mind some such possibility as this: that every language has, as Mr Wittgenstein says, a structure concerning which, *in the language*, nothing can be said, but that there may be another language dealing with the structure of the first language, and having itself a new structure, and that to this hierarchy of languages there may be no limit.[10]

Russell's proposal for avoiding Wittgenstein's inexpressibility-of-logic problem is in effect a double anticipation of (i) Hilbert's formalist approach to the foundations of mathematics (according to which mathematical systems are nothing but totalities of meaningless signs operationally defined in a distinct "metamathematical" language) and (ii) Tarski's hierarchy-of-languages strategy for avoiding the syntactic and semantic paradoxes. It is precisely this jointly formalist-cum-hierarchical strategy for grounding logic that Carnap later develops and explicitly implements in *Logical Syntax*.

Given its dual Hilbertian–Tarskian provenance, Carnap's general solution to Wittgenstein's inexpressibility-of-logic problem unsurprisingly has two parts. First, Carnap holds that the arbitrary choice of a set of rules governing well-formedness and inference for a logical calculus (i.e., formation rules and transformation rules) strictly determines the meaning or interpretation of the calculus. And second, he holds that the introduction, by means of arbitrary choice, of the formation and transformation rules must always occur in a metalanguage whose sole function it is to mention signs in the object-language calculus. The first part of the solution conveys *the thesis of conventionalism*: the meaning of expressions in a logical calculus is strictly determined by the free imposition of formal constraints on the manipulation of the various signs constituting the calculus. So logical meaning derives from a source outside of logic itself. And the second part of the solution conveys *the thesis of metalogic*: metalanguages or "syntax languages" are nothing but purely formal or presuppositionless devices for providing access to a totality of meaningless signs, upon which logical constraints can be freely imposed. So, in particular, logic requires no language-user or community of language-users in order to determine the set of symbols that collectively constitute a logic. In other words, logic has a strictly external and objective ground.

Nevertheless, the hardy Tortoise survives Carnap's subtle strategizing. A central doctrine of *Logical Syntax* is that logical truths (or analytic sentences) are contentless tautologous consequences of the arbitrarily and meta-linguistically adopted postulates or rules of a given calculus. But in "Truth by Convention" Quine presents a deep difficulty for conventionalism:

In the adoption of the very conventions . . . whereby logic itself is set up, however, a difficulty remains to be faced. Each of these conventions is general, announcing the truth of every one of an infinity of statements conforming to a certain description; derivation of the truth of any specific statement from the general convention thus requires a logical inference, and this involves us in an infinite regress. . . . In a word, the difficulty is that if logic is to proceed *mediately* from conventions, logic is needed for inferring logic from the conventions.[11]

This formulation is highly compressed; hence it is useful to look at how Quine redescribes the same point twenty years later in "Carnap and Logical Truth":

The linguistic doctrine of logical truth is sometimes expressed by saying that such truths are true by linguistic convention. Now if this be so, certainly the conventions are not in general explicit. Relatively few persons, before the time of Carnap, had ever seen any convention that engendered truths of elementary logic. Nor can this circumstance be ascribed merely to the slipshod ways of our predecessors. For it is impossible in principle, even in an ideal state, to get even the most elementary part of logic exclusively by the explicit application of conventions stated in advance. The difficulty is the vicious regress, familiar from Lewis Carroll, which I have elaborated [in "Truth by Convention"]. Briefly the point is that the logical truths, being infinite in number, must be given by general conventions rather than singly; and logic is needed then to begin with, in the metatheory, in order to apply the general conventions to individual cases.[12]

As I read him, Quine is saying in both texts (but more explicitly in the second) that if every member of the infinite class of logical truths of a given calculus is to be generated as a contentless tautologous consequence of its defining rules, then logic is required to show that the relevant sentences *follow from* those very conventions. If there were only a finite number of such truths, they could be antecedently provided in a nonlogical way and listed one by one. But conventions are essentially general and require logic (and in particular, the principle of universal instantiation) for their application. So paradoxically, and also precisely in the manner of Carroll's vicious regress argument, it follows that a logic cannot be constituted by conventions without presupposing and using a logic that is not itself constituted by conventions.[13]

Is there any way of escaping Quine's objection? It is arguable that Quine's anticonventionalist argument is triply ambiguous, as between what might be called (i) a thoroughly metaphysical reading, (ii) a partially epistemological reading, and (iii) a thoroughly epistemological reading.[14] On the thoroughly metaphysical reading (which is the one I have adopted in the previous paragraph) conventionalism about logical truth is a thesis about the *nature* of logical truth, to the effect that logical truths are generated by conventions; and Quine's worry is that logical truths cannot be generated by conventions without presupposing and employing logic. On the partially epistemological reading, conventionalism about logical truth is still a thesis about the nature of logical truth, but Quine's worry is now that logical truths cannot be *known* to be generated by conventions without presupposing and employing logic. And on the thoroughly epistemological reading, conventionalism about logical truth is an epistemological thesis to the effect that logical truths are known to be generated by conventions; and Quine's worry is once again that logical truths cannot be known to be generated by conventions without presupposing and employing logic.

Now whether we adopt the thoroughly metaphysical or the thoroughly epistemological reading, Quine's argument comes out logically cogent. But things are different for the partially epistemological reading. It is obvious that one could consistently hold (i) that logical truths are in fact generated by conventions and (ii) that they cannot be known to be so generated without presupposing and employing logic. So on the partially epistemological reading, the metaphysical thesis of conventionalism for logical truths comes out unscathed and Quine's argument is fallacious.

Since I have, in effect, offered the thoroughly metaphysical reading as my favored interpretation of Quine's texts, it is obvious that I am strongly inclined to adopt it as a genuine instance of the logocentric predicament: conventionalism about logical truth presupposes and employs logic in order to explain logic. But in view of the partially epistemological reading of Quine's argument, it is also not unreasonable to wonder whether he has left open a way to avoid his objection. This motivates us to explore Prior's distinct, and I think ultimately even more powerful, argument against conventionalism.

3.3 Prior's Runabout

I have just said that I think that Prior's argument against conventionalism is, at the end of the day, perhaps even more powerful than Quine's. It is therefore regrettable that Prior's argument is presented in a subtly but crucially ambiguous way that has misled many interpreters—perhaps including Prior himself[15]—and therefore needs to be sorted out before we can reach the bottom line. To do this, I will introduce the concept of the *inferential role* of a linguistic term. The inferential role of some term *T* is how *T* functions in inferences leading to or from sentences containing *T*. In "Investigations into Logical Deduction," Gentzen argues that the inferential role of a logical constant constitutes its meaning. Gentzen also argues that the inferential role of a constant, in turn, is strictly determined by the arbitrary choice of metalinguistically expressed inference rules governing the use of that constant. So inferential role logically strongly supervenes on syntactic conventions.

Prior sharply disagrees with Gentzen.[16] What Prior explicitly says in "The Runabout Inference Ticket" is that he is attacking the theory according to which "there are inferences whose validity arises solely from the meanings of certain expressions occurring in them." He illustrates this target theory as follows:

One sort of inference which is sometimes said to be in this sense analytically valid is the passage from a conjunction to either of its conjuncts, e.g., the inference 'Grass is green and the sky is blue, therefore grass is green'. The validity of this inference is said to arise solely from the meaning of the word 'and'. . . . [I]f we are asked what is the meaning of the word 'and', at least in the purely conjunctive sense . . . the answer is said to be *completely* given by saying that (i) from any pair of statements P and Q we can infer the statement formed by joining P to Q by 'and' (which statement we hereafter describe as 'the statement P-and-Q'), that (ii) from any conjunctive statement P-and-Q we can infer P, and (iii) from P-and-Q we can always infer Q. Anyone who has learned to perform these inferences knows the meaning of 'and', for there is simply nothing more *to* knowing the meaning of 'and' than being able to perform these inferences.[17]

For reasons I will get to shortly, let us call the theory that Prior has described *Theory X*. The rest of his argument is devoted to showing that Theory X is false, because by hypothesis it generates "inferences whose validity arises solely from the meanings of certain expressions occurring in them," yet some of these inferences lead from true premises to a false conclusion. Otherwise and more long-windedly put, Theory X is false because it is contradictory, in that it allows for at least one *invalid* inference despite the fact that by hypothesis this inference is *valid* by virtue of the meaning of a logical constant whose meaning is constituted by its inferential role, which in turn is strictly determined by the arbitrary choice of metalinguistically expressed inference rules governing the use of that constant.

Prior elegantly shows this by, first, conventionally defining a new constant, 'tonk', in terms of two rules that jointly constitute the new logical operation of "contonktion":

(Rule I) From P, derive P-tonk-Q.

(Rule II) From P-tonk-Q, derive Q.

Then, second, he demonstrates that the tonk rules are unsound in the sense that they allow us to prove sentences that are not logical consequences of their premises, for example:

{1} (1) $2 + 2 = 4$ (Premise)
{2} (2) $2 + 2 = 4$ tonk $2 + 2 = 5$ (From (1) by Rule I.)
{1,2} (3) $2 + 2 = 5$ (From (2) by Rule II.)

It is quite clear, I think, that Prior has successfully refuted Theory X by reductio ad absurdum. The difficulty lies in properly interpreting this refutation. This is because Theory X is not in fact a monolithic item, but instead a complex theory composed of two logically independent elements. The first element of Theory X is the thesis that the meaning of a logical constant is constituted by its inferential role (call this *the inferential role thesis*); and the second element of Theory X is the thesis that the inferential role of a logical constant is strictly determined by the arbitrary choice of metalinguistically expressed inference rules governing the use of that constant (this is of course our old friend *the conventionalist thesis*). The interpretive difficulty arises because the inferential role thesis could be true even though the conventionalist thesis is false. You can consistently assert that the meaning of a logical connective is constituted by its inferential role and also deny that its inferential role is strictly determined by conventions.[18] Indeed, as far as I can tell, nothing in Prior's argument against Theory X turns specifically on problems intrinsic to the inferential role thesis.[19] That logical constants are *either* meaningful by virtue of inferences leading to or from sentences in which those constants occur, *or* else they are not meaningful in that way but instead in some other way, seems to have nothing directly to do with the issue of how to explain or justify valid deductions. That is because any theory of valid deduction will assume that the logical constants are meaningful, *no matter how* this meaningfulness is accounted for. In other words, the semantics of logical constants cancels out as an issue directly relevant to theories of valid deduction. So it seems that the clearest and cleanest and most correct interpretation of Prior's argument is that it is a successful reductio of the conventionalist thesis. Yet this leaves the inferential role thesis untouched.

That, I claim, is the bottom line on Prior's runabout inference ticket. Now I need to link this bottom line explicitly with my larger argument. The fundamental worry about the nature of logic raised by Carroll, early Wittgenstein, and Sheffer is that there is no epistemically noncircular theoretical standpoint from which logic can be legitimately explained or justified, because every attempt to explain or justify logic presupposes and employs logic. So logic is not merely circular, but indeed viciously circular or groundless. Carnap offers conventionalism about logical truth as a way of explaining and justifying logic from a strictly external and objective standpoint. But Quine (at least on one reading—the thoroughly metaphysical reading) refutes conventionalism about logical truth by showing that it entails a version of Carroll's vicious regress. Gentzen then offers conventionalism about valid deduction together with an inferential role thesis about the meaning of logical constants, as a two-part way of explaining and justifying logic in an epistemically noncircular way. But Prior refutes (at the very least) conventionalism about valid deduction by showing that it is self-contradictory. We are left with the unrefuted thesis that logic is explanatorily and justificatorily circular. So there is no epistemically noncircular theoretical standpoint from which logic can be adequately explained or justified. Hence logic is viciously circular, that is, groundless. I conclude that no form of conventionalism is correct, and that no form of conventionalism can avoid the logocentric predicament.

3.4 Dummett, Haack, and the Justification of Deduction

From the fact that no form of conventionalism can avoid the logocentric predicament, of course it does not follow that the logocentric predicament is unavoidable *period*. One still has the option of rejecting conventionalism wholly or in part, and then, by partially or wholly nonconventionalist means, perhaps still being able to explain or justify logic without presupposing or employing logic.

In "The Justification of Deduction" Dummett argues that it is indeed possible to avoid the justificatory circularity of logic. His overall argument is dense and subtle.[20] But for our purposes it can be reduced to four distinct moves.

His first move is to claim that the very need for a justification of deduction is typically mispresented by philosophers by means of a false parallel with the need for a justification of induction. Whereas we *do not* antecedently believe induction to be justified, we *do* antecedently believe deduction to be justified. Moreover, metalogical soundness and completeness proofs for rules of inference are the obvious or prima facie plausible candidates for the justification of deduction.

Dummett's second move is to argue that the classical circularity objection to the justification of deduction (an objection he attributes to Goodman[21]) is misguided, for two reasons. First, according to Dummett, although the worry about circularity correctly draws attention to the fact that if I were trying to persuade someone that deduction is justified it would not be cogent to presuppose or use deduction in my justificatory argument, nevertheless it overlooks the fact that an adequate explanation of something typically works *backward* from the assumed fact of an explanandum (as conclusion) to its explanans (as premises): so it is quite legitimate to appeal to the thing to be explained (in this case, valid deduction) in the construction of its explanation. This distinction between "suasive" and "explanatory" arguments also picks up on an asymmetry Dummett finds between the justification of induction and the justification of deduction: whereas only suasive arguments are appropriate for the justification of induction (since we need to be persuaded that induction is justified), by contrast only explanatory arguments are appropriate for the justification of deduction (since we already believe that deduction is justified). Second, according to Dummett, the classical circularity objection to the justification of deduction is misguided because the adoption of a holistic semantics for natural and formal languages makes the circularity objection irrelevant, by preemptively building circularity into the nature of logic itself at the more fundamental level of the theory of meaning.

Dummett's third move is to set aside semantic holism for the purposes of his argument, thus reinstating the relevance of the problem of the justificatory circularity of deduction, and to assume the existence of a "molecular" (that is, sentence-based, as opposed to a word-based or atomic) compositional semantics.[22] Against this backdrop, he then distinguishes between three levels of approach to the justification problem. The first two levels correspond to the problem of the justificatory circularity of deduction. The first level says that derived rules of inference are justified in terms of primitive rules of inference. But this is trivial because it leaves open the problem of the justification of the primitive rules. The second level says that there are metalogical soundness and completeness proofs for the primitive inference rules used in object-language deductions. This, unlike the first level, is to the point. And not only *that,* says Dummett: it is also something we are antecedently inclined to believe. The third level expresses an attempt to show "how deduction is possible." This corresponds to the problem of how logical deduction can be at once necessary or truth-guaranteeing and also informative. Dummett orders these levels by increasing degree of philosophical significance. Hence he takes the issue addressed by third level approach to be more basic and important than the second-level issue.

Dummett's fourth and final move is to assert that the second-level metalogical strategy of giving soundness and completeness proofs for the primitive inference rules used in object-language deductions, despite its not being a response to an issue of highest philosophical significance, still avoids justificatory circularity. This, according to him, is because such proofs adequately explain how and why carrying out a certain deduction according to a certain primitive inference rule will be truth-preserving (soundness), or how and why a certain truth-preserving deduction is determined by one of the primitive inference rules (completeness), simply by exploiting the semantic powers of a logical language that is distinct from the language in which an object-language deduction is expressed. In other words, for Dummett *the semantics of the metalanguage* is the external and objective source of the justification of deduction. This turn toward metalogical semantics is clearly opposed to conventionalism, which as we have seen concentrates instead on logical syntax.

Haack also published a paper called "The Justification of Deduction." In it she offers a synoptic objection to all attempts to justify deduction, whether conventionalist or nonconventionalist, nonmetalogical or metalogical. Her argument is pithy:

(1) All justification is either nondeductive (for example, inductive)[23] or deductive.

(2) But on the one hand a nondeductive (for example, inductive) justification of deduction is *too weak*, since it will never absolutely guarantee that the conclusion of an argument is true whenever the premises are true.

(3) And on the other hand deductive justifications of deduction are *circular*, since valid deduction must be presupposed and employed in order to show that deductions are truth-preserving or valid.

(4) Therefore, deduction cannot be justified.

In my opinion Dummett's dense and subtle argument has not the slightest adverse critical effect on Haack's pithy argument, for five reasons. First, the putative asymmetry that Dummett finds between the justification of induction and the justification of deduction is irrelevant to Haack's steps (1) and (2). Dummett does not deny that all justification is either nondeductive or deductive; nor, presumably, does he deny that an inductive justification of deduction (as an example of a nondeductive sort of justification) would be too weak. Second, Dummett's distinction between suasive and explanatory arguments, when applied to the justification of deduction, at best shows that *we antecedently believe* it to be cogent to use metalogical soundness and completeness proofs to justify deduction by an explanatory argument, not that it *really is* cogent to do so. Third, although it is true that adopting a holistic semantics would absorb the worry about circularity, since Dummett himself explicitly opts for a molecular compositional semantics, the worry about justificatory circularity still remains in force. Fourth, even if Dummett is right that the issue of "the possibility of deduction" is of greater philosophical significance than the worry about justificatory circularity, it does not follow from the latter's lesser significance, even when taken together with our antecedent belief that metalogical soundness and completeness proofs justify deduction, that metalogical soundness and completeness proofs *really do* justify deduction. Fifth and last, but not least, although an appeal to the semantics of the metalanguage in which soundness and completeness proofs are constructed as the ground of the justification of deduction does indeed avoid the difficulties of conventionalism, such an appeal is still wide open to the following objection: the semantics of the metalanguage both presupposes and employs deduction. So, in order to justify the deductions carried out in the semantics of the metalanguage, a meta-metalanguage and *its* semantics are also required. In other words, Dummett's semantic version of the strategy of appealing to metalogical soundness and completeness proofs only pushes the problem of justification one step further up a regressive ladder of ordered semantic theories for higher and higher metalanguages.[24] I conclude that Haack's pithy argument stands: there is no way out of the justificatory circularity of logic.

3.5   How I Learned to Stop Worrying and Love the Predicament

The conclusion of the previous section was that Haack is absolutely correct in asserting the existence of the logocentric predicament in its justificatory version. Dummett's metalogical semantic strategy for justifying deduction has the useful advantage of avoiding the confusions of conventionalism; yet, if I am correct, Dummett's strategy too is ultimately subject to the justificatory circularity of logic. When Haack's result is taken along with Carroll's, Wittgenstein's, Sheffer's, Quine's, and Prior's support for the logocentric predicament, it is natural to conclude that the predicament is *an intrinsic feature of logic*. Hence, it seems, we must also accept the groundlessness of logic. For convenience, I will henceforth call the thesis that the logocentric predicament is an intrinsic feature of logic *acknowledging the predicament*.

There are at least six ways of acknowledging the predicament. The first way is Carroll's: in the face of the explanatory and justificatory circularity of logic, accept the groundlessness of logic and then opt for *logical prudentialism*. According to logical prudentialism, we mitigate the groundlessness of logic by appealing to its personal utility. Thus the logical enterprise, although in itself groundless, is still worth pursuing by virtue of the fact that it enables one to talk and think rings around those who (unfortunately for them!) do not study logic.[25]

The second way of acknowledging the predicament is a more nuanced and interesting strategy, deriving from the later Wittgenstein: in the face of the explanatory and justificatory circularity of logic, accept the groundlessness of logic and then opt for *logical communitarianism*. Logical communitarianism says that our acceptance of the groundlessness of logic forces us to recognize that logic, like all human institutions, is based radically and solely on a mass of more or less coordinated desires and decisions, silently or explicitly adopted social conventions, and historically entrenched communal practices:

241 "So you are saying that human agreement decides what is true and what false?"—It is what human beings *say* that is true and false; and they agree in the *language* they use. This is not agreement in opinions but in form of life.
242 If language is to be a means of communication there must be agreement not only in definitions but also (queer as this may sound) in judgments. This seems to abolish logic, but does not do so.[26]

The third way of acknowledging the predicament is, in a sense, a combination of logical prudentialism and logical communitarianism and is best articulated by Michael Resnick and Crispin Wright:[27] in the face of the explanatory and justificatory circularity of logic, accept the groundlessness of logic and then opt for *logical nonfactualism* or *expressivism*. The idea here is that logic is essentially normative and practical, not cognitive and theoretical. My grasp of, for example, logical necessity is deeply analogous to my finding something funny within the well-entrenched normative human practice of humor, and *not* analogous to my knowing some empirical fact. So logical discourse is prescriptive, not descriptive, and thus its epistemic circularity is irrelevant to its actual nature.

The fourth way of acknowledging the predicament is articulated by Quine,[28] Nelson Goodman,[29] and the Hilary Putnam of the 1980s and '90s:[30] in the face of the explanatory and justificatory circularity of logic, accept the groundlessness of logic and then opt for *semantic and epistemic holism* about logic. According to this view, the groundlessness of logic is a direct consequence of the deeper dual fact that the nature of logic (a) is determined by our whole conceptual scheme and (b) consists in the coherence (that is, the mutual consistency, mutual implication, and mutual reinforcement) of all the individual members of the total web of concepts and beliefs, including logical beliefs, nonlogical natural scientific beliefs, and empirical beliefs.

The fifth way of acknowledging the predicament is defended by the 1970s Putnam[31] and by Haack:[32] in the face of the explanatory and justificatory circularity of logic, acknowledge the groundlessness of logic and then opt for *logical instrumentalism* or *pragmatism*. According to this view, logic is groundless because its *only* defining or intrinsic or a priori feature is its tendency to generate the logocentric predicament. Beyond that, logic is nothing but an empirical theory whose overall character is determined by human interests and, like other theories, wholly revisable in the light of experience.

I think that each of these five ways of acknowledging the predicament has serious problems. In the first place, logical prudentialism, logical communitarianism, logical expressivism, and logical pragmatism are often taken to be forms of scientific naturalism about logic in that they explicitly assert, or at least assume, the logical strong supervenience of logic on the natural facts. But scientific naturalism about logic is self-refuting (see section 1.4).

One obvious response to this objection would be to give up scientific naturalism, perhaps in favor of some nonreductive form of naturalism. In the second place, then, and more decisively, logical prudentialism, logical communitarianism, logical expressivism, and logical pragmatism all fail to explain why logic lies not merely accidentally but necessarily at the foundation of all the sciences, or why it is that logical principles are built into the very structure of all rational discourse and rational inquiry.[33] Another way of putting this is that none of them adequately explains our robust intuition that logical discourse is (to use Quine's phrase) "obvious, actually or potentially."

In the third place and finally, logical holism cannot guarantee that what it sets up as logical truths by virtue of its coherent web of concepts and beliefs are in any sense really and independently true, that is, in any sense true not only inside but also outside the web, *in the world*.[34] This is because holism is committed to coherentism, and coherentism is a form of antirealism.[35] But as Benacerraf points out (see section 1.5), the "standard" or Tarskian semantics of theoretical discourse of any sort, including logic, is realistic. So logical holism is inconsistent with the semantic realism of logical discourse.[36]

This leaves us with the sixth and last way of acknowledging the predicament. This way is sharply distinct from the other five, because while (like the others) it asserts that the logocentric predicament is an intrinsic feature of logic, it *rejects* the further step that logic is thereby groundless. Every treatment of the logocentric predicament that we have looked at so far assumes that all legitimate explanations and justifications must be epistemically noncircular. Given that assumption, the line of reasoning to the groundlessness of logic is airtight: (i) if the logocentric predicament is an intrinsic feature of logic, then logic cannot have an epistemically noncircular explanation or justification; (ii) every legitimate explanation and justification must be epistemically noncircular; (iii) so logic has no legitimate explanation or justification: that is, logic is groundless.

One way of responding to this line of reasoning would be to try to stop the inference to (iii) by giving up the implicit assumption that logic as a whole *needs* a legitimate explanation or justification. In other words, someone could try to defend a thoroughgoing "localism" about explanation and justification, such that only partial explanations and partial justifications are possible, and not *global* explanations or justifications. The main worry I have about this response, however, is that it gives up far too quickly on the project of global explanations and justifications. Surely, other things being equal, global explanations and global justifications are rationally preferable to partial ones. So as long as there is still a possibility of the former, it seems to me that we should fully explore it and not yet settle for the Blue Monday of diminished rational expectations.

Indeed, it seems to me that the best way to avoid the groundlessness of logic in the face of the logocentric predicament is simply to give up premise (ii), which asserts that every legitimate explanation and justification must be epistemically noncircular. For there are two (or anyhow at least two[37]) sharply distinct sorts of epistemic circularity: (1) begging the question—that is, an argument whose conclusion is to be found among its premises—which is informally fallacious, and therefore epistemically illegitimate; and (2) a *presuppositional argument*, which is not epistemically illegitimate. My proposal, then, is that the correct account of the logocentric predicament is that it involves a presuppositional argument.

A sentence X is a presupposition of a sentence Y if and only if the truth of X is a necessary condition of the truth of Y and also a necessary condition of the falsity of Y.[38] For example, the sentence 'John has some children' is a presupposition of the sentence 'All John's children are asleep'. A presuppositional argument for a sentence S_1, as I will understand it, involves an inference from S_1 to a presupposition of S_1.[39] Call the sentence that expresses a presupposition for S_1, 'S_2'. Then the conclusion of a presuppositional argument is not S_1, but instead S_2. So a presuppositional argument *for* S_1 is an argument *to* S_2, the sentence that expresses S_1's presupposition. The conclusion S_2, as a presupposition of the premise S_1, partially or wholly explains S_1 because it states an otherwise merely implicit necessary condition for the truth conditions of S_1. In some cases, S_2 can also be a sufficient condition for the truth conditions of S_1. For example, 'John has some children, and none of them are awake, and John and his children are all living humans, and other things being equal there is no intermediate living human condition between waking and sleeping' is also a sufficient condition for the truth conditions of 'All John's children are asleep'. Here S_2 is not merely *a* presupposition of S_1, but also *the* presupposition of S_1.

Whether the conclusion of a presuppositional argument is only *a* presupposition or *the* presupposition of the argument's premise, however, presuppositional arguments are epistemically circular only to the minimal extent that they introduce no new truth-conditional information into the argument over and above that which is already contained in the premise. But crucially, they make explicit some information that is otherwise merely *implicitly* contained in the premise. The conclusion of a presuppositional argument thus partially or completely *unpacks* truth-conditional information implicitly contained in the premise. So presuppositional arguments are epistemically legitimate and not question-begging because they render explicit some information that is otherwise merely implicit, and because the conclusion of the argument is not to be found among its premises.

Now, if the relevant statement or proposition expressed by S_1 is the statement or proposition that there are some logics structurally distinct from classical or elementary logic, then it follows that my argument in chapter 2 for the logic faculty thesis is, in effect, a presuppositional argument for the existence of nonclassical logics. This is because my argument for the logic faculty thesis says that assuming that some nonclassical logics exist, a single universal unrevisable a priori protologic is innately contained in a multiply embodiable modular constructive cognitive capacity for logical representation (the logic faculty), and as a consequence every logical system whatsoever, whether classical or nonclassical, is cognitively constructed by rational animals. But even if it were in fact *false* that some nonclassical logics exist— that is, even if there were no nonclassical logics, and only classical or elementary logic existed—then the logic faculty thesis would still be true, because it explains the existence of classical logics and nonclassical logics alike. So, if sound, this argument shows that whether it is true or false that some nonclassical logics exist (so whether S_1 is true or S_1 is false), the logic faculty thesis is true, which in turn shows that S_1 has a presupposition in S_2, the sentence that expresses the logic faculty thesis. Indeed, if sound, the argument for the logic faculty thesis shows that this thesis is *the* presupposition for the claim that some nonclassical logics exist, and the same goes for the claim that some classical logics exist.

In this way, it seems to me that the correct conclusion to draw from the logocentric predicament is *not* that logic is groundless, but rather that logic, whether classical or nonclassical, has a legitimate *presuppositional* explanation in the logic faculty thesis and thereby in logical cognitivism. For obvious reasons, this is what I call *the cognitivist solution* to the logocentric predicament.

Suppose we adopt the cognitivist solution to the logocentric predicament. Then, like a Gestalt shift, the background of our entrenched philosophical picture becomes the foreground and everything takes on a new look. Indeed the logocentric predicament turns out to be *just what we would expect if the logic faculty thesis were true*. For if the logic faculty thesis were true, we would thereby have both to constructively and epistemically presuppose and employ the protologic in explaining or justifying logic, because by hypothesis rational humans are animals who construct, analyze, and evaluate all logics by means of the logic faculty, which has the protologic innately contained in it. In other words, logic does not require an external (nonlogical) and objective (nonmental) ground of explanation and justification: rather, it requires only an internal and mentalistic ground, that is, a *logico-psychological* ground. Logic is both globally explained and globally justified by the fact that rational animals possess a logic faculty. So the previously highly disturbing fact that we must presuppose and employ logic in order to explain or justify logic turns out to be merely a superficial token of the much deeper and entirely nondisturbing fact that logic is cognitively constructed by rational animals. In other words, logic is globally explained and globally justified by *our cognitive constitution*. The explanatory and justificatory buck for logic stops right at the fundamental cognitive architecture of our rational human nature. Or, in still other words, if we adopt the cognitivist solution, then the logocentric predicament is A Big Easy, not A Kill-Ease.

Now is the right time to recall the claim—for which I argued in chapters 1 and 2—that the logic faculty thesis not only coherently but also apparently uniquely resolves the psychologism and *e pluribus unum* difficulties, both individually and when taken as a package deal. This provides a strong case for the logic faculty thesis by means of an inference to the best explanation. Let us then add to this the two claims for which I have argued in this chapter: (1) that the logocentric predicament (the explanatory and justificatory circularity of logic) is an intrinsic feature of logic, and (2) that the logic faculty thesis, and apparently the logic faculty thesis alone, both accounts for the logocentric predicament and also avoids the groundlessness of logic. The obvious conclusion is that we now have a very strong case for the logic faculty thesis, because it, and apparently it alone, provides a coherent triple resolution of the psychologism, *e pluribus unum*, and logocentric predicament problems. But the logic faculty thesis is the first of two parts of logical cognitivism. So we now have a very strong case for half of logical cognitivism too.

4 | Cognition, Language, and Logic

"Really, now you ask me," said Alice, very much confused, "I don't think—"
"Then you shouldn't talk," said the Hatter.
—Lewis Carroll[1]

4.0 Introduction

The upshot of chapters 1 to 3 is that we can best account for the nature of logic by appealing to the notions of *the logic faculty*, *the protologic*, and *rational animals*. The general theory that incorporates these three notions is *logical cognitivism*. My basic claim, so far, is that the logical faculty thesis both coherently and also apparently uniquely solves the problem of psychologism, the *e pluribus unum* problem, and the logocentric predicament. If correct, this establishes the first half of logical cognitivism. I now turn from the nature of logic back to the nature of human rationality.

A rational animal, as I proposed in the introduction, is an animal that is a normative-reflective (i.e., a rule-following, conscious, intentional, volitional, self-evaluating, self-legislating, reasons-giving, reasons-sensitive, and reflectively self-conscious) possessor of strict modal concepts, and more specifically a normative-reflective possessor of the concepts of necessity, certainty, and unconditional obligation. The upshot of chapters 4 through 7 will be that *rational human animals are essentially logical animals*, in the sense that a rational human animal is defined by its being an animal with an innate constructive modular capacity for cognizing logic, a competent cognizer of natural language, a real-world logical reasoner, a competent follower of logical rules, a knower of necessary logical truths by means of logical intuition, and a logical moralist. This is *the logic-oriented conception* of human rationality. If the overall argument of these four chapters is sound, then it establishes the second half of logical cognitivism.

The logic-oriented conception of human rationality stands in contrast to the traditional conception, going all the way back to Descartes,[2] but amusingly incarnated by Carroll's fictional Mad Hatter, according to which the human capacity for thought (by which I mean specifically, in this context, the human capacity for rational cognition[3]) and the cognitive capacity for natural language are strongly equivalent, thus according to which a rational human animal is essentially a talking animal. I do not want to deny that rational human animals are *necessarily also* talking animals: I want to deny only that rational human animals are *nothing but* talking animals.

The present chapter thus deals with the deep and manifold connections between human rationality, cognition, language, and logic. I will argue (1) that what I call *the standard cognitivist model* of the mind, as developed by Noam Chomsky and Jerry Fodor, is, with a few important critical refinements, correct; (2) that what I then call *the refined standard cognitivist model* of the mind implicitly includes the conception of a *logic of thought* that is presupposed by what Fodor calls a "language of thought"—that is, a *lingua mentis,* or mental language; (3) that the protologic is to the logic of thought as the universal grammar or UG in Chomsky's sense is to the language of thought, and hence the principles of the protologic (whatever they turn out to be) are in a substantive sense "the laws of thought"; and finally, (4) that although on the one hand it is inconceivable and therefore impossible for there to be rational human animals who lack linguistic competence (that is, the ideal speaker-hearer's knowledge of her own natural language), on the other hand it is not merely conceivable but also a matter of fact that there are linguistically competent yet nonrational human animals. The general conclusion to draw from these four theses is that human cognition is shot through with rationality, and thereby counts as *human thought*, precisely to the extent that it is shot through with a capacity for cognizing logic. Moreover, while it is true that rational human animals are necessarily also linguistic animals, nevertheless not all linguistic humans are rational, nor are all linguistic animals rational. So rational human animals are essentially logical animals, but *not* essentially talking animals.

4.1 The Standard Cognitivist Model of the Mind

What I am dubbing the standard cognitivist model of the mind is a comprehensive doctrine of human cognition and thought that has its early origins in Kant's transcendental psychology,[4] but has been more recently and quite fully developed in the writings of Chomsky and Fodor and is shared more or less explicitly by most or at least a great many contemporary cognitive psychologists, philosophers of mind and language, and cognitive neuroscientists.[5] With due allowance made for the lack of fine-grained detail imposed by broad-stroke generalization, we still can, I think, capture the main thrust of the standard cognitivist model of the mind by conjoining the following five theses:

(1) *Representationalism or intentionalism*   Human cognition consists primarily in the generation, manipulation, and transformation of mental representations, which in turn are neurophysiologically realized, typically nonconscious, and essentially Turing-computational functional mental states with object-directed or self-directed contentfulness, or intentionality.

(2) *Innatism or nativism*   Fundamental aspects of human cognition and thought are intrinsic to the human mind, strictly determined genetically by the human brain, and neither derived from (by mere generalization) nor strictly determined by any set or sort of sensory, behavioral, and environmental inputs to the human animal plus capacities to generalize from it, even though it is always responsive to such inputs.

(3) *Constructivism*   The human mind–brain is spontaneous, active, or dynamic in the sense that while in order to function properly its cognitive activity must (on the whole and other things being equal[6]) be triggered by appropriate, relevant, and real external experiential stimuli or inputs, nevertheless its representational outputs inevitably embody and express both formal and material contributions that uniquely derive from or are uniquely determined by the mind–brain, and these outputs can be of infinite complexity despite their finite generative basis (this feature is also known as "creativity" or "productivity").

(4) *Modularity*   The representational and constructive ability of the human mind–brain is to a significant extent, and perhaps even massively, organized as a network of mental modules or *cognitive faculties*. Cognitive faculties, in turn, are cognitive capacities that are (i) "dedicated" or operationally specialized, (ii) "fast" or reflex-like, (iii) "domain-specific" or informationally

specialized, and (iv) "encapsulated" or informationally isolated from one another and from central processes or systems. Cognitive faculties are typically innate. And the network of innate faculties in humans properly constitutes the deepest level of our cognitive architecture, that is to say, properly constitutes the *nature* of the cognitive mind–brain.

(5) *The mental language thesis* Human knowledge of natural language in particular, but also human cognition and thought more generally, must occur in a *lingua mentis* or mental language. Such a mental language, more precisely, is a semiotic or sign-based, symbolic or meaningful subjective system of mental representations, or even more precisely an internal, individual, and intensional[7] code, that is shared in common by the several cognitive capacities and faculties.

To be sure, these five theses are not by any means self-explanatory or self-justifying, even for committed cognitivists. So I will need to unpack each of them at least briefly.

(1) Representationalism or intentionalism As Fodor has been tirelessly pointing out since the mid-1970s,[8] if we are to treat the human animal as a cognizer capable of knowing a natural language, of inference, of problem solving, of theorizing, of science, of ordering its preferences, of assessing utilities, of evaluative judgment, and of decision making, there seem to be few coherent or defensible alternatives to the idea that the basic vehicles and basic elements of such cognitive activities are mental representations. Mental representations are mental states intrinsically characterized by "aboutness" or intentionality, that is, a mental state's *object-directedness* or *self-directedness* and *contentfulness*.

A mental state is *object-directed* or *self-directed* to the extent that (1) the animal in that state is to some appreciable degree attentively focused[9] on some or another individual thing or property or state of affairs, or on itself; and (2) the animal is capable of self-consciously contextually individuating either its intentional object, or itself. An animal's object-directedness or self-directedness can be perceptual, imaginational, conceptual, doxic (that is, belief-based, propositional-attitude-based), desiderative, emotive, or volitional. As a consequence, the intentional target might or might not have causal relevance or efficacious causal powers, and in the case of intentional objects, it also might not actually exist.

Correspondingly, however, a mental state is *contentful* to the extent that (a) any individual thing or property or state of affairs or self that is a target of the mental directedness of the animal can be focused on in two or more cognitively significant or informative ways (also known as "modes of presentation," or MOPs) by that animal, and (b) the words or phrases that express these distinct MOPs are also not identical as regards their cognitive significance or informativeness (also known as their "sense"). This latter point is restated by saying that two sentences differing only in the uniform intersubstitution of words or phrases expressing distinct MOPs of the same thing do not have the same content (or do not have the same sense) just insofar as those words or phrases cannot be uniformly intersubstituted in every linguistic context while preserving the same truth-value, despite the fact that they ordinarily refer to the same object. This is, of course, a Fregean[10] point.

As the previous paragraph indicates, there is a close connection between intentionality and *natural language*, in that it is assumed, or at least hypothesized, that every intentional state is linguistically expressible. Even more precisely, there is a close connection between the *contentfulness* of intentionality and the fact of *referential opacity*,[11] whereby coreferential words or phrases cannot be uniformly intersubstituted in all linguistic contexts without change of truth-value. Correlatively, uniform intersubstitution that *does* preserve truth-value is *referential transparency*.[12] So, insofar as referential opacity, along with its Fregean semantics of senses or *Sinne*, is admitted, along with its correlate referential transparency, into the privileged class of things that really and truly exist, then intentionality is thereby admitted into the privileged class as well. And of course the reification of intentionality gains further force when it is noted that the range of linguistic contexts in which intersubstitution fails is heavily biased toward embeddedness within psychological verbs.

The most obvious function of intentionality is that it is directly implied by ordinary ascriptions of belief and desire, and by the more or less subtle and more or less accurate but still perfectly ordinary interpretations of human behaviors (whether one's own or someone else's) as agent-centered actions. For example:

A few days passed away, and Catherine, though not allowing herself to suspect her friend, could not help watching her closely. The result of her observations was not agreeable. Isabella seemed an altered creature. When she saw her indeed surrounded only by their immediate friends in Edgar's Buildings or Pulteney-street, her change of manners was so trifling that, had it gone no farther, it might have passed unnoticed. A something of languid indifference, or of that boasted absence of mind which Catherine had never heard of before, would occasionally come across her; but had nothing worse appeared, *that* might only have spread a new grace and inspired a warmer interest. But when Catherine saw her in public, admitting Captain Tilney's attentions as readily as they were offered, and allowing him an almost equal share with James in her notice and smiles, the alteration became too positive to be past over. . . . Isabella could not be aware of the pain she was inflicting; but it was a degree of wilful thoughtlessness which Catherine could not but resent.[13]

On the face of it, such direct appeals to intentionality are not only salient and necessary for individual human lives, human practices, and social institutions, but also scientifically intelligible and explicable. Folk psychology might or might not be a theory in the precise sense (assuming there really *is* a precise sense) in which theories in the natural sciences are theories.[14] But at the same time, folk psychology undeniably has an apparent or prima facie ontology, semantics, and epistemology: a prima facie ontology of mental representations and intentional states; a prima facie semantics of Fregean senses, referential opacity, and referential transparency; and a prima facie epistemology of belief/desire attributions and behavior-to-action interpretations. In other words, nonreductive explanatory appeals to intentionality and nonreductive explanatory appeals to rational human nature apparently stand or fall together, and in this sense, folk psychology possesses undeniable theoretical integrity and legitimacy.

If this is so, however, then there must be a general theory or science of mental representations and intentionality, that is, there must be a *cognitive science*. And the most promising and fruitful approach to cognitive science, if only because it is the *only* positive and moderately successful research program to emerge since the collapse of the behaviorist paradigm in the 1960s, has it that mental representations (1) are real (i.e., intersubjectively verifiable, irreducible, ineliminable) phenomena; (2) have an intrinsic syntax or sign design; (3) have their logical form and intentional content alike constituted by information-processing procedures defined over that intrinsic syntax (assuming a backdrop of sensory, behavioral, and environmental inputs to the animal); (4) are adequately formally modeled by the operations of universal Turing machines or digital computers;[15] (5) are type identical with specific causal roles within functional (i.e., multiply realizable, second-order physical, causally structured) organizations of organisms or machines;[16] (6) are token identical with causally efficacious realizations of those causal roles in the brain; and (7) are logically strongly supervenient on the brain's underlying neurophysiological (including neurobiological) properties, taken together with various local and nonlocal environmental factors affecting the brain.

(2) Innatism or nativism   An aspect *A* of an animal's mind is innate if and only if *A* is intrinsic to the animal's mind and *A* is neither derived from nor strictly determined by any set or sort of sensory, behavioral, or environmental inputs to the animal—or, as I shall say, *A* is neither derived from nor strictly determined by *the external experiential stimulus*—despite its being always sensitive to such an input. Otherwise put, something innately in the mind of an animal is a necessary or inherent part of that mind, not an accidental or extrinsic part, and it is underdetermined by the external experiential stimulus, even though it is always affected by that stimulus. Still otherwise put, an innate aspect of the mind is an a priori aspect of the mind. It is that part of the mind which, if you lost it permanently, would permanently make you into a different kind of animal; and it is not modally controlled by the empirical world, although it inevitably tracks the empirical world. So, while an innate aspect's character or operations must (on the whole and other things being equal) be triggered by an appropriate, relevant, and real external experiential stimulus, that character and those operations are not in any way strictly fixed or forced by that input. The best working hypothesis for explaining the presence of innate aspects in a human animal's mind is that they are genetically strictly determined by the human brain.

As Chomsky began pointing out in the late 1950s and early '60s, innatist or nativist models of the mind–brain are most effectively vindicated by an argument-strategy known familiarly as "the poverty-of-the-stimulus argument."[17] The nub of the poverty-of-the-stimulus argument is that innate components of an animal's mind are features intrinsic to an animal that best explain either (i) cognitive outputs *from* the animal or (ii) manifest cognitive traits *of* the animal, whenever those outputs or traits have structures or contents that are significantly underdetermined by external experiential inputs *to* the animal, plus its generalizing ability. But here is a more precise version:

The poverty-of-the-stimulus argument

(1) *M* is a mapping from an external experiential stimulus to some cognitive output, or some manifest cognitive trait, of animals of a certain kind *K*.

(2) *M* is such that the external experiential stimulus plus capacities to generalize from it significantly underdetermine the relevant cognitive output or the relevant manifest cognitive trait[18] for *K*-animals (= the "poverty" of the external experiential stimulus).

(3) There are only three possible factors that can plausibly determine mappings from cognitive inputs to cognitive outputs or traits: (i) the external

experiential stimulus alone; (ii) the external experiential stimulus together with some cognitive factor other than innateness, namely a capacity for generalization; or (iii) the external experiential stimulus together with innate organs or devices contained in animal cognizers. Appeals, for example, to sheer chance, divine preordination, or some nonprobabilistic and nontheological but otherwise unknown and mysterious X-factor are implausible and theoretically unhelpful.

(4) Therefore, the best explanation of the mapping M is that the minds of K-animals contain an innate organ or device sufficient to bring about the relevant cognitive output or manifest cognitive trait, given the external experiential stimulus.

Here is a familiar concrete application of the poverty-of-the-stimulus argument, originally offered by Chomsky. The external experiential stimulus for human animals who acquire mastery of natural languages (a stimulus that includes parental grammar training, parental vocabulary training, communal speech-act initiation, ostensive word–world pairings, etc.), plus a capacity for generalization, significantly underdetermines the syntactic and semantic/conceptual structures of those animals' outputs, whether in the form of language production or language understanding. In particular, inputs plus the capacity for generalization cannot account for the creativity or productivity of language production and language understanding (see the discussion of constructivism under (3) just below). So neither the relevant external experiential stimulus alone nor the relevant external stimulus plus the capacity for generalization can explain the mapping from that stimulus to the human mastery of natural languages. And it is both implausible and theoretically unhelpful to postulate that language acquisition occurs either by sheer chance, divine intervention, or some as yet unknown nonprobabilistic and nontheological X-factor. Hence the best overall explanation for the mapping from those external experiential inputs to those cognitive outputs is that humans contain an innate language acquisition device or organ that is sufficient to bring about linguistic outputs having those syntactic and semantic/conceptual structures. The alternative anti-innatist, or empiricist, explanation of language acquisition is eliminated because it appeals, quite inadequately and hand-wavingly, either to the bare external experiential stimulus plus some black-box-like generalizing psychological propensities or mechanisms (e.g., Humean association, stimulus–response arcs), or else to the bare external experiential stimulus plus possession by the creature of some black-box-like generalizing "multipurpose learning strategies."[19]

In the ensuing and famous "nativism versus empiricism" debate, it was correctly pointed out by Chomsky that it is in fact a conceptual mistake to frame the issue as an opposition between an appeal to innate aspects of the animal's mind–brain on the one hand versus a complete rejection of innatism on the other. On the contrary, viewing the debate in light of the poverty-of-the-stimulus argument shows clearly that the empiricist is, in a certain basic way, every bit as committed to innateness as the nativist. This is because the empiricist after all appeals to *innate* propensities or mechanisms such as Humean association, or to *innate* general multipurpose learning strategies, in order to explicate the capacity for generalization. Quine, for example, explicitly appeals to "innate quality spaces" in *Word and Object*.[20] Thus the supposedly fundamental difference between the innatist and empiricist positions boils down to just what *sort* of underlying structure is assigned to the innate component of cognition by the two hypotheses respectively, given the poverty of the stimulus. The Chomskyan innatist ascribes a comparatively rich or maximal set of structures to the innate component, while the empiricist ascribes a comparatively thin or minimal set of structures to it. But the need for an appeal to *some* sort of innate component in cognition is conceded by *both* parties to the debate.[21] So in that sense there really is no deep or fundamental nativism versus empiricism controversy: *everyone* who is not a radical skeptic about rational cognition[22] is an innatist of some kind.

(3) **Constructivism** As Kant first argued, the human mind is inherently spontaneous, active, or dynamic, and not passive, static, or inert. Indeed, the mind is the same as the *life* of a rational human animal. This means that, given the appropriate inputs, the mind is *self-determining*. Its operations, while always triggered by external inputs, are also always in certain respects unprecedented and underdetermined by those inputs. This is as true of the capacity for cognition (also known as "the understanding") as it is of the capacity for desire and volition (also known as "the will"). According to this activist-cognitivist model of the mind, the basic operation of the capacity for cognition is the *construction* of mental representations. But at this point an obvious critical question arises:

Q Fine. But according to the activist-cognitivist model of the mind, how does the construction of mental representations actually differ from the classical empiricist's account of the origins and genesis of our ideas by means of association, or indeed from any more sophisticated empiricist account in terms of general multipurpose learning strategies?

A   I can handle that worry. Construction implies the special nonassociationist features of (i) *cognitive generativity* and (ii) *cognitive creativity or productivity*. Let me tell you a little bit about them.

(3.i)   Cognitive generativity   To account for the animal mind's processing of mental representations, the cognitivist postulates that such a mind contains potentially or actually explicit formal procedures[23] for assigning determinate features to representational outputs. These determinate features will be syntactic or semantic/conceptual structures of a familiar sort in the case of mental representations that are either linguistic or at least require language. But in the case of nonlinguistic mental representations, the features can be syntactic or semantic in different ways. For example, as Ray Jackendoff has argued, spatial or temporal syntax, and spatial or temporal semantics, are assigned to all human sensory representations, whether or not they are also combined with language.[24] Whether linguistic or nonlinguistic in character, however, all processing of mental representations is based on formal procedures for assigning determinate features to outputs. An animal's mind thereby *generates* its outputs precisely by *implementing* these formal procedures. A generative theory of X is, perforce, a formal procedural theory of X.

(3.ii)   Cognitive creativity or productivity   This feature of the constructive activity of the animal's mind is often confused with generativity. This confusion is not entirely unjustified, because the creative or productive feature implies the generative feature, although the converse is not the case. Creativity/productivity is usually characterized informally as the ability of an animal to "make infinite use of finite means," according to Von Humboldt's apt gloss, famously resuscitated by Chomsky in *Aspects of the Theory of Syntax*. Essentially the same idea appears at least as early as Kant's first *Critique*, under the awkward label "the epigenesis of pure reason."[25]

Kant's label, gallumphing as it is, is actually more informative than Humboldt's more elegant slogan. 'Epigenesis' is a technical term in classical biology. According to the theory of epigenesis, every organism has an inherent self-originating and self-organizing vital force (hence a natural anticipation of spontaneity) whereby it gradually develops from some relatively simple seed into an open-endedly complex state of the same organism in an orderly and step-by-step way, by virtue of the fact that the original constitution of its seed includes an inherently reusable, self-applicable (that is, recursive) mechanism for converting elements of its environment into proper parts of itself.[26] This is to be contrasted with the "preformation" theory of biological development, according to which an organism is fully formed in its seed-state and merely acquires greater bulk over time. Kant's deep insight is that the operations of our cognitive capacity, our volitional capacity, and our overall rational capacity are all significantly analogous to epigenesis.

This appeal to epigenesis leads directly to a notion, shared by Kant, Humboldt, and Chomsky alike, to the effect that an animal is cognitively creative/productive if and only if it can cognitively generate infinitely many or infinitely complex representational outputs by operating on finite sets of relatively simple inputs, by virtue of the fact that the mind of the animal contains some inherently reusable, self-applicable device or organ for doing so. Even more precisely put, the creativity/productivity of the animal's mind is equivalent to its containing some representational organ or device equivalent to a *discrete combinatorial system,* and this in turn is equivalent to what can be computed by a universal Turing machine (that is, any recursive function). So, in the hands of the cognitivist, the profound but rather vague Kantian–Humboldtian conception of the mind as a creative/productive vital force or subjective agency takes on the crisp form of Turing computability. The creative/productive *agent* is a computational *engine.*

This combination of creativity/productivity and computability brings out a further important point about the cognitivist conception of innateness. What is innate need not be, strictly speaking, *ideas,* or mental representations as such, whether they are concepts or beliefs. On the contrary, what is necessarily innate for the cognitivist are the generative and creative/productive capacities, or *powers,* of the animal's mind for constructing mental representations, given finite and relatively simple inputs, in certain very specific but also possibly infinitely many or infinitely complex ways.[27] And this thesis answers an important worry of the empiricist, namely, that it is exceedingly unlikely that the human mind could ever innately contain a stock of such highly specific and highly internally structured concepts as *carburetor, bureaucrat,* and *quantum potential,* because, given their high degree of specificity and internal structure, it seems that such concepts could be acquired only from experience.[28] The cognitivist's reply is that what the mind innately contains is not those concepts as such, which is to say that it does not innately contain fully formed and fully developed concepts under precisely those labels. What the mind innately contains, instead, is a highly versatile generative and creative/productive mental power for constructing these highly specific and highly internally structured concepts and infinitely many others (including Fodor's favorite, the highly humble concept *doorknob*) under relevant, appropriate, and real, although still underdetermining, external triggering conditions.[29] The world supplies the right raw materials and the right occasions for construction, and the innate powers of our mind do all the rest of the work.

(4) Modularity Just as generativity is often confused with creativity/productivity (because the latter requires the former but not conversely), so too is modularity often confused with innateness. Every innate cognitive faculty of the mind is modular, but cognitive modularity does not in and of itself entail innateness. A cognitive module can, both in principle and in fact, be acquired through experience.[30] Granting that, the thesis of modularity is then a claim about how the mind is designed or structured. More specifically, the modularity thesis holds that many (let us call this "moderate modularity") and perhaps most (let us call this "massive modularity") of our cognitive capacities are *dedicated, fast, domain-specific,* and *encapsulated.*

A *dedicated* cognitive capacity is one that is set up to perform a certain cognitive task. Good examples are the visual recognition of shapes,[31] face recognition,[32] linguistic syntax recognition,[33] and subitizing (i.e., immediate recognition of numbered collections without counting).[34]

A *fast* cognitive capacity is one that not only works more quickly, relatively speaking, than other capacities, but also requires relatively fewer cognitive resources. So in this regard (if not in absolutely every regard) a fast cognitive capacity is like a reflex. For example, our ability to recognize faces is fast, while our ability to recognize elm trees is slow.

A *domain-specific* cognitive capacity is one that is highly sensitive to one sort of thing, normally applied only to that sort of thing, and highly resistant to other sorts of inputs including even inputs that are superficially quite similar. This is manifest, for example, in the contrast between our strong ability to recognize schematic faces (e.g., "Mooney faces," the famous chiaroscuro drawings of faces that were used in recognition experiments in the late 1950s by cognitive psychologist Craig Mooney) that are right-side up, and our weak ability to recognize schematic faces that are turned upside down.

Finally, an *encapsulated* cognitive capacity is one that neither shares its characteristic sort of information processing with other cognitive faculties nor interacts directly with the explicit (i.e., conscious or self-conscious), implicit (i.e., nonconscious or preconscious), or culturally mediated beliefs, desires, and volitions of the cognizing animal. For example, there is no empirical evidence for direct or lateral communication between our capacity to recognize faces and our capacity to parse phrases or sentences; and again, we continue to see the famous Necker cube phenomenon as one of spontaneously reversing three-dimensional aspects even when we believe that the figure is flat or two-dimensional, and even when we *want or will to see* one aspect only.[35]

Generally speaking, the best evidence for the existence of cognitive modularity is twofold: first, the introspective or phenomenological fact that the operations of a cognitive capacity are partially or wholly independent of our explicit, implicit, or culturally mediated theories, judgments, beliefs, desires, and volitions; and second, the fact that a cognitive capacity can break down (and here I mean primarily aphasias and agnosias) autonomously, that is, without materially affecting the functioning of other capacities. A good example of the first kind of evidence is that you did not have to study English grammar and did not have to will the parsing of that last sentence in order to parse it; nor could you stop yourself from parsing it once you read it. And a good example of the second kind of evidence is the phenomenon of prosopagnosia, the inability to recognize faces.

Although modularity is, strictly speaking, logically independent of innateness, the combination of the two provides a powerful explanatory tool in cognitive science. Cognitivists concede that certain important and characteristically human cognitive activities known as "central processes" or "central systems" (for example, theoretical judgment, belief fixation, problem solving, preference forming, desire, emotion, evaluative judgment, volition, and decision making) are, or at least seem to be, nonmodular and hence noninnate.[36] Nevertheless, it remains true that precisely to the extent that any activities of the mind *can* be studied as innate modules, they are susceptible of a representationalist, apriorist, constructivist analysis, and also of cognitive psychological explanation more generally. Given the assumption of innate modularity, the computational representational rules of the operation of the relevant faculty can then be articulated and understood; and to the extent that generative innate modules are also creative/productive, infinite sets of infinitely complex representational outputs can be comprehended in terms of relatively simple Turing-computation-style information-processing schemata. And all of this can be done while paying relatively little attention to the role of external experiential stimuli. So the seeming fact of the existence of nonmodular, noninnate central processes or systems is no inherent barrier to cognitive science. On the contrary, it is an open-ended opportunity to extend the modularity and innateness theses to mental phenomena that previously were taken to be explanatorily intractable from the standpoint of cognitive psychology.[37]

(5) **Mental language** Suppose that representationalism, innatism, constructivism, and modularity are all true. What, then, is the nature of human cognition? The cognitivist holds that in virtue of representationalism, cognition must be intentional, intensional, and systematic, so cognition must be language-like, and more specifically, Turing-computational. In virtue of innatism, cognition must be internal and individual. In virtue of constructivism, cognition must be spontaneous, generative, and productive. And in virtue of modularity, cognition must be dedicated, fast, domain-specific, and encapsulated. Combining all these features leads to the general thesis that human cognition and thought must occur in a system of rule-governed, internally structured, relationally ordered, transformable, meaningful, and computable mental signs: a *lingua mentis* or mental language.[38] Chomsky calls this the "I-language," and Fodor calls it the "language of thought" or LOT.[39]

The meaningful signs (hence symbols) of the mental language must be more than merely referentially or extensionally meaningful. Otherwise put, the mental language must contain more than names (or other singular referring terms, such as indexicals) for individual things, or names of properties. Since the main purpose of the mental language is to carry and incorporate (hence to be the vehicle of) an animal's intentionality, the mental language's meaningfulness must also capture all the nuances of modes of presentation and more generally capture all the nuances of referential opacity, with its attendant semantics of Fregean senses. Therefore the mental language must also be *fine-grained descriptive* or *intensional*: that is, it must also express *concepts*.

It seems clear, too, that the mental language is not strictly identical to any natural language. This is shown by the phenomenon of global aphasia, or the total breakdown of the cognitive capacity for understanding and speaking a natural language. There is credible empirical evidence that global aphasia is consistent with the continued existence of thought and logical reasoning.[40] So, since according to the LOT hypothesis all thought and logical reasoning necessarily occur in a mental language, and since the operations of mental language can occur in the absence of the capacity for cognizing a natural language, it follows that the mental language is not a natural language.

According to Fodor, moreover, the mental language is universal and sui generis: there is one and only one mental language for all cognizers, and it is syntactically distinct from every natural language.[41] Why so? Fodor says that in order for all elements of the human cognitive capacity to be representationally interactive within a given human animal (were it otherwise, the several parts and operations of the individual mind–brain would be psychologically incommensurable with one another), the mental language must be written in a single code for that animal. And in order for cognition and thought to be the same across a given species (were it otherwise, the several members of that species would be psychologically incommensurable with one another), the mental language must also be written in a single code for all conspecifics. And in order for cognition and thought to be essentially the same across different species (were it otherwise, the capacities for cognition and thought would be the exclusive possessions of humans), the mental language must also be written in a single code for all cognizing animals: but since, as a matter of fact, some species of cognizing animals (for example, cats and dogs) do not acquire natural languages, it follows that the mental language cannot be syntactically equivalent to any natural language. Taking together all these requirements, it follows that the mental language or the language of thought is, as Fodor puts it, *Mentalese*.

Mentalese is necessarily a logical language in the sense that it is the medium of the human animal's theoretic and inferential activities. For the cognitivist, this directly implies that Mentalese must be a Turing-computable language. And that is because (i) nothing will count as the medium of theory and inference unless it has at least the structure of sentential logic; (ii) logical truth and theoremhood or provability are recursive functions in sentential logic; and (iii) a universal Turing machine can compute any recursive function. To the extent that the mental language is Turing-computable, human cognition is not merely logical in nature but again, quite precisely, *computational* in nature.

So far I have presented the standard cognitivist model of the mind in an entirely positive and sympathetic light. Now I want to shift gears somewhat. My own opinion, held on behalf of logical cognitivism, is that the standard cognitivist model of the mind is *largely* but *not completely* correct. For the standard cognitivist model of the mind to be completely correct, or at least to be somewhat closer to being completely correct, it needs a few critical refinements. I will briefly propose six of these critical refinements in sections 4.2 through 4.7. The upshot will be that the cognitivist conception of the human mind that underlies logical cognitivism is not the standard cognitivist model of the mind as such, but instead must be a *refined* standard cognitivist model of the mind.

4.2 Refining Cognitivism I: Mental Signs and Mental Symbols

The mental language, like all languages, is a system of signs, and hence is a system of *mental signs*. As a system of mental signs, it necessarily has a syntax, or more precisely put, a rule-governed formal sign-design. This sign-design comprises both the relational ordering and the internal structural features of mental signs. And it also comprises both the formation of atomic or molecular mental signs, as well as the transformation of mental signs under various operations on the atomic or molecular mental signs.

Mental signs are the vehicles of mental representation. A mental sign becomes a *mental symbol*, and thus a mental representation, when it is invested with meaning (this includes both reference/extension and sense/intension). But a mental sign does not have meaning on its own, purely by virtue of its syntax. If this is so, then no *tokens* of mental signs have meaning on their own, purely by virtue of their syntax, even together with all the causal relations into which such tokens enter. More generally, as many philosophers of mind have noted,[42] it is crucial not to confuse the *vehicle* of a mental representation with the *intentional target* and *intentional content* of a mental representation.

A mental symbol necessarily has a semantics, both referential and intensional. But a mental sign as such, including its syntax, does not itself have a semantics. The meaningfulness of mental symbols is the same as the object-directedness or self-directedness and contentfulness of mental states. In turn, the meaningfulness of mental symbols requires that the corresponding syntactic system of mental signs be embodied in an animal and incorporate the activities of that animal. This is because the object-directedness or self-directedness and contentfulness of conscious mental states is largely determined by the animal who has those conscious mental states. The animal largely determines the object-directedness or self-directedness of a given mental state by attentively focusing to some appreciable degree on *this* (sort of) object as opposed to *that* (sort of) object, or on *itself* as opposed to something *else*; and the animal largely determines the contentfulness of a given conscious mental state by being able to focus attentively on the same object or itself under *this* mode of presentation (MOP) as well as under *that* MOP.

But the very same mental sign or syntactic system of mental signs can in principle be realized or tokened in something other than an animal, even in something that plays the very same functional role as the animal, and yet fail to be the vehicle of conscious mental states that are object-directed or self-directed and have content. For example, to adapt a thought experiment employed by Ned Block,[43] all the members of the entire nation of China, who in turn are causally connected to a humanoid robot, might be compelled to implement a certain syntactic system of mental signs (say, the system of mental signs in my mind–brain that corresponds to my conscious grasp of English) and also perform the very same functions that occur in my body when I hear and consciously understand some English sentence (say, 'The quick brown fox jumps over the lazy dog'). But no one would seriously hold that the entire nation of China, plus its humanoid robot, thereby consciously understands English and has the object-directedness and contentfulness of the mental states of a suitably alert English speaker who says "The quick brown fox jumps over the lazy dog." How could a physical realization that is made up mostly of monolingual speakers of Chinese ever possibly consciously understand English? Of course a few individual members of the Chinese nation will be able to consciously understand English. Even so, *surely* the Chinese nation considered as a single unit, even a unit that is functionally equivalent to the syntactic system of mental signs that is the vehicle of my grasp of English, cannot consciously understand English. But without object-directedness or subject-directedness and contentfulness, mental signs cannot be meaningful, and hence cannot be mental symbols. So neither a mental sign nor a functionally defined syntactic system of mental signs suffices for the semantics of mental symbols.

This sharp distinction between mental signs and mental symbols, such that the former, even when taken together with its functionally defined syntax, does not necessitate the latter, implies that the representationalism or intentionalism thesis of the standard cognitivist model of the mind must be logically detached from representational functionalism, or metaphysical functionalism about intentionality. Metaphysical functionalism in general holds that mental properties are type-identical to functional properties, token-identical to first-order physical realizations of those functional properties, and either locally or globally logically strongly supervenient on first-order physical properties. Representational functionalism in particular holds that the intentional or semantic properties of mental symbols are the same as the functionally defined syntactic properties of their corresponding mental signs. But if the intentional or semantic properties that belong to mental symbols can fail to be instantiated when the same functionally defined corresponding syntactic system of mental signs is differently realized, then representational functionalism is false.[44] Later, in section 4.6, we will see that the failure of representational functionalism also carries with it the failure of *computational* representational functionalism.

4.3 Refining Cognitivism II: Why There Must Be Lots of LOTs

The overarching claim of the cognitivist is that mental representation or intentionality must occur in a mental language or language of thought (LOT) that is not strictly identical to any natural language, that is the medium of the various innate faculties that severally and jointly construct mental representations, and that is also the medium of mental processing more generally. But is the LOT really universal and sui generis? Is there really one and only one mental language for all cognizers, and is it really syntactically distinct from every other language? Must the LOT be Fodor's *Mentalese*? Here we must look more closely at the arguments used to justify this very strong thesis.

The first argument says that in order for all elements of the human cognitive capacity to be representationally interactive within a given human animal (otherwise, the several parts and operations of the individual human mind–brain would be psychologically incommensurable with one another), the mental language must be written in a single code for that animal. That seems correct. But obviously it does not entail that there is one and only one mental language for all cognizers, since even if each animal's mental language is written in a single code, each animal might still have its own distinctive mental code. Nor does it entail that the mental language within a given animal is syntactically distinct from every other language, since even if each animal's mental language is written in a single code, the given animal's single code might still be syntactically identical with some other languages, including natural languages.[45]

The second argument says that in order for cognition and thought to be the same across a given species (otherwise, the several members of that species would be psychologically incommensurable with one another), the mental language must also be written in a single code for all conspecifics. The conclusion of this argument is a non sequitur. Obviously any constraint on mental language that made the several members of the same species psychologically incommensurable with one another would have to be rejected. But the claim that in order to avoid psychological incommensurability, every animal within the species has to have *exactly the same mental language*, seems to be logical overkill, that is, a thesis much stronger than is required. Otherwise put, there appears to be nothing that stands in the way of our claiming that psychological commensurability is perfectly consistent with *similarity well short of identity* across the mental languages possessed by the members of a given species. Just as I can verbally communicate adequately with someone who speaks a different dialect or idiolect of English, there seems to be no good reason why I could not communicate or otherwise psychologically resonate adequately with someone whose personal mental language is as different from my personal mental language as Cockney English or Canadian English is from the so-called Queen's English, or as my twenty-year-old daughter's version of English is from my version of English. A sufficient similarity of mental languages is all that is required for commensurability.

The third argument is that in order for cognition and thought to be essentially the same across different species (otherwise, the capacities for cognition and thought would be the exclusive possessions of humans), the mental language must also be written in a single code for all cognizing and thinking animals. But since some species do not or cannot acquire natural languages, it follows that the mental language cannot be syntactically identical to any natural language. This argument contains not one but two non sequiturs.

First, although it seems obviously true that the capacities for cognition and thought actually are (in the case of cognition) or at least can be (in the case of thought) found in nonhumans, so that cognition and thought are by no means the exclusive possessions of humans, it does not follow that the mental languages of all species must be written in the same code, just as it did not follow from the obvious fact that members of the human race are not psychologically incommensurable that the mental languages of all humans must be written in a single code. The presence of cognition or thought in nonhumans is perfectly consistent with a sufficient similarity well short of identity across mental languages.

Second, supposing then that we grant what seems obvious, namely (i) that some actual nonhumans are cognizers, and that there actually are or anyhow conceivably *could be* nonhuman thinkers,[46] and (ii) that some nonhumans who are clearly also cognizers (for example, cats, dogs, and horses) do not or cannot acquire natural languages. Even granting all of that, it still does not follow that the mental languages of all cognizers *and* thinkers must be written in a single code distinct from any natural language. On the contrary, if we combine the thesis that cognition and thought actually do occur or in principle could occur in a nonhuman species, with the thesis of mental language, and with the thesis that some cognizing species are not or cannot be natural language speakers, this seems to lead to just the *opposite* conclusion: namely, that there is a cross-species *diversity* of mental languages. At the same time, moreover, this cross-species diversity of mental languages is perfectly consistent with the possibility that some mental languages—namely, those found in the species that can or actually do acquire natural languages—are syntactically identical with some other languages, including natural languages.

The conclusion we are left with for the purposes of refining the standard cognitivist model of the mind is this: the thesis that cognition and thought must occur in a mental language or LOT is unexceptionable and correct; but, at the same, time it is also plausible to hold that there are lots of different LOTs, some of which are syntactically identical with natural languages, and some of which are not syntactically identical with natural languages.

4.4 Refining Cognitivism III: Innateness and Instinct

According to the cognitivist, the human mind–brain contains various innate mental powers or faculties. And since the medium of the operations of the several mental faculties is the mental language, in this dispositional sense the mental language is also innate. Innateness, in turn, is the presence of something intrinsic and a priori in a mind–brain. Finally, the presence of something intrinsic and a priori in the human mind–brain is to be explained by the fact that the human brain genetically strictly determines it.

It seems to me that the generalized poverty-of-the-stimulus argument sketched in section 4.2 vindicates the cognitivist's doctrine of innateness *right up to* the thesis that innateness is genetically strictly determined by the human brain, but *not including* that thesis. In other words, not only does it seem to me that the thesis of genetic strict determination by the human brain is logically independent of the thesis of innateness, it also seems to me that the innate component of the human mind is *not* genetically strictly determined by the human brain. Or, in still other words, it seems to me that cognitive innateness is *not* nothing but cognitive instinct.

To say that the human brain genetically strictly determines the innate component of the human mind is to say that the innate component of the human mind logically strongly supervenes on the genetic makeup of the human brain. To be friendly to my opponents, I will grant that the genetic makeup of the human brain is indeed strictly determined by its underlying biological properties, whether or not these are strictly determined by Darwinian evolution. On the other side, the defenders of the view I am criticizing (which, again, is that innateness is genetically strictly determined by the brain) will presumably also grant *me* the friendly assumption that the innate faculties of the mind–brain are multiply realizable in the sense that they can conceivably and therefore possibly occur in different biological species, and more precisely in different sorts of brains. So they will grant me that actual human biology is not strictly speaking logically or metaphysically necessary for our innate faculties. But what they are *not* prepared to grant is that the innate component of the human mind can vary independently of the constitution of the human brain; or more precisely, they are not prepared to grant that the innate component of the human mind can vary independently of the first-order physical properties of the species-specific constitution of the brains (no matter what they are made of) that realize our cognitive functions. So, according to the defenders of the view I am criticizing, it is logically or metaphysically impossible for a brain genetically organized like ours (no matter what it is made of) not to have the innate mental faculties it actually has, and it is equally logically or metaphysically impossible for innate mental faculties to have any features that do not correspond directly to the basic genetic features of the human brain.

But these last two claims seem false. Take for example what Chomsky calls the "science forming faculty."[47] (I am not asserting that there *actually is* such a faculty, but only that there conceivably *could be*, and if there *could be*, then the genetic strict determination thesis is in trouble.) If the genetic strict determination thesis is correct, then logically or metaphysically there could not be a brain that was both genetically organized like the normal human brain and yet not also capable of doing science.

David Chalmers has argued, following in the large footsteps of Descartes and Kripke, that it is conceivable and therefore logically and metaphysically possible for there to be purely physical (that is, both first-order physical and functional) duplicates of actual conscious humans that either lack all phenomenal consciousness (these are "zombies" in the philosophical sense), or partially lack phenomenal consciousness, or have some sort of inverted phenomenal consciousness.[48] If this is so, then consciousness does not logically strongly supervene on the physical facts.

Now the doing of science, whatever else it involves, must involve the capacity for "abductive" inference, or inference to the best explanation. In the general spirit of modal arguments against the logical strong supervenience of the mental on the physical, we can then ask: Must every possible purely physical duplicate of an actual human scientist, even if it is phenomenally conscious, and even if it has some innate faculties, *also* be capable of abductive inference? Couldn't there be a logically possible world physically identical to our actual world in 1906, containing a physical duplicate of Einstein's body, Einstein's consciousness, and some of his innate faculties, which nevertheless lacks an abductive capacity? Otherwise put, isn't it at least *logically and metaphysically possible* that in some worlds physically identical to this one, Einstein's body is not a *scientist's* body? Is the structure of Einstein's brain so fine-grained that it alone (or in conjunction with the environmental factors affecting the brain) strictly fixes *all* of his innate mental powers in *all* their specificity?

The critical point I am trying to make consists in a measured or qualified denial of the thesis that the genetic makeup of the human brain strictly determines the innate component of the human mind. For the purposes of my critical argument, I do not even need to deny that the genetic makeup of the human brain strictly determines the *existence* of an innate component in the mind. But innateness, it seems to me, is intrinsically more fine-grained than are human brains. Similarly, it seems wholly implausible to me that the precise *character* of a given phenomenally conscious state (its being experienced by the conscious subject in just *this* way or just *that* way) logically strongly supervenes on the physical world: so phenomenal consciousness is intrinsically more fine-grained than is the physical world. Therefore, by analogy, just as the physical world is too rough-grained or coarse in its underlying structure to specify precisely the "what-it's-like-to-be" for a given mental state of a given phenomenally conscious animal, so too the genetic makeup of the human brain is too rough-grained or coarse in its underlying structure to specify precisely the innate faculties of a given rational animal.

If this line of reasoning is sound, then it also follows that the innate language faculty is not genetically strictly determined by the human brain, any more than the innate science-forming faculty is. Just as, on the assumption that the notion of a science-forming faculty is intelligible, it is conceivable and therefore logically possible that some of Einstein's purely physical duplicates are not scientists, so too it is conceivable and therefore possible that some of his purely physical duplicates are not linguistically competent.

I need to emphasize that my rejection of biological scientific naturalism in respect of innateness is *not* a rejection of the thesis that there are natural laws that link our innate mental powers with human biology. Innate faculties are of course instantiated in the human brain, and of course there are discoverable lawlike connections between what human brains do and what our innate mental powers can do. My point is rather that the precise character of our innate mental powers is *logically and metaphysically underdetermined* by their biological basis and in particular is *logically and metaphysically underdetermined* by the genetic makeup of the human brain. Hence the natural laws linking the precise character of our innate mental powers with human biology, whatever those laws turn out to be, are never going to be merely logically or analytically or metaphysically yielded by the basic or fundamental natural laws governing the first-order physical properties of the world. In other words, our innate mental powers, though doubtless nomologically connected to the first-order physical properties of the world, are nevertheless not *explanatorily reducible* to them.

4.5 Refining Cognitivism IV: Modularity, Encapsulation, Promiscuity

As we have seen, the cognitivist's conception of modularity is a thesis about how the mind is basically organized as an interactive set of dedicated, fast, domain-specific, and encapsulated information-processing units (in a word, faculties), whether or not these faculties are also innate. According to the cognitivist, then, encapsulation is a *necessary* feature of modularity. Encapsulation, again, is the propensity of a cognitive capacity not to share its characteristic sort of information with other cognitive faculties, and not to interact directly with the explicit, implicit, or culturally mediated theories, judgments, beliefs, desires, and volitions of the cognizing animal.

I agree completely with the thesis that the clearest examples of modularity (visual shape recognition, face recognition, linguistic syntax recognition, subitizing, etc.) are all encapsulated. What I want to challenge is the idea that *every* cognitive module is encapsulated, and thereby also challenge the thesis that modularity *must* include encapsulation. Just to lay my cards on the table right from the start, my rationale for arguing this is that I believe that there is at least one, and apparently only one, cognitive capacity that is dedicated, fast, domain-specific, innate (that is, in other words, an innate modular capacity or a faculty) and *informationally promiscuous* in the sense that it not only shares its characteristic sort of information with every other faculty but also is directly interactive with the explicit, implicit, and culturally mediated beliefs, desires, and volitions of the cognizer. Otherwise put, what I want to argue is that there is at least one cognitive capacity that is *central* in that it belongs to the central processes or systems of cognition, yet also *modular*.

This cognitive capacity, you will probably have already guessed, is the innate capacity for the logical processing of information, or what I have dubbed the logic faculty. As we will see in chapter 5, there already exists a considerable body of empirical evidence for the logical capacity's being modular in the sense of its being dedicated, fast, and domain-specific. But at the same time, in view of the well-known "reasoning tests" first developed in the 1960s by Peter Wason[49] and replicated many times since, there is also considerable empirical evidence for our logical capacity's being cognitively penetrable by explicit, implicit, and culturally mediated theories, judgments, beliefs, desires, and volitions. Most important of all, however, is the fact that the very animals who possess the cognitive capacities that are the most likely candidates for modularity (visual shape recognition, face recognition, linguistic syntax recognition, subitizing, etc.) *are all logical animals or reasoners*. That is, despite the fact that the modular capacities just mentioned are all "peripheral" or immediately linked to external experiential input sources, and despite the fact that they are all encapsulated, they are also all immediately subject to logical control and logical processing, in the sense that not only do their outputs play a direct role in the reasoning processes of the animal, but also those outputs also are *relationally ordered and internally structured* so as to play this role. In rational cognizers, perceptual recognition implies the construction of perceptual concepts that enter directly into perceptual judgments, and these judgments in turn are necessarily constructed under logical constraints governing sentential formation and transformation, meaningfulness, truth-evaluability, and consequence. Yet how can it be that a cognitive capacity is *at once* dedicated, fast, domain-specific, cognitively penetrable by explicit or implicit judgments, beliefs, desires, or volitions, and also informationally ubiquitous in the peripheral modules?

My proposal has two parts. First, I believe that cognitivists have generally failed to notice a crucial distinction between two types of modular cognitive capacity or faculty: (1) peripheral cognitive faculties and (2) central cognitive faculties. And second, I believe that the logic faculty is not only a central faculty but, as far as can be determined by introspection and evidence from cognitive psychology, the *only* central cognitive faculty.

A peripheral cognitive faculty is a cognitive capacity that is dedicated, fast, domain-specific, and encapsulated. Moreover, it is encapsulated precisely because it is peripheral and at the front lines of cognition: its main job is to be directly responsive to external experiential inputs and *not* to be directly responsive to, or, otherwise put, to be sealed off from, information derived from the other cognitive capacities, including the central processes. A central faculty, by sharp contrast, is dedicated, fast, domain-specific, and *not* encapsulated but instead *informationally promiscuous*. It is promiscuous because its cognitive job is *precisely* to operate on the encapsulated information buried within the various peripheral modules and get that information out of the closet, that is, get that information appropriately structured and ordered for the purposes of central processing. This happens both in the periphery-to-center direction whereby the central faculty imposes or laminates its processing activities right onto the original activities of the peripheral modules, and also in the center-to-periphery direction whereby the outputs of the central module can reflect cognitive penetration by central processes that are not grounded in faculties.

More precisely, then, my thesis is that the logic faculty operates in both the periphery-to-center and center-to-periphery directions of information processing: on the one hand, it imposes logical form and content onto the characteristic mental representations of the various peripheral cognitive modules (for example, as cognitive psychologists Richard Gregory and Irwin Rock have argued, logical inference plays a direct constructive role in visual perception, at least to the extent that it is expressed in visual judgments[50]); and on the other hand, in actual reasoning contexts it can be affected by the explicit, implicit, or culturally mediated theories, judgments, beliefs, desires, and volitions of the animal (for example, as cognitive psychologists R. A. Griggs and J. R. Cox have shown, precisely how a deductive reasoning task is verbally presented to us substantially influences our success in completing it[51]). In other words, then, the special job of the logic faculty in the cognitive economy of the rational animal is precisely to *mediate* between the various peripheral modules on the one hand, and the theories, judgments, beliefs, desires, and volitions that make up the other central elements of the animal's cognitive life on the other.

4.6   Refining Cognitivism V: Construction and Computation

According to the cognitivist, the constructivity of the human mind–brain is its representational generativity and its creativity/productivity. Its generative character is the human animal's containing potentially or actually explicit formal procedures for assigning determinate features to representational outputs, and its creative or productive character is the human animal's ability to make infinite representational use of finite representational means. The basic model for these features of the mind is the universal Turing machine, which can compute not only any recursive function but also (according to the Church–Turing thesis) any effective procedure or algorithm, including (according to Turing's theory of computers and intelligence) all cognitive algorithms: hence cognitive constructivity is closely linked with *Turing computability*.

As John Searle has pointed out, there is an important ambiguity in the close linkage of cognition and computation.[52] According to the thesis of *strong AI*, the human brain is nothing but (that is, is explanatorily reducible to) a universal Turing machine or digital computer, and the human mind is nothing but (again, is explanatorily reducible to) a computer program that runs on a universal Turing machine. But according to a more carefully qualified version, *weak AI*, the human mind–brain is merely able to be accurately modeled to some nontrivial extent by universal Turing machines and their programs. Thus, if weak AI is true, the human brain might not be nothing but a digital computer and the human mind might not be nothing but a computer program. Strong AI is usually construed as another version of metaphysical functionalism, namely *machine functionalism,* according to which mental properties are type-identical to Turing-computational functional properties, token-identical to realizations of those Turing-computational properties in the brain or some other suitable physical realizer, and either locally or globally logically strongly supervenient on first-order physical properties. As is well known, Searle rejects strong AI although he accepts weak AI. I agree with his rejection and his acceptance alike.

But my reasons for doing so are somewhat different from his. Searle's basic argument against strong AI, the highly controversial Chinese Room argument,[53] is that computational mental syntax does not itself entail mental semantics or intentionality. This is because it is conceivable and therefore logically and metaphysically possible for a conscious monolingual speaker of English (alone inside a room, satisfying the basic constraints of Turing's famous behavioral test for the ascription of mental properties having to do with intelligence) behaviorally to simulate and implement a computer program for speaking Chinese, yet fail to have a conscious understanding of Chinese. Most criticisms of Searle's argument have consisted in pointing out that the functionalist argument-step from computational mental syntax to computational mental semantics is legitimate *if* one takes into account the total functional context of the realization or implementation of the program for speaking Chinese, and does not focus primarily on the conscious states of the monolingual English speaker. This attack on Searle's argument is effective, I think, *if* it is also assumed that intentionality does not necessarily include consciousness. For if intentionality does not necessarily include consciousness, then even if the monolingual English speaker in the Chinese Room has no conscious awareness of understanding Chinese, nothing in principle stands in the way of the entire computational functional system's nonconsciously understanding Chinese.

But, at the same time, this attack on the Chinese Room argument is effective *only if* intentionality does not necessarily include consciousness. For if intentionality necessarily includes consciousness, then Searle's argument is sound. If intentionality necessarily includes consciousness, then obviously there cannot be an understanding of Chinese without the subject's conscious awareness of understanding Chinese. But since by Searle's initial hypothesis the creature in the Chinese room is a monolingual English speaker, clearly she will not have a conscious awareness of understanding Chinese, no matter which computer program for talking Chinese she behaviorally realizes or implements. Hence the computational system as a whole will not understand Chinese.

So here is the crucial point about the Chinese Room argument: if the resolution of the debate about its soundness really turns on the thesis that intentionality necessarily includes consciousness, then since neither Searle nor the defenders of strong AI have offered conclusive considerations for or against that thesis,[54] the debate is a stalemate.

So what I propose is to bypass that thesis and end the stalemate by focusing on states that are, by hypothesis, both conscious *and* intentional. Strong AI is false if it cannot show that functional states logically and metaphysically suffice for intentional states of any sort, including of course conscious intentional states. Now if what I have argued in section 4.3 is correct, then there really *is* a logical and metaphysical gap between any functionally defined mental syntax (including of course a *computational* functionally defined mental syntax) and *conscious mental semantics*. This is, again, because (i) an individual animal can have conscious intentional states, yet (ii) every (computational) functionally defined mental syntax that corresponds precisely to the mental syntax of an animal who has a conscious intentional state (for example, the computational functionally defined mental syntax which corresponds to my conscious understanding of English, whatever it happens to be) has some possible realizers that are not individual animals and which thereby, intuitively, fail to have the appropriate mental semantics (for example, the Chinese nation taken as a unit, intuitively, cannot itself consciously understand English despite its being able to realize or implement the mental syntax that corresponds precisely to my conscious understanding of English); hence, (iii) functionally defined mental syntax does not alone entail conscious mental semantics. So strong AI is false.

4.7 Refining Cognitivism VI: Cognizers and Their Brains

I have just argued that strong AI is false. If my argument is sound, then the human brain is not nothing but a digital computer, and the human mind is not nothing but a digital computer program. The falsity of strong AI, however, is perfectly consistent with the truth of weak AI, that is, the thesis that human brains and human minds are to some nontrivial extent modelable as digital computers and digital computer programs respectively. In turn, the conjunction of the falsity of strong AI and the truth of weak AI leaves intact another important thesis of the standard cognitivist model of the mind: that the human mind, construed as a constructive representational network of innate modular capacities operating in the medium of a mental language, is logically strongly supervenient on the human brain. In section 4.4 I argued that the innate component of the human mind is not strictly determined by the genetic makeup of the human brain, which is tantamount to showing that the innate component of the human mind is not logically strongly supervenient on the human brain. But what about the rest of the mind?

That the human mind (including consciousnness, intentionality, and rationality) logically strongly supervenes on the human brain is an important thesis, because as Searle has pointed out, it is possible to reject strong AI and deny that mental properties are identical with physical properties including physical properties of the brain, and yet also accept the logical strong supervenience thesis. Indeed, Searle himself holds that strong AI is false and that consciousness and intentionality are irreducible to the physical, but holds also that the mind logically strongly supervenes on *the causal powers of the brain*, in the sense that (i) as a matter of conceptual and therefore logical and metaphysical necessity anything that realizes the causal powers of the brain is mental, and (ii) as a matter of conceptual and therefore logical and metaphysical necessity there cannot be a change in something's mental properties without a corresponding change in the causal powers of the brain.[55] Now it is possible to hold that the human mind logically strongly supervenes on the human brain without holding specifically that the human mind is *caused* by the human brain (for example, one could be a functionalist and hold that the fundamental microphysical constitution of the brain merely *logically and metaphysically fixes* all functional mental properties according to fundamental physical laws), so the generalized version of Searle's good point is that it is possible to reject strong AI, and hence reject metaphysical functionalism, and also reject type–type identity physicalism about the mental, yet also defend a causal logical strong supervenience materialist thesis.

Nevertheless it seems to me that just as cognitivism should be logically detached from strong AI, so cognitivism should be logically detached from Searle-style materialism. More precisely, I think that Searle's materialism about conscious intentionality is false: human conscious intentionality is not logically strongly supervenient on the human brain.[56] My justification for denying the logical strong supervenience of human conscious intentionality on the human brain is provided by what I will call "the Chinese brain argument."[57]

The crux of the Chinese brain argument is the following thought experiment. Suppose that as a child I learned Chinese in China as my first language, but then at a relatively young age (say, 9) I moved to a predominantly English-speaking country like Australia, Canada, the UK, or the United States, suffered from both culture shock and a nasty blow on the head (say, with a hockey stick) delivered by one of my new compatriots, then quickly learned English by total immersion in my newly adopted culture; and after the age of ten (a sufficient time for the formation of some permanent scar tissue in the neural pathways of the left cerebral hemisphere of my brain, say in Broca's area or Wernicke's area, as a result of the hockey stick incident), I never spoke another word of Chinese. Suppose further that as an adult in my mid-forties, utterly assimilated to my adopted culture, I cannot consciously remember any Chinese, no matter how hard I try. So I am now, in my midforties, a monolingual speaker of English. Then one day, as a result of a new and shocking stimulus (say, another blow on the head or some traumatic emotional event), suddenly a fluent stream of grammatically and semantically correct Chinese comes pouring out of my mouth, such that it would convince any monolingual speaker of Chinese who happened to be listening to me that I was a fluent and competent speaker of Chinese. But inside my own mind I experience the stream of sounds coming out of my mouth, in a manner most distressing to me, *as an involuntary, intrusive, and utterly foreign voice.* Not only do I have no control over my utterances, as in Tourette's syndrome, or in Wernicke's or "fluent" aphasia; and not only do the words seem to come from outside myself, as in dissociative personality disorder—but worst of all, I do not have the slightest idea what I am saying! That is, I still do not *consciously* remember a word of Chinese and in fact I do not even know that it *is* Chinese that is pouring out of my mouth.

Now this scenario seems fully conceivable and therefore logically and metaphysically possible, in light of the extensive empirical evidence from cognitive psychology. Indeed for my purposes here, I do not have to *explain* the etiology of fluent aphasia, the experience of intrusive voices, or distortions of the memory. Instead, I need only appeal to the *fact* that these phenomena actually exist and are therefore logically and metaphysically possible. But if they are logically and metaphysically possible, then it follows immediately that human conscious intentionality is not logically strongly supervenient on the human brain, since the Chinese brain example shows clearly that conscious intentional states can vary independently of the causal and neurobiological constitution of the human brain and its various functions. Some part of my brain is nonconsciously speaking Chinese, *but I, the conscious thinker, am not speaking Chinese*: at the level of conscious intentionality, I remain a monolingual speaker of English. Brains are necessarily causally implicated in conscious intentionality, but conscious intentionality is not strictly determined by the causal powers of the brain.

4.8 Why There Must Be Some Logics of Thought Too

Perhaps I should summarize the various twists in the plot of this chapter so far. On behalf of logical cognitivism I have accepted the standard cognitivist model of the mind, which consists in the conjunction of the theses of representationalism or intentionalism, innatism or nativism, constructivism, modularity, and mental language, subject to the following critical refinements:

(i) that functionally defined mental syntax does not alone entail mental semantics;

(ii) that there is no universal sui generis mental language, but instead there are many different mental languages, whether across individuals within the same species, or across species, some of which are syntactically identical with natural languages;

(iii) that innate components of the mind are not strictly determined by the genetics of the human brain;

(iv) that not all innate modular cognitive capacities, or faculties, are peripheral and informationally encapsulated, but on the contrary at least one (and apparently only one) such faculty, the logic faculty, whose role it is to mediate between the peripheral faculties and central processes, is central and also informationally promiscuous;

(v) that strong AI is false although weak AI is true; and

(vi) that human conscious intentionality is not logically strongly supervenient on the human brain.

Let us suppose now that, as the refined standard cognitivist model of the mind states, human cognition must occur in the medium of a mental language or a language of thought. What I want to argue is that this language of thought requires a *logic of thought*, a logic that is constructed by the logic faculty, which in turn innately contains a single universal unrevisable a priori protologic, which is used for the construction of every classical or nonclassical logical system. In section 2.5 I dubbed this *the logic of thought thesis*. The main point to recognize for the purposes of the present argument is that every basic consideration or reason that supports the mental language thesis *also* supports the logic of thought thesis.

If the human mind is inherently representational or intentional in the sense that it includes both a representational syntax and representational semantics, then since it is obvious that no syntax or semantics, whether linguistic or mental, could fail to be constrained by the protologic (given that the protologic is a single universal unrevisable a priori set of logical principles and concepts), it follows that the protologic is required by the representational architecture of the human mind.

If the human mind, as representational, has an innate linguistic component as per the poverty-of-the-stimulus argument, and if the protologic is required by the syntactic and semantic architecture of the representational mind, it follows that the protologic is innate.

If the human mind is essentially active in the constructivist sense that involves generativity and creativity/productivity, and if the protologic is required by the representational architecture of the mind, and if the protologic is innate, and if the mind includes a generative and creative/productive capacity for natural language, it follows that the human mind includes a generative and creative/productive cognitive capacity for representing logic.

If the generative and creative/productive cognitive capacity for language is innate, as per the poverty-of-the-stimulus argument, and if every innate cognitive capacity is modular and hence every innate cognitive capacity is an innate cognitive *faculty*, and if the protologic is innate, and if the mind includes a generative and creative/constructive cognitive capacity for representing logic, it follows that there is an innate cognitive faculty for logic, namely the logic faculty.

If the representational, innate, constructive, and modular features of the human mind jointly entail that human cognition occurs in a mental language or language of thought, and if there is an innate cognitive faculty for logic, it follows that there is a mental logic or logic of thought.

What more can be said about the logic of thought? In chapter 5 I will develop a picture of it, based on the psychology of reasoning. But in view of the various refinements I have made to the standard cognitivist model of the mind, four things can be noted right now.

First, since the language of thought can differ across individuals within the same species and also across different species (provided there is sufficient similarity to guarantee psychological commensurability between individuals within the same species, sufficient similarity across species to guarantee that there can be nonhuman cognizers and thinkers, and sufficient difference across species to allow for some cognizing species to lack natural language), then so too the logic of thought can differ across individuals and also across different species. By the same token, if there *must be* many LOTs, then there must be many logics of thought too.

Second, since no innate component of the human mind is strictly determined by the genetic makeup of the human brain, then so too the innate cognitive faculty that constructs the logic of thought, the logic faculty, is not strictly determined by the genetic makeup of the human brain.

Third, since the logic faculty is a central and informationally promiscuous faculty of the human mind (and apparently the only one), whose role it is to mediate between the peripheral faculties and the central processes of theory-formation, judgment, belief, desire, and volition, then it follows that the logic of thought is the mediating cognitive medium between the mental representations constructed by the peripheral faculties, and the mental representations utilized or penetrated by the central processes.

Fourth and finally, there is an important structural analogy between the language of thought and the logic of thought. As Chomksy has argued, if the language of thought is to be the medium in which we cognize natural language, then there must be an innate language faculty that contains the UG, the total set of a priori (although not necessarily *consciously* knowable a priori) grammatical principles and concepts of natural language. In a precisely analogous fashion, if the logic of thought is to be the medium in which we cognize logic, there must be an innate logic faculty that contains the proto-logic, the total set of unrevisable (hence consciously knowable) a priori logical principles and concepts. In this way, just as the UG contains the "laws of language," so too the protologic contains the "laws of thought."

4.9   Rationality, Language, and Logic

So far I have developed an argument for the logic of thought by accepting and critically refining the standard cognitivist model of the mind, as historically anticipated by Kant and as developed principally by Chomsky and Fodor. But we have now reached a fundamental issue on which the standard cognitivist model of the mind and indeed also the refined standard cognitivist model of the mind do not decisively legislate: the nature of the connection between human rationality and our cognitive capacity for natural language.

According to a traditional view initiated by Descartes and renewed by Chomsky and Donald Davidson,[58] our cognitive capacity for natural language and human rationality are essentially connected: rational human animals are essentially talking animals. But according to an alternative view initiated by Kant and at least implicitly promoted by Fodor,[59] it is the cognitive capacity for logic, and not the cognitive capacity for natural language, that is the essence of human rationality. I do not wish to deny that rational human animals are necessarily also talking animals, because I accept the Chomskyan view that the language faculty is innate in human cognizers. Nor do I want to deny the Fodorian view that every rational animal, whether human or not, has a mental language or language of thought. Moreover, since the refined standard cognitivist model of the mind entails that some mental languages are syntactically identical with natural languages, I also hold that rationality itself entails natural language. What I reject is the view that human rationality is *essentially (that is, necessarily and sufficiently) connected* with the cognitive capacity for natural language. In its place I want to put the compound thesis that rational human animals are essentially logical animals, and that although every rational human animal is necessarily also a cognizer of natural language, nevertheless not all linguistically competent human animals are rational and thus not all linguistic animals are rational. In other words, I want to defend the view that it is *our logical capacity* that is essentially connected with our rationality, and not our cognitive capacity for natural language.

My argument for this view has three parts. (A) First, I will argue that it is inconceivable and therefore logically impossible for there to be rational animals who lack a cognitive capacity for natural language. Thus rationality entails a cognitive capacity for natural language. (B) Second, I will argue that there are actual cases of human beings who are linguistically competent yet nonrational. Thus human rationality is logically distinct from the human cognitive capacity for natural language. (C) Third, I will argue that necessarily every animal that has a cognitive capacity for logic is rational, and also that it is inconceivable and therefore impossible for a rational animal to lack a cognitive capacity for logic. So a human animal is rational if and only if it has a cognitive capacity for logic, and although every rational human animal is necessarily also a cognizer of natural language, nevertheless not all linguistically competent human animals are rational, and thus not all linguistic animals are rational.

(A) Rational animals, I have proposed, are normative-reflective animals possessing concepts that express strict modality. And I have also accepted the Chomskyan thesis that human animals have an innate cognitive capacity for natural language. But is it conceivable and therefore logically and metaphysically possible that a race of *nonhuman* rational animals might arise and exist (say, on Mars) in a set of contingent biological and environmental conditions such that the members of this race have no need whatsoever for speech and consequently have no capacity for natural language? In other words, could there be a race of Martians who enjoy life, liberty, the pursuit of happiness, science, normativity, the capacity for reflection, and access to strict modal concepts, but in an entirely nonsocial framework and thereby quite apart from any systematic verbal medium of communication? Some philosophers (in particular, Peter Carruthers, Robert Kirk, and Robert Stalnaker[60]) have thought so. And if they are correct, then obviously rationality does not entail natural language.

But it seems to me that the Carruthers–Kirk–Stalnaker scenario is logically and metaphysically impossible. The crucial point here, I think, lies in my assumption that in order to *be* rational, a creature must not only be instrumentally rational and holistically rational but also possess *principled rationality*. Principled rationality, in turn, entails the possession of concepts of strict modality, including logical necessity. And logical necessity is bound up with the notion of consequence, which is defined in terms of such logico-linguistic notions as *schematizable language*, *nonlogical constant*, and *logical constant*. So even if the languageless Martians could exist, they would at best be instrumentally and holistically rational, not rational in the principled sense. Or in other words, the languageless Martians would at best be *partially* rational, not *fully* rational. Furthermore, it seems to me very unlikely that even instrumental rationality and holistic rationality are possible without at least a disposition to logical reasoning and thus a dispositional grasp of concepts involving logical necessity: how could a creature figure out how best to satisfy its desires, given its beliefs, without an ability to make valid inferences? And how could a creature construct coherent systems of cognitive and practical judgments without an at least implicit grasp of the concepts of logical consistency and logical consequence? Therefore, rationality entails natural language, or at the very least, a cognitive capacity for natural language.

(B) There is much empirical evidence to indicate that the human cognitive capacity for natural language is not only innate but also modular. And one of the most striking sources of evidence for modularity is the existence of autistic language savants:[61] human beings who are introverted to the point of severe psychological impairment, but whose linguistic competence remains intact. Autism is first and foremost a condition that adversely affects a thinking subject's ability to relate to other thinking subjects, and can perhaps be most adequately explained as a failure to possess a "theory of mind"[62] module. Be that as it may, many autists manifest severe impairment in their intellectual, practical, and emotional capacities, and hence are nonrational; and the autistic language savants who have been studied in fact fall into that class. So there are nonrational humans who are linguistically competent, and therefore human rationality and the human cognitive capacity for natural language are logically and metaphysically distinct.[63]

(C) Suppose that I am correct that rational animals are normative-reflective animals who possess concepts expressing strict modality. Now consider any animal who is not only able to think logically in the implicit or unreflective sense of being able to cognize according to the principles of some logic or another, but is also capable of *explicitly or reflectively doing logic*, that is, of self-consciously grasping the principles of the protologic or of some other logic, be it a classical or nonclassical logical system. Such an animal is able to recognize and obey (or disobey) norms; is able to reflect on herself, her own psychological capacities and states, and her own behaviors and actions; and is able to grasp basic logical concepts such as that of consequence, which of course expresses the strict modality of necessity. Therefore, every such animal must be a rational animal. Now consider on the other hand an animal who lacks a cognitive capacity for logic. Such an animal is not only incapable of doing logic but is also incapable of thinking logically even in the implicit or unreflective sense of merely being able to cognize according to the principles of some logic or another. In other words, such an animal's intentional states are not able to be ordered by that animal according to any conception of logical consequence—not even the universal unrevisable a priori conception of consequence contained in the protologic. The animal's intentional states are to that extent random, disorganized, and disconnected. The animal cannot in any sense deliberate about beliefs and desires or make plans. The animal cannot make up its mind, or decide, on the basis of reasons. It then seems to me obvious that such an animal is nonrational. So it is impossible for a rational animal to lack a cognitive capacity for logic. It follows that an animal is rational if and only if it has a cognitive capacity for logic.

Taking together the conclusions of (A), (B), and (C), I conclude that human rationality is essentially our cognitive capacity for logic, not essentially our cognitive capacity for natural language. The cognitive capacity for natural language is innate in humans and entailed by rationality itself. But because there are nonrational linguistic humans, rational humans are essentially logical animals, not essentially talking animals.

This may not, in and of itself, seem like a very important thesis outside the theoretical hothouse of the philosophy of rationality. But when we combine the logic-oriented conception of human rationality with the refined standard cognitivist model of the mind, a unified picture of human nature begins to emerge that is deeply and importantly different from both early modern (whether rationalist or empiricist) and postmodern (that is, skeptical antirationalist-*cum*-naturalist) conceptions of ourselves. In stark contrast with the latter, we are essentially logical animals, and thus it is *not* the case that "all is permitted" in the realm of thinking, just as not all is permitted in the realm of rational intentional action. But at the same time, in equally stark contrast with the former, the construction of all meaningful representations of the world and of ourselves is left up to us, as active cognizers, in the sense that the set of basic or primitive meaningful representations is not *given* to us either nonempirically (as innate ideas) or empirically (as ideas of sensation). And in this sense we can, to some considerable extent, be held *personally responsible* for cognizing the world and ourselves, just as we can, to some considerable extent, be held personally responsible for outer-directed and self-directed rational intentional actions. In other words, rational thinking is both *inherently normatively constrained by logic,* and also *the fundamental cognitive project* for creatures like us. As I explore some of the basic details of the logic-oriented conception of human rationality in the next three chapters, its double-sided contrast with both early modern and postmodern conceptions of human nature will be thrown into sharper relief.

5 The Psychology of Reasoning

All correct reasoning consists of mental processes conducted by laws of thought which are partly dependent upon the nature of the subject of thought.
—George Boole[1]

For Bacon, Hume, Freud, or D. H. Lawrence, rationality is at best a sometimes thing. On their view, episodes of rational inference and action are scattered beacons on the irrational coastline of human history. During the last decade or so these impressionistic chroniclers of man's cognitive foibles have been joined by a growing group of experimental psychologists who are subjecting human reasoning to careful empirical scrutiny. Much of what they found would appall Aristotle. Human subjects, it would appear, regularly and systematically invoke inferential and judgmental strategies ranging from the merely invalid to the genuinely bizarre.
—Stephen Stich[2]

5.0 Introduction

In the previous chapter I sketched a general conception of human rationality, *the logic-oriented conception*, according to which rational human animals are essentially logical animals. The logic-oriented conception of human rationality can then be added to the logic faculty thesis, which says that rational animals possess an innate faculty for logical representation which contains the protologic (a single universal unrevisable a priori set of metalogical principles and logical concepts used for the construction of every logical system). The logic faculty thesis, in turn, coherently and apparently uniquely solves three fundamental problems about the nature of logic: the problem of logical psychologism, the *e pluribus unum* problem, and the logocentric predicament. The conjunction of the logic faculty thesis with the logic-oriented conception of human rationality is the doctrine of logical cognitivism, which constitutes a broadly Kantian theory of rationality and logic. But as Monty Python's

Flying Circus's dinosaur expert Ann Elk would say: "it's *my* theory, and what it is too." In any case, in this chapter I want to show that logical cognitivism is well supported by the experimental or scientific psychology of human (deductive) reasoning.[3]

As I pointed out in the introduction and again in chapter 1, one of the many important socio-philosophical lessons of the late nineteenth- and early twentieth-century debate about logical psychologism is that the emergence of psychology as an autonomous empirical science was symbiotically connected with the simultaneous emergence of pure logic as an autonomous nonempirical science. In other words, the extreme purity of the a priori science of pure logic and the extreme empiricality of the experimental science of psychology are conceptually complementary, just as, in a moral or theological context, conceptually there cannot be "perfect saints" without "miserable sinners" and conversely. Indeed some empirical psychologists even coined a term, 'psychological logicism', for what they regarded as the fallacious introduction of logical laws into the description of causally efficacious mental processes.[4] Despite this extrusion of logic from mental processes, and despite Frege's equal and opposite "extrusion of thoughts from the mind,"[5] however, psychologists have retained a deep interest in the psychology of logic, by repeatedly and concertedly trying to answer this hard question: what is human deductive reasoning?

The upshot of my discussion will be that Boole's observation is correct. What defines human deductive reasoning is the human animal's cognitive tendency to follow, without fail, a few innate normative principles and concepts of logic, derived from the protologic, that in turn intrinsically constrain both what the human cognizer can know about logic (for more details, see chapter 6), and how she actually conducts her thinking (for more details, see chapter 7). This cognitive tendency is what I call *protological competence.*[6]

In section 5.1, I briefly spell out and then critically compare two sharply opposed classical accounts of human reasoning offered by William James and Jean Piaget. In a certain way, these accounts frame the space of possible theories of reasoning. Most contemporary work on the psychology of logic, however, takes its direct cue from the famous "reasoning tests" developed by Peter Wason in the 1960s and studied intensively by him and Philip Johnson-Laird. The results of these tests (together with the similar results of similar tests on probability judgments later developed by Daniel Kahneman and Amos Tversky[7]) are often taken to show that human cognizers are irrational, in the sense that humans of normal or higher intelligence regularly, systematically, and even flagrantly deviate from logical norms of truth, consistency, and validity in their reasoning.[8] This conclusion was later challenged by L. J. Cohen in 1981,[9] thereby triggering a vigorous and wide-ranging *fin de siècle* debate, centered on the psychology of human reasoning, about the nature and scope of human rationality.[10] In section 5.2, I take a preliminary look at the reasoning tests (or more precisely, at those tests that bear directly on *deductive* reasoning) and the rationality debate. Then in sections 5.3 through 5.6 I offer a critical survey of four contemporary scientific psychological theories of human reasoning: (i) the mental logic theory, (ii) the mental models theory, (iii) the heuristics-and-biases theory, and (iv) the minimal rationality theory. In section 5.7, I sketch and defend my favored alternative to those theories, the protological competence theory. Then finally in section 5.8, against the backdrop of the protological competence theory, I take a second look at the reasoning tests and the rationality debate.

5.1   James, Piaget, and the Psychology of Reasoning

The psychology of human reasoning owes its origins to the intellectual spadework of William James and Jean Piaget. As is well known, James works out a highly flexible or pluralistic conception of the human mind, a conception that is in equal measures phenomenological or introspective, pragmatic or human-interest-driven, behavioral or operational, and neurophysiological. As is equally well known, by contrast, Piaget's conception of the mind emphasizes formal cognitive functions and structures, a rigidly fixed developmental teleology, and a quasi-nativist conception of the cognitive architecture characteristic of the several developmental stages. These conceptions, in turn, carry over directly into their respective psychologies of reasoning. James's account consists in an attempt to distinguish introspectively, pragmatically, and behaviorally between thought processes involving "true reasoning" and other types of thought processes. Piaget's account on the other hand consists in an attempt to show how the intact human animal gradually, inevitably, and systematically works its way from a stage of prelogical cognizing, in infancy, to a more or less explicit grasp, in maturity, of a particular system of mathematical logic.

As I mentioned above, what I want to argue in this section is that in a certain way James's and Piaget's accounts frame the space of possible theories of the nature of human deductive reasoning. What I mean is this. For James, in the end, virtually *any* sort of human thinking counts as logical thinking. By contrast, for Piaget, cognition is logical just to the extent that by means of an invariant genetic process it conforms itself to the canons of a single system of mathematical logic. So James's theory takes us to the limit of letting in radically too much as human reasoning, whereas Piaget's theory takes us to the opposite limit of letting in radically too little.

James on human reasoning   James's psychology of logic is spelled out in the "Reasoning" chapter of his seminal *Principles of Psychology*. At first glance his account seems to cover an irrelevantly wide array of topics, including "the intellectual contrast between brute and man"; "different orders of human genius"; speculations on the brain-basis of reasoning; and some fairly invidious (and from our cultural-ethical standpoint, fairly scandalous) comparisons between the reasoning powers of Germans and Italians and between those of "men as a whole and women as a whole."[11] But James's seemingly scattergun and occasionally chauvinistic approach to human reasoning in fact helpfully provides us with a pithy, pungent anticipation of the deeply important, tangled, and highly controversial philosophical, psychological, cultural, and ethical issues that are at stake in the late twentieth-century and now twenty-first-century debate about human rationality (see sections 5.2 and 5.8, and chapter 7).

James's basic thesis is that human deductive reasoning is a special kind of orderly thinking, applicable to any subject matter (what later psychologists call "domain-general"), and essentially bound up with everyday believing and practical decision making (what later psychologists call a central or "higher-level" process), involving "general characters" (that is, concepts) and predication.[12] This is opposed to a merely associative thinking that involves random ordering or ordering by sheer similarity. In turn, James's account of conceptualization and predication is oriented heavily toward the Aristotelian logic of the syllogism. But what is crucial for him is a sharp contrast between reasoning and mere associative thinking, manifested introspectively, in deliberative processes leading to human action and in speech behavior. Associative thinking, contained in "trains of images suggested one by another . . . [in] a sort of spontaneous revery," *can* in fact be rational to the extent that it "leads to rational conclusions, both practical and theoretical."[13] But associative thinking deals only with "empirical concretes, not abstractions."[14] The production of abstract or general characters out of a stream of thoughts, together with the manipulation of these general characters, yields a new element in cognition and therefore isolates the phenomenon of reasoning: *"Let us make this ability to deal with NOVEL data the technical differentia of reasoning."*[15]

Within the domain of thought-orderings essentially involving concepts and their predicative manipulation, however, James does not adequately distinguish between sequences of thoughts that are linked to one another in a merely temporal or causal way, and sequences (such as those of the categorical syllogism) that include specifically *logical* linkages. Indeed he is led into inconsistency in the very examples he uses to illustrate his main thesis. He identifies the inferences of "true reasoning," as opposed to the hops, skips, and jumps of associative thinking, by using the notion of a thought process that is altogether underdetermined by sensory experience. Here he explicitly contrasts "reasoned thought" with "empirical thought."[16] But if he were correct that reasoning is essentially the imposition of conceptual and predicative structures onto the stream of thoughts, then he would not be able to distinguish between (i) experientially imposed associative sequences involving conceptual and predicative structures that are nevertheless formally fallacious (for example, 'This cat over here is a four-legged animal; that cat over there is a four-legged animal; therefore all cats are four-legged animals'), and (ii) nonexperiential valid deductive sequences (for example, 'All cats are quadrupeds; all quadrupeds are four-legged animals; therefore all cats are four-legged animals'), as he himself concedes: "association by similarity and true reasoning may have identical results."[17]

James disguises this problem somewhat by saying that the reasoned conclusion is the one that bears a more "obvious" or "evident" relation to its preceding thoughts than any of the infinitely many different thoughts that might otherwise have been suggested by them.[18] Obviousness or self-evidence here is clearly a stand-in for intuitions about logical consequence. But his appeal to the relative generality and relative simplicity of the conclusions of reasoning,[19] as opposed to those of empirical thinking, plainly will not do: these features are neither necessary nor sufficient to explain even the examples he gives, since equally simple or equally general hypotheses, as well as simpler and more general hypotheses, are perfectly consistent with the data given to the cognizer, although by James's own reckoning some of these hypotheses would constitute formally incorrect or fallacious conclusions from the data or premises given to the cognizer.

James implicitly acknowledges this further difficulty by directly appealing to "sagacity"[20] or the capacity to pick out the *right* or essential characters from the data in order to guarantee that the inference is reasoned and not merely empirical: "in reasoning, we pick out essential qualities."[21] But ultimately he has no way of distinguishing between the right and wrong choice of characters, except by covertly assuming some sort of special cognitive access to purely logical notions and principles. So, *excluding* this covert assumption (for which he has not argued and to which he is not entitled without invoking a great deal more theoretical machinery than he is prepared to allow), James is left with the uncomfortable result that virtually any thought process involving concepts and predication, whether formally correct or fallacious, will count as "true reasoning" or logical thinking, whenever it *seems sagacious* to the thinking or acting subject, or to the subject's interlocutors.

Piaget on genetic logic A notable feature of James's treatment of human reasoning is his uncritical reliance on the Aristotelian syllogism as the paradigm of logic,[22] and corresponding to that, his evident lack of interest in the leading work in nineteenth-century logical theory: after all, *Principles of Psychology* was published thirty-six years after Boole's *Laws of Thought*, and eleven years after Frege's *Begriffsschrift*. Just the reverse is true of Piaget, whose first book was a study in formal logical theory and who quite explicitly sets his theory of the human mind against the backdrop of a Boolean conception of logic that is further informed by the logical framework of *Principia Mathematica*. What for convenience I will call *Piagetian logic*, developed in Piaget's 1949 *Traité de logique* is a classical bivalent propositional calculus plus an algebraic logic of classes, hence a conservative mathematical extension of classical propositional logic.

Assuming Piagetian logic as a theoretical backdrop, Piaget holds that the mind of the intact human animal is an essentially dynamic phenomenon, continuous with ecologically embedded biological systems more generally, whose dynamism takes the specific form of an irreversible and continuous four-stage developmental process:

1. the sensorimotor period (0–2 years),
2. preoperational thought (2–7 years),

3. concrete operations (7–11 years), and finally
4. propositional or formal operations (from 11–12 to 14–15 years).[23]

With the exception of the sensorimotor period, which contains a bare minimum of logical structure and is therefore "prelogical," each cognitive stage contains significantly more logical structure than the next lowest stage; and each cognitive stage is an incomplete or gappy adumbration of the next highest stage.

In all stages prior to the last stage, the cognitive architecture of the human animal is in some respects in "equilibrium" with its external experiential inputs, which is to say that there is an isomorphism between the logical structure of the animal's mind and the logical structure of its surrounding objective world, and also a practical matchup between the coping abilities of the animal and the goading urgencies of its local and global environment. But in these cognitively lower or immature stages it is also in some respects *not* in equilibrium, which is to say that the cognitive framework of the animal is in certain salient ways both formally and practically inadequate to the richer logical framework of its world. In any cognitive stage, the human animal's mind becomes "operational" to the extent that it constructs language-like mental representations (each of which belongs to a "structured whole," or a relatively complete formal system of such representations) and attempts to apply them directly to its world. The notion of a psychological operation, in turn, corresponds directly to the notion of a logical operation. The last or fourth stage of propositional or formal operations, in which the cognizer finally achieves an adequate mind–world equilibrium, constitutes an implicit or explicit grasp of "adult logic," that is, Piagetian logic.[24]

Leaving aside Piaget's developmentalism, there is obviously a significant overlap between his conception of the human mind and what in the previous chapter I called the standard cognitivist model of the mind. Indeed, if we take Piaget's stage 4, the stage of formal or propositional operations, to provide an account of the cognitive structure of the human mind per se, then it is easy enough (1) to see a human cognizer's system of Piagetian formal or propositional operations as quite similar to a Chomskyan or Fodorian mental language, in that like a mental language it imposes significant constraints on mappings from sensory inputs to cognitive outputs; (2) to see Piagetian logic as the logical analogue of Chomsky's universal grammar or UG; and (3) to see the ideal stage 4 cognizer's mastery of Piagetian logic as the analogue of Chomsky's linguistic competence.

Looked at in these three ways, however, Piaget's psychology of reasoning has two obvious problems: (i) Piagetian logic clearly does *not* stand to all other logics as the UG stands to all natural languages, since it is obvious that Piagetian logic is neither implicit in, nor generative of, all the other logics; and (ii) there is little or no empirical evidence to suggest that when left to their own devices, intact humans will ever naturally develop an implicit or explicit grasp of Piagetian logic and reach stage 4, the level of formal or propositional operations—in fact, just the contrary seems to be the case. As we will see in the next section, there is a significant amount of empirical evidence that seems to show that when cast back upon their own resources in experimental test situations involving deductive tasks, humans of normal or above-average intelligence naturally employ reasoning strategies *not* in accordance with normative logical principles derived from Piagetian logic or from any other classical or extended nonclassical logical system. It appears that human beings are irrational animals!

But even if humans are not irrational and have instead a reasoning competence that matches the normative principles of some logic, why should those normative principles derive specifically from *Piagetian logic*? More precisely, why should the structure of the human mind in its ultimate stage of cognitive development necessarily mimic the structure of Piagetian logic in particular and not the structure of some *other* conservative extension of classical or elementary logic, or for that matter, the structure of some deviant logic? Piaget is a diehard (albeit conservative) nonclassicist and therefore subject to worries we have already rehearsed about diehard nonclassicism as a solution to the *e pluribus unum* problem (see section 2.5). But now we can also see that the basic problem with Piaget's psychology of reasoning is its excessively narrow commitment to Piagetian logic as both the paradigm of logic and the determiner of the structure of adult human cognition.

5.2 Reasoning Tests and the Rationality Debate I

Broadly speaking, until the mid-1960s, the psychology of human deductive reasoning tended to run in two somewhat divergent directions. On the one hand it followed the lead of James, with his emphasis on the introspective, pragmatic, or behavioral contrast, built into his notion of "sagacity," between formally correct and fallacious reasoning. And on the other hand it followed the lead of Piaget, with his emphasis on "genetic epistemology," which is the doctrine that the cognitive architecture of the adult human mind is biologically predetermined and also essentially configured by Piagetian logic. Following *both* James and Piaget, in the same period psychologists of human reasoning also by and large (1) assumed that reasoning is a domain-general, central cognitive process, and (2) concentrated narrowly on deductive reasoning according to canons of syllogistic logic or classical propositional logic, and thus almost universally avoided the study of reasoning according to canons of extended logic or deviant logics, not to mention the study of inductive, probabilistic, or practical reasoning.

In the Jamesian direction, M. C. Wilkins in 1928 argued on the basis of empirical studies that the semantic content of syllogisms tends to affect the performance of human subjects whose cognitive task is to discriminate between valid and invalid syllogisms.[25] And in 1935 and 1936, R. S. Woodworth and S. B. Sells argued on the basis of further empirical studies that the variability in human performance on the syllogism task noted in the Wilkins study could be explained to some extent by the "atmosphere effect."[26] The atmosphere effect is that the logical mood of the premises of a syllogism tends to make it easier for the subjects to draw conclusions in the same mood but harder to draw conclusions in converse, contrary, or contradictory moods, because propositions or sentences in such moods tend to cancel the atmosphere of the propositions or sentences in the converse (etc.) moods.

But by contrast, in the Piagetian direction, Mary Henle argued in 1962 that the frequent failure of subjects to complete the syllogistic task correctly in the earlier tests could be traced to their irrelevantly focusing on the semantic features of syllogisms and failing to "accept" the purely logical task.[27] When the logical task was properly accepted, she suggested, the errors would disappear and reveal the underlying competence of the intact mature human thinker to reason in accordance with normative syllogistic principles.

The contrast between the basically Jamesian strategy of Wilkins, Woodworth, and Sells on the one hand, and the basically Piagetian strategy of Henle on the other, carries over into an even more salient and important contrast between two fundamentally different approaches to human reasoning. It is a striking and repeatedly confirmed empirical fact that most human subjects of average or even above-average intelligence perform poorly on deductive tasks under experimental test conditions. Furthermore, even when subjects perform well, they appear to do so by following procedures that have little or nothing to do with normative logical principles. As Stich aptly puts it: "Human subjects, it would appear, regularly and systematically invoke inferential and judgmental strategies ranging from the merely invalid to the genuinely bizarre." But this robust empirical fact can be taken to show either (1) that humans are irrational animals, or (2) that despite their bad test results, humans are rational animals after all. For convenience, let us call the first approach *irrationalism* and the second approach *rationalism*.

The line of reasoning that sharply diverges in these two ways can be expressed more explicitly. Given the robust fact of highly variable human performance on deductive tasks under experimental test conditions, and assuming

(α) that there are normative logical principles;

(β) that human rationality at least partially consists in reasoning in accordance with these normative logical principles;

(γ) that these normative logical principles express the inference rules, axioms, logical truths, or basic logical concepts of Aristotelian logic, Piagetian logic, or some other system of classical or nonclassical logic; and

(δ) that successfully or unsuccessfully performing deduction tasks (more specifically, successfully or unsuccessfully performing the "Wason selection task"—which will be described in some detail immediately below—and other similar reasoning tasks) under experimental test conditions adequately reveals human reasoning ability;

then

(ϵ) *if* it is true that humans do not naturally reason in accordance with normative logical principles but instead naturally use intrinsically nonlogical strategies to deal with the deduction tasks, then it follows that humans are irrational[28] animals; but

(ζ) *if* on the other hand it is true (a) that some test-induced nonlogical factor introduces a "cognitive illusion" that determines performance on the original tests by effectively impeding the natural capacities of humans to reason in accordance with normative logical principles; and also (b) that at least in principle, this nonlogical factor can be experimentally manipulated so as to reveal, on some other versions of the tests, our underlying competence to reason in accordance with normative logical principles, then it follows that humans are rational animals after all.

And that is the crux of the rationality debate.

The rationality debate, to a surprising extent, grows out of a single psychological experiment: the *selection task,* invented by Wason and subsequently studied in detail by Wason and Johnson-Laird.[29] Here is the task, as crisply described by Lance Rips:

> Suppose you have in front of you four index cards. On the first card is the letter E, on the second K, on the third the numeral 4, and on the last 7. The experimenter tells you that each card contains a numeral on one side and a letter on the other, although you are not allowed to look at the flip sides of the cards. The experimenter also informs you that you are to "decide which cards [you] would *need* to turn over in order to determine whether the experimenter was lying in the following statement: If a card has a vowel on one side, then it has an even number on the other side" (Wason, 1966, p. 146). (In later versions of the problem, subjects were asked to "name those cards and only those cards, which need to be turned over in order to determine whether the rule is true or false"; see Wason and Johnson-Laird, 1972, p. 173)[30]

Most often, people say either that the E and the 4 cards should be turned over (46% in Wason and Johnson-Laird's study), or that the E card alone should be turned over (33%). But unfortunately both of those answers are wrong. To determine whether the logical rule is true or false, only the E and 7 cards should be turned over. This is because the rule is in the form of a material conditional of classical propositional logic. A material conditional of classical propositional logic is true just in case either its antecedent is false or its consequent is true; and it is false just in case its antecedent is true and its consequent is false. So both the K card (which makes the antecedent false) and the 4 card (which makes the consequent true) are consistent with the rule, whether or not they are turned over. Only the E card (which makes the antecedent true) and the 7 card (which makes the consequent false) are such that when they are turned over, they will *determine* the truth or falsity of the rule.

In Wason and Johnson-Laird's study, only 4 percent of the people who tried the selection task (all of them university undergraduates of normal or higher intelligence) actually completed it successfully. But the plot thickens. Wason and Johnson-Laird also discovered that by simply changing the verbal formulation of the task, they could significantly improve success rates, while holding the basic logical structure fixed. Indeed, R. A. Griggs and G. R. Cox later found that the success rate jumped to a whopping 74 percent when (i) the original rule was replaced by the rule, *If a person is drinking beer, then the person must be over 19*; (ii) the new rule was said to refer to four drinkers sitting around a table; and (iii) the original four cards were replaced by new ones according to the following scheme:

E-card a card that says "drinking beer" on one side.

K-card a card that says "drinking Coke" on one side.

4-card a card that says "22 years of age" on one side.

7-card a card that says "16 years of age" on one side.[31]

It has also been shown, however, that not *every* change in the experimental test materials has the same effect of radically improving success rates. For example, K. I. Manktelow and J. Evans discovered that the success rate was a mere 7 percent (i.e., closely comparable to the original results) when (a) the original rule was replaced by *If I eat haddock, then I drink gin;* (b) the new rule was said to refer to what I ate at a particular meal; and (c) the original four cards were replaced by these new ones:

E-card a card that says "I eat haddock" on one side.

K-card a card that says "I eat macaroni" on one side.

4-card a card that says "I drink gin" on one side.

7-card a card that says "I drink champagne" on one side.[32]

Taken together, these results lead directly to two quite different questions that have sometimes been run together in the massive subsequent literature:

(1) Precisely *which* nonlogical factor determines the variability in human performance on the selection task and other similar reasoning tasks?

(2) Given the fact that *some* isolable nonlogical factor determines the variability in human performance on the selection task and other similar reasoning tasks:

(2i) do humans naturally deal with the selection task and other similar reasoning tasks by using strategies not in accordance with normative logical principles, that is, by using intrinsically nonlogical strategies? Or, alternatively,

(2ii) does the relevant nonlogical factor in the selection task and other similar reasoning tasks constitute only a cognitive illusion that determines poor performance on the "standard abstract selection task" (i.e., the original E, K, 4, 7 version of the selection task), without affecting our underlying competence for reasoning in accordance with normative logical principles?

It is question (2), and *not* question (1) on its own, that yields the debate about human rationality. Discovery of the correct answer to (1) will not, in and of itself, cut one way or the other. Irrationalists and rationalists would *both* like to know what actually causes highly variable human performance on the selection task and its analogues, and they are not (usually) antecedently committed to any specific sort of account. By contrast, however, a positive answer to option (2i) entails the irrationalist conclusion that humans are irrational, whereas a positive answer to (2ii) entails the rationalist conclusion that humans are rational after all. Nevertheless, we can conclude to either irrationalism or rationalism *only if* we assume, as spelled out under the four enabling assumptions listed above: (α) that normative logical principles exist; (β) that human rationality at least partially consists in reasoning in accordance with these normative logical principles; (γ) that these principles express the inference rules, axioms, logical truths, or basic concepts of Aristotelian logic, Piagetian logic, or some other system of classical or nonclassical logic; and finally, (δ) that successfully or unsuccessfully performing the selection task and other similar reasoning tasks under experimental test conditions adequately reveals human reasoning ability.

That is the gross architecture of the rationality debate. Let's look now at some of the fine grain.

The basic empirical data deriving from the Wason selection task are these:

(1) that a large majority of people of ordinary or higher intelligence fail to complete the standard abstract selection task successfully; but

(2) when certain nonlogical factors are added to the experimental materials of the standard abstract selection task (thereby yielding some or another version of the "thematic selection task"), then a sizable majority of people of normal or higher intelligence successfully complete the selection task; but

(3) not every addition of nonlogical factors to the experimental materials of the standard abstract selection task induces a successful completion of the task: in fact, the success rate for some versions of the thematic selection task is approximately as low as that of the standard abstract selection task.

What is the difference between the standard abstract selection task and the thematic selection task? In the standard abstract selection task, a set of (relatively speaking) *content-free* and *value-neutral* experimental materials is presented to the subject, whereas in the thematic selection task, certain semantically *contentful* or *value-laden* factors have been added to the experimental materials. The irrationalist argues that in the case of successful performance on any version of the thematic selection task, the added semantically contentful and/or value-laden factors trigger an intrinsically nonlogical cognitive competence of the subject (be it on the one hand a semantic competence, or on the other hand a competence bound up with "heuristics and biases"[33]) thereby making it possible for her to complete the selection task successfully. The irrationalist then infers that human cognitive competence for deductive reasoning tasks is not in accordance with normative logical principles, and concludes that humans are irrational.

This in turn highlights one crucial difference between the irrationalist and the rationalist: their treatment of the standard abstract selection task. The irrationalist takes the dismal showing of most test subjects on the standard abstract selection task to be an accurate indicator of the inherently unsound logical reasoning capacities of humans of normal or higher intelligence. By sharp contrast, the rationalist believes that some nonlogical factor in the experimental materials of the standard abstract selection task effectively suppresses the inherently sound logical reasoning capacities of humans of normal or higher intelligence. So the irrationalist thinks that the low success rate on the standard selection task together with a high success rate on some versions of the thematic selection task shows that our cognitive capacity for reasoning is intrinsically nonlogical and thus irrational, whereas the rationalist thinks that the low success rate on the standard abstract selection task together with a high success rate on certain versions of the thematic selection task shows that our cognitive capacity for reasoning is intrinsically logical and thus rational. Stalemate!

Or is it? At this point the rationalist, and here I specifically have in mind Cohen,[34] attempts to break the deadlock by using an a priori argument:

(1) There is a basic issue, not addressed by irrationalists, about how we advance from the plurality of distinct logical systems, to the normative criteria of deductive reasoning that are used by psychologists to evaluate performance on the reasoning tests and to construct their empirical theories about human reasoning.

(2) Only pretheoretic human intuitions about deduction, gradually refined and justified in the process of seeking a narrow reflective equilibrium[35] across logical beliefs, and then later systematized as a model of *idealized competence in the Chomskyan sense*, will mediate the advance from the plurality of distinct logical systems to the normative criteria of deductive reasoning. This is because there is no other relevant source of data, no other adequate method of belief justification, and no other acceptable

way of constructing the normative cognitive theories on the basis of which psychologists of reasoning build their empirical theories of deductive reasoning.

(3) For these reasons, the conclusion to draw from the robust fact of highly variable performance on the reasoning tests *cannot* coherently be that humans are irrational, since the normative cognitive framework behind the reasoning tests is derived entirely from pretheoretic human intuitions about deduction, by way of the method of reflective equilibrium and the idealized Chomskyan competence–performance distinction. On the contrary, then, it is *presupposed* by the reasoning tests that humans are rational, at least at the level of idealized deductive competence.

(4) Moreover, given the distinction between idealized deductive competence and deductive performance, the high variability of performance in the reasoning tests can be adequately explained by appealing to a cognitive illusion (analogous to the Müller-Lyer and Ponzo visual illusions) induced by the standard abstract selection task, an illusion that can be made to disappear in at least some versions of the thematic selection task by manipulating the relevant nonlogical factor in the experimental materials.

Cohen's important argument has been much discussed and much criticized by cognitive scientists and philosophers. Not surprisingly, philosophical criticism has focused mainly on step (2), and in particular on Cohen's crucial appeals to (a) reflective equilibrium, and (b) the idealized Chomskyan competence–performance distinction. In particular, Steven Stich and Edward Stein have argued that neither reflective equilibrium nor the idealized Chomskyan competence–performance distinction taken individually, nor the two of them taken together, suffices to guarantee a priori that humans are rational, given the empirical evidence from the reasoning tests.[36] Stich and Stein conclude that the rationality or irrationality of human reasoners is, at least in part, irreducibly an empirical question, which leaves wide open the possibility that irrationalism is true.

Assuming that Stich's and Stein's worries are cogent, then Cohen's a priori gambit fails and we are back at the condition of stalemate in the debate about rationality. But there is another way out. Remember that the irrationalist and rationalist alike accept the four assumptions (α) through (δ). Later I will argue that although the first and second assumptions (α) and (β) are true, the third and fourth assumptions (γ) and (δ) are false. This blocks the entailment from *either (2i) or (2ii)* to substantive conclusions about human rationality or irrationality. Furthermore, I will argue that the empirical evidence from the reasoning tests in fact supports *neither irrationalism nor rationalism*, but instead supports my nonirrationalist, nonrationalist doctrine, the protological competence theory. So, if I am correct, it turns out, contrary to the assumptions of both of the opposing sides in the debate about human rationality, that the basic connection between the empirical evidence from the reasoning tests and the nature of human rationality needs to be significantly reconceived.

In any case, the crucial irrationalist and rationalist cards are now on the table. In order to evaluate the cogency of their views, I will turn in the next four sections to a critical survey of the leading contemporary psychological theories of human reasoning.

One caveat before I do that, however. As we have seen in chapter 1, logical cognitivism entails the denial of logical psychologism. By contrast, the four scientific psychological reasoning theories I will consider are all, implicitly or explicitly, psychologistic.[37] That issue has already been adequately covered in chapter 1, so for the purposes of the present chapter I will leave it aside. Nevertheless, it is important to keep in the back of one's mind that no matter what significant similarities there might be between logical cognitivism and any of the psychological theories of reasoning, they all differ sharply from logical cognitivism on that crucial point.

5.3 The Mental Logic Theory

The mental logic theory was anticipated in the 1940s and '50s by Piaget's account of the fourth genetic cognitive stage of "formal operations," and again in the 1960s by Henle's Piaget-inspired early defense of rationalism. The philosophical foundations of the mental logic theory were adumbrated by Fodor in 1979,[38] under the joint influences of Chomsky's psycholinguistics and the computational theory of mind. But, presumably owing to the vicelike grip of the dying hand of behaviorism on mid-twentieth-century experimental psychology, with the exception of a seminal paper published by Martin Braine in 1978,[39] the mental logic theory was not explicitly developed or seriously defended by psychologists until the 1980s and '90s. The leading proponents of the mental logic theory are Braine and his collaborator David O'Brien,[40] John Macnamara,[41] Willis Overton,[42] Lance Rips,[43] and Norman Wetherick.[44]

The mental logic theory says that there is a logic internal to the human mind–brain and that intact, mature human beings reason by spontaneously and systematically applying, to the representations of natural-language sentential inputs, a set of content-free, syntax-sensitive inference rules (also known as "inference schemas") that are severally preencoded in an informationally encapsulated mental module. The inference rules or inference schemas fit the format of a Gentzen-style natural deduction system. The preencoding of the inference rules in the mental module may or may not be innate. But in any case the set of inference rules or schemas directly corresponds, in a proof-theoretic way, to some particular classical or nonclassical logical system.

At this point I should concede explicitly what the alert reader will already have noticed, namely that the mental logic theory is somewhat similar in motivation and formulation to logical cognitivism, and in particular somewhat similar to the logic of thought thesis I developed in section 4.8. Both theories trace their historical lineage back to Kant and Boole. And both theories are thoroughly cognitivist in character. More precisely, both theories say that there is a logic internal to the human mind–brain in the sense that the human mind–brain contains a mental module with some science of the necessary relation of consequence already specified in it, and that this mentalistic fact about us partially or wholly explains human reasoning.

Despite their gross or high-level similarity, however, the mental logic theory and the logic of thought thesis must be sharply distinguished. So I will present the basic elements of the mental logic theory by explicitly bringing out the significant differences between the two doctrines. And to keep confusion to a minimum, from now on I will use the capitalized term 'Mental Logic' whenever mental logic being in the sense specific to the mental logic theory is being discussed.

Here are seven important differences between the mental logic theory and the logic of thought thesis.

(1) *A difference concerning the treatment of modularity* The mental logic theory assumes that all mental modules or faculties, including the logic module, are of the same cognitive order. This means that all mental modules are peripheral or lower-level, triggered by sense perception, and in the specific case of the logic module, triggered by the perception of certain sequences of natural language sentence-inscriptions or utterances. But then the mental logic theory has trouble explaining why humans of ordinary or

higher intelligence do so badly on the standard abstract Wason selection task. It is, on the face of it, natural to think that the abstract nature of the materials (hence the lack of semantic content or value-laden factors) would guarantee that the logic module would be smoothly and appropriately triggered by the standard abstract selection task. So it is a genuine problem for the mental logic theory that the logic module actually operates so poorly under what should be optimal conditions.[45] By contrast, the logic of thought thesis postulates the cognitive existence of a fully nonperipheral or central module (namely, the logic faculty) that *mediates* between the peripheral modules and central cognitive processes. This means that human reasoning involves an extra step of cognitive construction upon, or interpretation of, the perceptual input and therefore is inherently open to the possibility of various cognitive errors creeping in at that second stage. (For more on this last point, see section 5.7.)

(2) *A difference concerning informational encapsulation* The mental logic theory assumes that all mental modules, including the logic module, are informationally encapsulated. But then the mental logic theory has trouble explaining the experimentally well-substantiated thematic materials effects for the selection task. This is because informational encapsulation conceptually implies domain-specificity (since obviously there is no informational encapsulation without informational specialization), and then domain-specificity in turn entails that all or most thematic information that is irrelevant to the logic module's modus operandi is automatically screened out. So the informational encapsulation of the logic module should imply a high resistance of the module to thematic materials effects: but that contradicts the robust experimental evidence. By contrast, the logic of thought thesis says that the logic module is informationally promiscuous. This means, among other things, that human deductive reasoning is open to penetration by the central processes of theorizing, judgment, belief, desire, and volition, which are of course highly sensitive to thematic materials.

(3) *A difference concerning the scope of mental logic* The mental logic theory postulates a unique Mental Logic shared by all intact, mature human reasoners. But in the face of the manifest plurality of logics, and in view of the corresponding *e pluribus unum* problem, a special reason needs to be offered why *that* very logical system, and not some other one, is the unique Mental Logic of the human species. This is obviously another very heavy burden of proof for the mental logic theory. By contrast, the logic of thought

thesis says that there are many different logics of thought across rational human animals, which obviously is much easier to square with the empirical evidence than the contrary thesis.

(4) *A difference concerning the formal characteristics of mental logic*   The mental logic theory assumes that the Mental Logic is cognitively basic and takes the form of some particular classical or nonclassical formal system. But this raises the *e pluribus unum* problem again. What I mean is that this two-part assumption leads the mental logic theory's defenders to be quite naturally and (as far as I can tell) almost invariably[46] committed to either diehard classicism or diehard nonclassicism. But both of these are problematic, as we saw in section 2.5. By contrast, the logic of thought thesis says that the logic of thought presupposes, as cognitively basic, the protologic. The protologic, again, is a special set of schematic logical structures: a single universal unrevisable a priori set of metalogical principles and logical concepts used for the construction of all logical systems, whether classical or nonclassical. So the protologic itself is neither a classical or elementary logic, nor a nonclassical logic, nor strictly speaking even a *logical system* as such, but is instead an *all-purpose logical tool-kit* used for cognitively constructing, analyzing, and comparatively evaluating logical systems. The notion of an "all-purpose logical tool-kit" is obviously only an illustrative and suggestive metaphor: so earlier on I proposed that we adopt Chomsky's "principles and parameters" theory of the UG as the best working model of the protologic (see section 2.6). I also proposed that the precise structural description of the protologic is a new and important collaborative empirical project for logicians and cognitive psychologists. In any case, the crucial philosophical point for my present purposes is the argument for the *existence* of the protologic, for a commitment to the existence of the protologic avoids commitments to diehard classicism, diehard nonclassicism, and logical pluralism alike.

(5) *A difference concerning the treatment of logical constants and logical consequence*   To the extent that the mental logic theory is committed to the thesis that the Mental Logic is a set of content-free, syntax-sensitive inference rules or schemas (a Gentzen-style natural deduction system) preencoded in an informationally encapsulated mental module, there is thereby a strong tendency within the mental logic theory toward the adoption of (1) an inferential-role theory of the meaning of the logical constants, and (2) a proof-theoretic or syntactic account of logical consequence, as opposed to

a model-theoretic or semantic account. In the next chapter I will express some worries about inferential-role semantics for the logical constants.[47] But perhaps more importantly, it is arguable that the commitment to a purely syntactic account of logical consequence runs into a Gödelian wall. Gödel's second incompleteness theorem shows that classical logical systems, plus the axioms of Peano logic, will contain some sentences that are true or valid but unprovable. The true unprovable sentences are clearly semantic consequences of the axioms in the sense that necessarily they are true if the axioms are true. So semantic consequence outruns provability, and therefore a *purely* proof-theoretic account of logical consequence cannot be right. By contrast, the logic of thought thesis implies (i) that logical consequence will always be at the very least semantic consequence, and (ii) that every logic of thought or mental logic will include some content-free, syntax-sensitive inference rules *and* some irreducibly semantic factors (whether in the form of semantic rules, directly referential terms, intensions, associated models, or whatever). This is because logics of thought are *interpreted* logical systems, insofar as they are the logics that competent speakers associate with their native languages. So unlike the mental logic theory, the logic of thought thesis can be used to explain human deductive reasoning in contexts involving the cognition of logical systems for which logical truth is strictly speaking unprovable.

(6) *A difference concerning the treatment of innateness* Not every version of the mental logic theory is committed to the innateness of the Mental Logic. All that is required is that the Mental Logic be internal to the human mind–brain, in the sense that it exists in an informationally encapsulated mental module. And that is perfectly consistent with the thesis that the creation of this encapsulated module occurred adventitiously. But on the other hand, many versions of the mental logic theory *are* committed to nativism. To the extent that a version of the mental logic theory is nativist and furthermore posits innate *knowledge* of inference rules, it is also committed to some form of the classical Cartesian rationalist thesis of innate ideas (see section 4.1, under heading (3)). But this version of nativism is subject to familiar and apparently cogent empiricist worries, now extended from innate concept-possession to innate logical knowledge, to the effect that it seems exceedingly unlikely that the intact, mature human mind could, from the beginning of its appearance on earth, ever innately contain a stock of such highly specific and highly internally structured contents as the primitive rules

'or'-introduction, 'or'-elimination, and the like, together with an indefinitely large number of derived rules based on the stock of primitive rules. By contrast, the logic of thought thesis rejects the innate ideas thesis as a defensible form of nativism and instead affirms merely the innate existence of some constructive logical capacities, or *innate logical powers*.

(7) *A difference concerning the treatment of human rationality* Although not every version of the mental logic theory is nativist, as I just pointed out, all versions of the mental logic theory entail rationalism:[48] if humans reason by using a Mental Logic, then the human reasoning capacity is intrinsically logical and some nonlogical factor in the experimental materials of the standard abstract selection task effectively suppresses the inherently sound logical reasoning capacities of humans of normal or higher intelligence. By contrast, the logic of thought thesis is consistent with the denials of both rationalism and irrationalism alike. (For more on this crucial point, see section 5.8 below.)

5.4   The Mental Models Theory

The mental models theory was inspired by the Scottish psychologist Kenneth Craik's prescient idea in the 1940s to the effect that perception-based reasoning and decision making operate by constructing small-scale models of reality.[49] That idea was systematically developed as a theory of deductive reasoning by Philip Johnson-Laird in the early 1980s, and then later by Johnson-Laird in collaboration with Ruth Byrne in the early '90s.[50] The mental models theory begins with what appears to be a thoroughgoing rejection of the mental logic theory. Indeed Johnson-Laird even coined the term 'mental logic' precisely in order to pick out the view he was most concerned to reject. So the mental models theory and the mental logic theory would seem to be radically opposed to one another. But closer inspection reveals a somewhat subtler situation.

The mental models theory, to be sure, explicitly denies the defining thesis of the mental logic theory to the effect that intact, mature human beings reason by spontaneously and systematically applying, to the representations of natural-language sentential inputs, a set of content-free, syntax-sensitive inference rules or inference schemas preencoded in an informationally encapsulated mental module. The dismal performance results of human reasoners of ordinary or higher intelligence on the standard abstract selection test and other similar reasoning tests, say the defenders of the mental models theory, clearly show that this cannot be the case.

On the contrary, according to the mental models theory, humans reason by spontaneously and systematically constructing and manipulating special imagelike or maplike (that is, nondescriptive, nonpropositional, shape-sensitive or spatially formatted, world-directed) mental representations of the relevant natural-language sentential inputs to the thinking subject, namely the premises and conclusions of deductive arguments.[51] These special imagelike or maplike mental representations are *mental models*. Mental models are thoroughly semantic. Each model represents a possible situation in the world, and an argument to a conclusion is valid if it holds true in all models of the premises. Mental models are the psychological correlates of Euler diagrams, Venn diagrams, C. S. Peirce's "existential graphs,"[52] truth tables, and semantic tableaux.[53] The crucial difference between these familiar logico-semantic representations and mental models is that the latter but not the former obey the constraint that the human mind's memory and general knowledge capacities are inherently finite, embodied, localized, and limited by actual-world conditions. For present purposes and for later purposes as well, this inescapable, definitive fact about us is what I will call *the cognitivist existential predicament*, or *the consequences for rationality of our animality*. Given the cognitivist existential predicament, mental models are always partial in the sense that they do not capture all the semantic information conveyed by the corresponding sentences; and ordinary reasoners rarely consider even all relevant possible models of the premises, not to mention the impossibility of their considering all possible models, period.

This partiality, however, naturally raises the possibility, as E. J. Lowe has persuasively argued, that the supposedly many differences between the mental models theory and the mental logic theory are mostly differences in a Pickwickian sense: namely, differences that don't make a real difference.[54]

First, at least implicitly, the mental models theory and the mental logic theory are *both* committed to theses to the effect (i) that the human mind–brain contains a science of the necessary relation of consequence already specified in it, and (ii) that this mentalistic fact about us partially or wholly explains human reasoning. According to the mental models theory, then, there *is* a logic internal to the human mind–brain, namely an axiomatic model theory.

Second, although it is true that the axiomatic model theory adopted by the mental models theory is not equivalent or reducible to any standard axiomatic model theory because of its reliance on partial models and restricted domains, this is only to say that the logic of the mental models theory is either a *fragment,* a *conservative extension,* or else a *deviant* of standard axiomatic model theory, not that it is *not* a logical theory.

Third, obviously there is a significant theoretical overlap between proof-theory and axiomatic model theory. Unless they are extended to arithmetic contexts or other complex domains, standard proof theory and standard axiomatic model theory are equivalent. So for a significant range of reasoning contexts, the logical force of the mental models theory and the mental logic theory is exactly the same.

In these ways the supposedly many differences between the mental models theory and the mental logic theory, it seems, boil down to a single authentic difference: the mental models theory says that in human reasoning we process linguistic information by constructing mental models *of* sentences, and the mental logic theory says that in human reasoning we process linguistic information by applying content-free, syntax-sensitive inference rules *to* sentences.

That is not to say that this authentic difference is negligible, however. Even quite apart from the reasoning tests, it is both introspectively and a priori quite plausible that all or at least most of our conscious logical reasoning operates by the construction and manipulation of mental models (especially when these models take the form of natural language imagery[55]), and not by applying content-free, syntax-sensitive inference rules.[56] The extension of the mental models theory to nonconscious or preconscious cognitive processes is thus a prima facie reasonable hypothesis. Furthermore, the claim of the mental models theory to be able to explain variable performance on the reasoning tests is based squarely on its hypothesis that conscious and non- or preconscious human reasoning alike is carried out in the processing medium of mental models. The root idea is that successful and unsuccessful performance on the selection task and other similar tasks are generally well correlated with the presentation of materials to the test subject that respectively promote or suppress the use of mental models. In view of the worries already expressed about the mental logic theory in section 5.3 under headings (1) and (2), both having to do with the role of nonlogical factors in the experimental materials of the selection task, it seems again quite plausible that the presence or absence of semantic contents in the reasoning materials is at least an important feature. To be sure, the mental models theory has trouble explaining the fact that success rates on some versions of the thematic materials selection task are almost as low as on the standard abstract task. But at least it remains true that high success rates are always associated with the presence of semantic content in the experimental materials, and that absence of semantic content is always associated with low success rates.[57]

Nevertheless, even granting the plausibility of the thesis that mental models are bound up with all or most preconscious and conscious human reasoning, and also the plausibility of the further thesis that mental models can at least partially explain the fact of variable performance on the selection task and other similar tasks, it still does *not* follow that humans *do not* naturally reason by means of a mental logic. And this is because the cognitive use of mental models is "nonlogical" only in a speciously narrow sense: it is "nonlogical" *only if* we construe logic as narrowly restricted to proof theory and classical logical systems. But on the perfectly reasonable assumption that an axiomatic model-theoretic semantics,[58] even if based on partial models and restricted domains, is *also* a logic, albeit a highly specialized kind of nonclassical logic, then despite its official theoretical animus toward the very idea of a Mental Logic, the mental models theory in fact entails the thesis that humans naturally reason by means of a mental logic. In other words, even though the mental models theory vigorously rejects the idea of a Mental Logic, if we assume a broader-minded view about what counts as a logic, the mental models theory remains a mental logic theory just the same.

To this extent, the mental models theory is wide open to the difficulties for the mental logic theory mentioned under headings (3) and (4) in section 5.4, namely, worries deriving directly from the *e pluribus unum* problem. If there is a unique logical system of mental models for all human reasoners, then we will want to know just *how* it accounts for the manifest plurality of logical systems, and *why* we should favor this logical system and not some other.

Moreover, to the extent that the logical system of mental models is held to be innate, then the mental models theory also falls afoul of the criticism of the mental logic theory mentioned under (6) in section 5.3: the questionableness of the doctrine of innate ideas. It seems exceedingly unlikely that the intact, mature human mind could, from the beginning of its appearance on earth, ever innately contain a stock of highly specific and highly internally structured beliefs about the primitive axioms of the logic of mental models, together with an indefinitely large number of beliefs about theorems based on the stock of primitive axioms.

Finally and perhaps most surprisingly, it follows from all of the above that despite its official endorsement of irrationalism, the mental models theory is in fact covertly a form of rationalism. On the one hand, the mental models theory is explicitly committed to the claim that human beings do not naturally reason according to normative logical principles:

[S]ubjects are not impeccably rational. Because they lack systematic inferential principles (rules of inference), they make genuine mistakes in reasoning.[59]

Nevertheless, on the other hand, despite their natural and all-too-human deviance from proof-theoretic logical rationality, according to the mental models theory human thinkers *do* naturally reason in accordance with the "metaprinciple" to the effect that "an inference is valid provided that there is no model of the premises in which its conclusion is false."[60] This is not to say that human reasoners never make mistakes; frequently, their performance is adversely affected by various nonlogical factors. But "[i]ndividuals who have no training in logic appear to have a tacit grasp of this metaprinciple," and therefore human reasoners are inherently competent to employ the metaprinciple of the logic of mental models:

[T]his meta-principle is defensible as a rational requirement for any system of deductive inference. It provides a rational meta-competence.[61]

I conclude that the mental models theory has some serious problems. Perhaps the most salient problem, which I have mentioned in passing already, is that despite the plausibility of the thesis that all or most preconscious and conscious human reasoning employs mental models, and despite the plausibility of the thesis that human reasoning is inherently sensitive to semantic content, the mental models theory on the face of it cannot account for the fact that success rates on many versions of the thematic selection task are almost as low as on the standard abstract task. Beyond its rejection of the mental logic theory, the next theory of reasoning I will look at grounds its claim to truth precisely on its ability to explain successful and unsuccessful performance on *all or almost all* versions of the selection task.

5.5   The Heuristics-and-Biases Theory

As we have just seen in sections 5.3 and 5.4, the mental logic theory and the mental models theory both say that human cognizers come to deductive problems already equipped with a logic internal to their minds–brains (be it either a system of content-free, syntax-sensitive inference rules on the one hand, or a nonclassical system of model-theoretic axioms on the other), and that this fact explains their reasoning activities. But we have also seen that the mental logic theory and the mental models theory each have significant problems. Suppose, then, that we reject the mental logic theory and the mental models theory alike. Does it follow that where human reasoning is concerned, everything is up for grabs? No. We can still adopt the heuristics-and-biases theory.

'Heuristics and biases' is a useful semitechnical term, coined by Kahneman and Tversky,[62] that stands for the total collection of characteristically human attitudes, emotions, desires, interests, goals, and short-term strategies or long-term plans, whether these be evolutionary, individual-prudential, or more broadly social-ethical. The heuristics-and-biases theory says that humans do *not* reason in accordance with normative principles of logic, but instead reason according to heuristics and biases. Deductive reasoning is practical and affect-driven, not logical. In this sense, the heuristics-and-biases theory is irrationalism *par excellence*.

One fairly obvious way of defending the heuristics-and-biases theory is to say that human reasoners deal with deductive problems on an ad hoc or case-by-case basis, depending on the differential availability, via long-term memory, of prior experience with factors the same as or similar to those found in the content domain of the relevant deductive problem.[63] But this pure associationist or empiricist version of the heuristics-and-biases theory (sometimes called "the availability theory"[64]) fails immediately, in view of these empirical facts: (i) on some versions of the thematic selection test, subjects perform very poorly despite being well acquainted with the content expressed by the experimental materials; and (ii) on other versions of the thematic selection test, subjects perform very well despite being unacquainted with the content expressed by the experimental materials. So the actual availability via long-term memory of prior experience with the content expressed by the thematic materials, while obviously having *some* nonnegligible impact on deductive reasoning, is neither sufficient nor necessary for successful performance on the selection task.[65]

This leaves two seemingly more plausible versions of the heuristics-and-biases theory: (A) the theory of *pragmatic reasoning schemas*, and (B) the theory of *social contract schemas*. I will briefly sketch each of them separately, then criticize both of them together.

(A) **Pragmatic reasoning schemas** The theory of pragmatic reasoning schemas was developed in the mid-1980s by Patricia Cheng and Keith Holyoak.[66] This version of the heuristics-and-biases theory says that humans do not reason in accordance with normative principles of logic, but instead reason according to generalized context-sensitive rules having essentially instrumental or pragmatic import, that is, rules that essentially reflect our short-term strategies or long-term plans to achieve our individual-prudential or social-ethical goals:

Our approach to reasoning implies that the schematic structures that guide everyday reasoning are primarily the products of induction from recurring experience with classes of goal-related situations. Reasoning rules are fundamentally based on our pragmatic interpretation of situations, rather than on syntactic interpretation of sentences.[67]

According to the theory of pragmatic reasoning schemas, there is strong experimental evidence to suggest that many versions of the thematic selection task invoke a particular pragmatic reasoning schema, "the permission schema," that coincides with knowledge of the material conditional of classical propositional logic. The permission schema "describes a type of regulation in which taking a particular action requires satisfaction of a certain precondition."[68] In other words, the permission schema is essentially *deontic*, or bound up with the concept of duty. By and large, it appears, variable performance on the selection task depends on whether the reasoner's recognition of the permission schema is supported by the experimental materials or suppressed by them. For example, in the original "unfacilitated" (or low-success-rate) abstract selection task, the content-free and value-neutral rule "If a card has a vowel on one side, then it has an even number on the other side" effectively suppresses recognition of the permission schema, whereas in the "facilitated" or high-success-rate version of the thematic selection task, the contentful and value-laden rule "If a person is drinking beer, then the person must be over 19," effectively supports recognition of the permission schema.

Nevertheless, the coincidence of a permission schema with logical knowledge of the classical *if . . . , then . . .* is more or less accidental, since the permission schema is not strictly equivalent with the material conditional. The permission schema does indeed rule out the deontic analogues of the fallacies of denying the antecedent (that is, arguments of the form $\varphi \rightarrow \psi, \sim\varphi \mid \sim\psi$) and affirming the consequent (that is, arguments of the form $\varphi \rightarrow \psi, \psi \mid \varphi$).

But unlike the material conditional of classical propositional logic, the permission schema is context-sensitive and involves the deontic concepts *must* and *may*. And by the same token, the permission schema does not map precisely onto the conditional of any nonclassical logic either. More generally:

evocation of a pragmatic schema will not necessarily lead to selection of the "logically correct" cases, both because different schemas will suggest different relevant inferences, and because the inferences based on any particular schema will vary depending on the mapping between the stated rule and those associated with the schema.[69]

So the reasoner who follows the permission schema in dealing with the selection task is using an intrinsically nonlogical and formally fallacious cognitive strategy. Human reasoning is rule-following, but not *logical* rule-following; it is instead *pragmatic* rule-following.

Moreover, and perhaps most interestingly, there is also good empirical evidence to suggest that recognition of the permission schema facilitates performance on a deontic analogue of the *abstract* selection task. This is because reliable facilitation occurs whenever the rule in the task is presented as a permission schema (or some other deontic schema such as obligation) in a content-free format.[70] In other words, according to the theory of pragmatic reasoning schemas, although humans appear to lack a mental *logic* module, they nevertheless seem to possess a mental *deontology* module.[71]

(B) Social contract schemas   The theory of social contract schemas was proposed in the late 1980s by Leda Cosmides.[72] This version of the heuristics-and-biases theory says that humans reason not according to normative principles of logic but instead according to social contract algorithms operating in contexts involving cooperative enterprises ("social exchange") that are adaptive under the laws of natural selection and thereby essentially *evolutionary* in character. Furthermore, these adaptive algorithms are pre-encoded in an innate mental module. Here is how Cosmides summarizes the view:

Human reasoning . . . has been considered essentially domain-general: the innate processes hypothesized—whether "logical," "inductive," or associationistic—have been thought of as operating consistently, regardless of content, with content-dependent performance attributed to differential experience. If this and other empirical studies establish that even human reasoning is not unitary and domain-general, but instead governed by an array of special-purpose mechanisms, this will provide substantial support for a modular approach to cognitive psychology. . . . Social contract theory [holds] that: (1) humans have algorithms specialized for reasoning about social exchange; (2) these algorithms will have certain structural properties, predicted by natural selection theory; and (3) these algorithms are innate, or else the product of experience structured by innate algorithms that are specialized for reasoning about social exchange.[73]

Two features of the theory of social contract schemas are especially worthy of note.

First, the theory's attack on rationalism explicitly proceeds via nativism and the modularity thesis, which one might otherwise unreflectively expect to be the special property of the mental logic theory and other forms of rationalism. Against this expectation, the theory of social contract schemas seeks to undermine rationalism not by denying that humans have an innate faculty for the sort of deductive reasoning isolated by Wason's selection task, but on the contrary, precisely by insisting that humans *do* have an innate faculty for the sort of deductive reasoning isolated by Wason's selection task. The critically crucial point is that for the theory of social contract schemas, the cognitive capacity located in that faculty is an intrinsically *nonlogical* capacity, by virtue of its being a capacity for recognizing and manipulating social contracts.

Second, although the theory of social contract schemas shares with the theory of pragmatic reasoning schemas the general idea behind the heuristics-and-biases theory, to the effect that it is practical information processing and not logical information processing that drives human reasoning, it (i.e. the theory of social contract schemas) accounts for the ground of practical cognition in a way sharply different from that of the theory of pragmatic reasoning schemas. According to the theory of pragmatic reasoning schemas, pragmatic reasoning schemas are generalized from everyday goal-oriented human behavior. But according to the theory of social contract schemas, social contract schemas are strictly determined by underlying evolutionary processes. The aim of an evolutionary process is for an organism to survive long enough to reproduce sexually and then leave as many copies of its genes in the world as possible. This is, as it were, the *evolutionary imperative*. Natural selection is the historical mechanism by which organisms are individually (ontogenetically) and species-specifically (phylogenetically) sorted and molded by the evolutionary imperative; and all behavioral and cognitive strategies that satisfy the evolutionary imperative are adaptive. For humans engaged in social exchange and driven by the evolutionary imperative, it is absolutely crucial not to let "cheaters" undermine contractual transactions by enjoying the benefits of a transaction without paying the cost. So in social exchange contexts, humans naturally look at cases in which benefits are accepted in order to check whether the cost is also paid, and also at cases in which costs are not paid in order to make sure that benefits have not been antecedently accepted.

This, according to the theory of social contract schemas, maps nicely onto the structure of the Wason selection task, which requires checking the card that represents the antecedent of the rule (the "benefits accepted" card), and the card that represents the denial of the consequent (the "cost not paid" card), in order to determine the truth or falsity of a classical material conditional. So, according to the theory of social contract schemas, successful and unsuccessful performance on the selection task is to be explained in terms of whether the content expressed by the experimental materials effectively supports or effectively suppresses the recognition of the adaptive "look for cheaters" procedure. Experimental results in fact indicate that the cost-benefit structure of social contract mental representations is what is cognitively salient for the various thematic materials effects in the Wason selection task, rather than the availability heuristic favored by the availability theory or the action-precondition structure of the permission schema favored by theory of pragmatic reasoning schemas.[74] In any case, it is manifest, as the theory of social contract schemas states, that this sort of cognitive processing has little or nothing to do with following the normative principles of a logic.

So those are my explicative glosses on the pragmatic reasoning schemas theory and the theory of social contract schemas, as exemplars of the heuristics-and-biases theory. But here are two worries about them specifically, and by implication, two worries about the heuristics-and-biases theory more generally.

The first worry concerns the fact that both theories invoke an intrinsically nonlogical mental module in the sense of the standard cognitivist model of the mind, the one for deontology, and the other for social contracts. Given this fact, we should *not* find a high success rate on thematic deductive reasoning tasks which *lack* experimental materials that express deontic notions or social exchange, whether or not they involve the selection task. Also, we should *not* find a high success rate on *abstract* deductive reasoning tasks, whether or not they involve the selection task. But in fact the mental logic theory literature shows that both sorts of cases, predicted not to exist by the pragmatic reasoning schemas theory and by the social contract schemas theory, have significant experimental support.[75] As D. P. O'Brien observes: "Because so much of the empirical work in the literature has reported errors on complex reasoning tasks, it has been easy to overlook that there is a body of simple sound inferences that people make routinely."[76] This strongly suggests that both the pragmatic reasoning schemas theory and the social contract schemas theory are too narrowly tied to the framework of the Wason selection task, and that they overgeneralize from their ability to account more or less well for the thematic materials effects.

The second worry concerns the question of whether the pragmatic reasoning schemas theory and the social contract schemas theory are really so radically different from the mental logic theory and the mental models theory after all. A crucial step in the irrationalist argument mounted by both the pragmatic reasoning schemas theory and the social contract schemas theory alike is that the cognitive competence posited by each theory is intrinsically nonlogical in character. But we saw in the case of the mental models theory that its claim to be a form of irrationalism depends heavily on a speciously narrow conception of the logic built into any Mental Logic, which restricts it to proof theory and classical systems. But if, as most working logicians are prepared to do, we allow model theory and nonclassical systems to count as logic, then the mental models theory too is a form of the mental logic theory, in the sense that (i) the human mind–brain contains a science of the necessary relation of consequence already specified in it, and (ii) this mentalistic fact about us partially or wholly explains human reasoning. Therefore, the mental models theory could be legitimately construed as a *slightly exotic* brand of rationalism, but as a brand of rationalism nevertheless.

An exactly analogous point can be made about the pragmatic reasoning schemas theory and the social contract schemas theory. Suppose we take the deontic principles of the pragmatic reasoning schemas theory, or the social contract algorithms of the social contract schemas theory, and consider them to be the axioms of some nonclassical logic that also explicitly and systematically allows for the indexicality or context-dependence of some of its nonlogical constants.[77] Then both the pragmatic reasoning schemas theory and the social contract schemas theory are committed to a version of the mental logic thesis, according to which the explanation of human reasoning is that humans naturally reason according to the normative principles of some highly specialized, context-sensitive, deontic or social contractarian nonclassical logic internal to the mind–brain. In other words, both the pragmatic reasoning schemas theory and the social contract schemas theory can be legitimately construed as *fairly exotic* brands of rationalism, but as brands of rationalism nevertheless.[78]

This line of criticism clearly generalizes to every psychological theory of reasoning that purports to prove irrationalism. As long as the supposedly intrinsically nonlogical cognitive competence isolated by the theory can be formulated as a set of rules or computational algorithms, then it can be formulated as a mental logic of some sort. That the logic is nonclassical, or even deviant, is irrelevant. The salient point is that the reasoner can, consistently with the empirical evidence, be taken to be following the normative principles of *that logical system.* Hence the psychological theory can be converted into a brand of rationalism. Leaving aside the obvious differences in the specific hypotheses offered by the several psychological theories of reasoning in order to account for the puzzling Wason selection test results, we can now see that the opposition between theories that support rationalism and theories that support irrationalism effectively boils down to the relatively trivial question of whether one is broad- or narrow-minded about what counts as a logical system for the purposes of postulating a mental logic. But chapter 2 offered a strong case for broad-mindedness. So, to that extent, *all* psychological theories of human reasoning support rationalism.[79] This points up the very important fact that all psychological theories of reasoning share a crucial nucleus of methodological and substantive assumptions. In particular, they all accept the standard cognitivist model of the mind[80] described in section 4.1, and also the assumptions (α) through (δ) listed in section 5.2 above.

This in turn leads to an important intermediate conclusion in the overall argument of this chapter. Since irrationalism reduces to rationalism, the claim of psychological theories of reasoning to be able to prove the truth of either irrationalism or rationalism ultimately depends *only* on those theories that make a case for rationalism. So I will wrap up my critical survey of contemporary psychological theories of reasoning by looking at a version of rationalism that is interestingly different from the others.

5.6   The Minimal Rationality Theory

The *minimal rationality theory* was worked out by Christopher Cherniak in the mid-1980s.[81] In effect, the minimal rationality theory starts from what in section 5.4 I called "the cognitivist existential predicament," or "the consequences for rationality of our animality." This is the inescapable, definitive fact that human cognizers always operate under limitations jointly determined by their finite embodied nature and by the contingent conditions of the actual world into which they have been originally thrown. The basic idea behind the minimal rationality theory is that a cognizing creature capable of deductive reasoning necessarily adopts the logical system that works best for it, given its real-world situation together with its beliefs and desires, and uses this logic as efficiently as it can, given its limited cognitive powers and the demands imposed on it by the world. So humans reason deductively by following the normative principles of the logical system most suited to the individual creature's predicament, although, even granting the inevitable adjustment of the logic to the creature, it is not automatically entailed that the reasoner will follow its minimal normative logical principles infallibly:

[I]t is true only in some cases that if p implies q and a person believes p he ought to infer q, in that this is required for rationality. . . . In determining whether the agent ought to make the inference from p to q in order to be minimally normatively rational, we must take into account not only (1) the soundness of the inference but also (2) its feasibility and (3) its apparent usefulness according to the agent's belief-desire set. Even in those cases where the believer of p ought to infer q in order to be minimally rational, there is no implication that a believer of p will actually do this. . . . What is the relation of the minimal normative thesis to the descriptive thesis, the minimal rationality condition, which actually predicts what a believer of p will infer? The blurred set of inferences required in a particular case for minimal rationality is only a proper subset of the set of inferences then required for minimal normative rationality. . . . Thus, it is a fact of our actual belief-attributing practices that minimal rationality is weaker than even minimal normative rationality.[82]

In this way, the minimal rationality theory offers us an essentially *humble* or *modest* rationalism by focusing on "bounded rationality" or "nonideal rationality." And by stressing the inherently suboptimal nature of human rationality, the minimal rationality theory is able to absorb much of the initial impact of the robust empirical fact that originally instigated the debate about rationality (namely that humans perform so poorly on the Wason selection task and other similar deductive reasoning tasks) by simply accepting that fact at face value but also denying that it entails human irrationality.[83] So most or at least much of what looked to be sheer irrationality from the standpoint of all those versions of rationalism and irrationalism that assume an "ideal rationality" model of human reasoners ("Human subjects, it would appear, regularly and systematically invoke inferential and judgmental strategies ranging from the merely invalid to the genuinely bizarre"), turns out to be just *authentic human rationality* under the minimal rationality theory's framework of systematically lowered expectations. In order to be rational, and in particular in order to be logically rational, humans need not always succeed on all deduction tasks. Instead, they need only *sometimes* succeed on *some* deduction tasks.[84]

Moreover, the minimal rationality theory expands the very idea of a deduction task. It now includes any situation in which a creature's beliefs and desires interact to produce behavior *that can be interpreted, by some third person who is capable of framing belief-report sentences, as the creature's adding a belief to the original set according to the normative principles of some logical system or another.* So, the rational creature, according to the minimal rationality theory, need not be able to use a natural language, or introspectively recognize or evaluate what it itself is doing, and hence need not be able to conceive of itself as a reasoner or as an agent. In other words, a deductive reasoner need be neither human, nor linguistically competent, nor normative-reflective.

What about the logical system that the interpreter ascribes to the creature? It need be neither classical propositional logic, nor elementary logic, nor even a conservative extension of classical or elementary logic. It need not be universal across reasoning creatures. And it need not "save logical truth": that is, it need not include any absolutely incorrigible sentences. The logical system ascribed to the creature need only be rich enough, determinate enough, general enough, and true enough to support a set of "feasible inferences" relative to some creature and its situation.[85] This is the *practical adequacy* of the logic, as opposed to its *metatheoretic adequacy*: its soundness, consistency, and completeness. Indeed, the logical system ascribed to the creature need not be metatheoretically adequate.[86]

Here is a two-part critical evaluation of the minimal rationality theory.

The first part is an endorsement. I think that there is something deeply right about the strategy of systematically incorporating the cognitivist existential predicament into the theory of human reasoning. The minimal rationality theory rightly acknowledges the fact that human logical abilities need not be infallible in order to be acceptably sound or inherently rational. It also rightly asserts the existence of a plurality of logical systems, not just within the domain of publically available logical systems, but even across individual cognizers. And it rightly takes a suitably broad or open-textured view of the nature of the logic from which humans derive their normative principles of reasoning. So in these respects, three cheers for the minimal rationality theory.

The second part, however, is a worry. My basic worry about the minimal rationality theory is that it sets the standards of rationality and logical ability essentially *too low*. According to the theory, to be a deductive reasoner a creature need only have a set of beliefs and desires according to the lights of some charitable and pragmatically minded interpreter,[87] and be able to generate some behaviors on the basis of the original belief set in such a way that the interpreter can ascribe a further belief and an inference to the creature. This criterion is radically overinclusive. According to the minimal rationality theory, cats, dogs, horses, and more generally every animal capable of having some beliefs and desires will be a logical cognizer, or thinker, and thus a rational animal. This is somewhat plausible if rationality is construed as basically instrumental.[88] But at the same time it cannot adequately capture either holistic rationality (the rationality of coherence or reflective equilibrium) or modal rationality (the rationality of necessity, obligation, and certainty), both of which, I am assuming, are capacities possessed by all and only rational animals, including rational human animals. Among other things, it follows from the logic-oriented conception of human rationality that an animal cannot *be* a logical animal or deductive reasoner unless it can represent itself *as* a logical animal or deductive reasoner.[89] In other words, although there is something deeply correct about a theory of rationality that systematically acknowledges the consequences for rationality of our animality and therefore is minimalist, humble, or modest to a nontrivial extent, it also seems to me that the conceptions of rationality and of logical reasoning offered by the minimal rationality theory are *excessively* minimalist. Of course we are crooked timbers. But the rationality of rational animals still exceeds minimal rationality.

5.7   The Protological Competence Theory

As we have just seen, the minimal rationality theory says that in order to be rational animals, humans need not reason perfectly in accordance with the normative principles of some universally accepted classical logic, but rather only minimally in accordance with the normative principles of the logical system that is practically adequate for that creature. By contrast, the proto-logical competence theory that I wish to defend says that in order to be rational animals, humans must reason *perfectly* in accordance with the normative principles of a single universal unrevisable a priori repertoire of meta-logical principles and logical concepts (namely, the protologic) although only *minimally* in accordance with the normative principles of their own mental logics or logics of thought. So, to the extent that the protological competence theory places only minimal requirements on human cognitive competence with respect to the mental logic or logic of thought, it is similar to the minimal rationality theory. But the protological competence theory differs sharply from the minimal rationality theory in requiring ideal compliance with respect to the innate protologic.

More explicitly, the protological competence theory is the two-part doctrine that humans reason deductively (1) by perfectly following the normative logical principles of the innate protologic, and (2) by using the innate protologic to construct a mental logic or logic of thought, whose normative principles they follow minimally. The first part of the protological competence theory is derived from the logic faculty thesis, as spelled out in section 2.7, and the second part is derived from the logic of thought thesis, as spelled out in section 4.8, together with what we have learned from our critical survey of the psychological theories of reasoning.

The theoretical advantages of a two-factor theory of human reasoning, as opposed to a single-factor theory, should be fairly obvious. A single-factor theory will always be burdened with a basic need to explain the nature of competence in human reasoning (which is, by hypothesis, the source of every successful deductive cognitive performance) in the face of another seemingly opposed basic need to explain the robust experimental fact of highly variable cognitive performance on the reasoning tests. This produces the dialectical tension between rationalism and irrationalism. But a two-factor theory, like the protological competence theory, can explain the nature of our reasoning competence by means of one factor and also explain variable performance on the reasoning tests by means of another distinct (although obviously not wholly unrelated) factor.

The universal human possession of an innate protologic explains the nature of our reasoning competence. What I mean is that human reasoning *just is* following and deploying the normative metalogical principles and logical concepts of the protologic, whatever these may turn out to be. If we were not doing this, then we would not be *reasoning*, but doing something else instead. Try to conceive of a creature that is actually "reasoning deductively" but has not the slightest determinate conscious conception of what it is for a conclusion to follow validly from a set of premises: this is patently absurd. This creature would be a "deductive reasoner" that could not (even with copious amounts of friendly extra tutoring, untimed online exercises and tests, and an essay) pass an introductory logic class. So to the extent that we meet the minimal standards of human rationality, following the protologic *constitutes* our reasoning activities. I will come back to this crucial point again shortly below in relation to the deductive reasoning tests.

At the same time, we can also explain the fact of highly variable performance on the reasoning tests (not to mention in everyday life) by appealing to the cognizer's use of the innate protologic to effect a cognitive construction of a logic of thought, via her cognitive construction of a language of thought (LOT), via the process of first-language acquisition, in the face of the cognitivist existential predicament. The explanation has three elements, each of which counts as a proper part of a cumulative explanation.

First, in section 4.3 I argued that there must be a plurality of LOTs, and in section 4.8 this was extended to the claim that there must also be a plurality of mental logics or logics of thought. Therefore, beyond the innate and thus universally shared protologic, each reasoner's mental logic will be only more or less similar to the mental logic of any other reasoner. Moreover, each reasoner's mental logic will be only more or less similar to the underlying logical structure of the natural language she speaks in common with other reasoners. And each reasoner's mental logic will be only more or less similar to any existing classical or nonclassical logical system. These three points together explain some errors or other variations in human reasoning in terms of obvious clashes or more subtle incommensurabilities between different logics.

Second, each reasoner is limited by his own finite embodiment, and by the demands of his actual world situation. That explains some errors or other variations in human reasoning in terms of brute contingent information-processing limitations and happenings over which the reasoner has little or no control.

Third, according to the account given in section 4.5 and touched on again in section 5.3, the cognitive function of the logic faculty in each reasoner is precisely to mediate between the information delivered by the peripheral mental modules and the information at large in the central processes of belief, desire, emotion, and decision. This explains some errors or other variations in human reasoning in terms of the inherent openness-to-penetration of logical reasoning processes by higher-level theoretical, practical, and affective attitudes, activities, and commitments of the more or less well-integrated rational human agent.

Taken together, these three partial explanations seem to me to be jointly sufficient to explain *all* the errors and variations in human reasoning.

The protological competence theory is also strongly supported by the deductive reasoning tests, in two ways.

First and most obviously, the tests strongly support the cumulative explanation of highly variable performance that is provided by the second factor of the theory.

But second and more surprisingly, the tests also strongly support the explanation of the nature of human reasoning competence that is provided by the first factor of the theory. Here I need only draw attention to the fundamental but easy-to-overlook fact that it is simply *assumed* by the experimenters and designers of the reasoning tests that all test subjects are capable of perfectly understanding, at the very least, what *counts* as a logical or deductive task in the first place. Correspondingly, it is also simply assumed that all test subjects can, at the very least, perfectly understand *some* logical principles and concepts. Otherwise the subjects would not be able to follow the test instructions; they would not be able to be significantly primed for the standard abstract selection task by preliminary clarification of the rule or by being asked to justify their decisions on related tasks; and they would not be able to comprehend the right answer when it is explained to them in debriefing sessions. And in fact the test subjects *uniformly and reliably can* do all of these things.[90] Indeed the test subjects *must be* uniformly and reliably competent in this sense, for otherwise the tests would be utterly pointless and void, like the mock psychological test in the famous Monty Python's skit, conducted by Dr. Peaches Barkowitz of the Rod Laver Institute, in which non-English-speaking humans and penguins (both placed in the same penguin pond and asked the same skill-testing questions in English) are experimentally judged to have exactly the same IQ. In other words, the very idea of *an experimentally feasible deductive reasoning test* assumes that humans of normal or higher intelligence can at the very least deductively reason by perfectly following the normative principles of the protologic, and this is obviously confirmed by the available empirical evidence.

5.8 Reasoning Tests and the Rationality Debate II

I pointed out in section 5.2 that rationalism and irrationalism, the two opposing parties to the rationality debate, share four enabling assumptions:

(α) that there are normative logical principles;

(β) that human rationality at least partially consists in reasoning in accordance with these normative logical principles;

(γ) that these normative logical principles express the inference rules, axioms, logical truths, or basic logical concepts of Aristotelian logic, Piagetian logic, or some other system of classical or nonclassical logic; and

(δ) that successfully or unsuccessfully performing the Wason selection task and other similar reasoning tasks under experimental test conditions adequately reveals human reasoning ability.

What can we now say about these assumptions?

If the protological competence theory is correct, then it is false that the normative principles of logic followed by human reasoners must be derived from some classical or nonclassical system. That is because the protologic is a set of schematic logical structures, in the form of a coherent repertoire of metalogical principles and logical concepts, that is used for the construction of all logical systems, and is not itself a classical or nonclassical logical system.

It also follows from the truth of the protological competence theory that there is a definite sense in which the Wason selection task and other similar reasoning tests do not properly test human reasoning competence. The Wason selection task properly tests only the knowledge of the principles of classical propositional logic within the class of mature, healthy humans of normal or higher intelligence. But a proper test of human reasoning competence would in fact test the knowledge of the principles and concepts of the protologic across the human species as a whole, and *not* test the knowledge of the principles of classical propositional logic within a speciously narrow class of humans.[91] As I have just argued in the last section, our competent knowledge of the principles and concepts of the protologic is in fact *assumed* by the Wason selection test and other similar reasoning tests, but not properly tested by them.

Furthermore, if the logic faculty thesis is correct, then the logic of thought or mental logic that is constructed by each cognizer is inherently open to penetration by information from central processes of theorizing, judgment, belief, desire, and volition. So the logic faculty thesis together with the logic of thought thesis *jointly entail* that there will be high variability in performance on the Wason selection test and other similar reasoning tests, depending on the character of the experimental materials and the individual subjects tested.

So rationalism and irrationalism alike are false. This can be restated by explicitly denying assumptions (γ) and (δ), as follows:

~ (γ) the normative logical principles implicit in all human reasoning express the principles and concepts of the protologic, not those of any classical or nonclassical logical system; and

~ (δ) successful or unsuccessful performance on the Wason selection test and other similar reasoning tests under experimental test conditions is largely irrelevant to adequately revealing human reasoning competence, because (a) this reasoning competence involves only the minimal capacity for understanding what *counts* as a logical or deductive task in the first place (i.e., a grasp of the principles and concepts of the protologic), and this is in fact presupposed by the reasoning tests, and (b) the high variability in performance on the reasoning tests is predicted by the logic faculty thesis together with the logic of thought thesis.

That leaves us with assumptions (α) and (β),[92] the robust empirical data from the psychology of reasoning, and the crucial recognition that rationalism and irrationalism do not exhaust the logical space of possible views on the nature of human rationality. And this in turn makes it possible for us to recognize that the protological competence theory, which is triply based on the logic faculty thesis, the logic-oriented conception of human rationality, and logical cognitivism, is in fact strongly supported by the psychology of human deductive reasoning.

This was our paradox: no course of action could be determined by a rule, because every course of action can be made out to accord with the rule. The answer was: if everything can be made out to accord with the rule, then it can also be made out to conflict with it. And so there would be neither accord nor conflict here.

—Ludwig Wittgenstein[1]

"But doesn't it follow with logical necessity that you get two when you add one to one, and three when you add one to two? and isn't this inexorability the same as that of logical inference?"—Yes! it is the same.—"But isn't there a truth corresponding to logical inference? Isn't it *true* that this follows from that?" The proposition: "It is true that this follows from that" means simply: this follows from that. And how do we use this proposition?—What would happen if we made a different inference—*how* should we get into conflict with truth?

—Ludwig Wittgenstein[2]

As an account of our knowledge about medium-sized objects, in the present, this is along the right lines. It will involve, causally, some direct reference to the facts known, and, through that, reference to those objects themselves. . . . [C]ombining *this* view of knowledge with the "standard" view of mathematical truth' makes it difficult to see how mathematical knowledge is possible. If, for example, numbers are the kinds of entities they are normally taken to be, then the connection between the truth conditions for the statements of number theory and any relevant events connected with the people who are supposed to have mathematical knowledge cannot be made out.

—Paul Benacerraf[3]

[I]t is tempting to press this line of reasoning one step further and suppose that once the machinery for the simulation of spatial operations had attained a critical degree of computational power and autonomy, it could, by analogical extension, provide aid in the solution of intellectual problems far removed from the exigencies of everyday life—including, possibly, some of the most profound and far-reaching achievements of the human mind.

—Roger Shepard[4]

6.0 Introduction

The previous chapter completes my five-pronged cumulative argument[5] for the logic faculty thesis in particular, and for logical cognitivism in general. If that argument is sound, then along the way we have also acquired good reasons, both a priori and empirical, for accepting the logic of thought thesis, the logic-oriented conception of human rationality, and the protological competence theory of human reasoning to boot.

The present chapter applies logical cognitivism to the theory of logical knowledge. Ironically enough, given the fundamental role of logic in the analytic tradition, the epistemology of logic is a surprisingly underdeveloped field.[6] So my route into it will be initially indirect, by way of two outstanding difficulties in the epistemology of mathematics: (i) Saul Kripke's "plus–quus" version of Wittgenstein's rule-following paradox, and (ii) Paul Benacerraf's dilemma about how the causal inertness of the abstract[7] objects we are committed to by virtue of our accepting a "standard" semantics of mathematical truth directly contradicts the further assumption of a "reasonable epistemology" to the effect that the objects of human knowledge must be causally related to our cognitive capacities. Both of these worries can be smoothly extended to logic.

Furthermore, I hold that there is a comprehensive solution to both the extended Kripke–Wittgenstein paradox and the extended Benacerraf dilemma. This solution is contained in the thesis that the rational human animal, by virtue of possessing the logic faculty, *is also an animal with an innate capacity for logical knowledge by means of logical intuition.* How can this thesis do the double job? I argue, first (sections 6.1–6.3), that adding a capacity for logical intuition to the innate logic faculty solves the extended rule-following worry, and as an extra bonus helps with two problems about the inferential role theory of the meaning of logical constants. And second (sections 6.4–6.5), I argue that the extended Benacerraf dilemma can be solved by combining (i) logical structuralism,[8] (ii) realism about logic generally and about logical necessity more specifically, and (iii) a theory of logical intuition based on our cognitive ability for the conscious scanning and manipulation of linguistic mental imagery.

My theory of logical intuition develops two important ideas briefly mentioned in section 2.3:[9] C. I. Lewis's idea that the ideographic compactness and precision of a symbolic logic is closely connected with our cognitive capacity for apprehending and retaining mental images, and the Tractarian Wittgenstein's idea that a properly sign-designed logical symbolism is itself the very medium of our a priori knowledge of logical truths and proofs.[10]

The theme of realism about logic and logical necessity brings out another important dimension of the logic faculty thesis. The logic faculty thesis says that logic is cognitively constructed by rational animals. This explains logic in terms of the innate abilities of a special class of animals. Hence the logic faculty thesis is naturalistic in the sense that it explains logic in terms of the innate abilities of a certain special class of sentient organisms, which in turn belong to the natural world if anything does. But at the same time, it is a *nonreductive* naturalistic explanation of logic. There are three reasons for this. First, the logic faculty is multiply embodiable and thus its essential properties are not identical to first-order physical properties. Second, since rational animals in general and rational human animals in particular are defined in terms of their possession of the logic faculty, the logic faculty thesis will obviously fail to account for logic solely in terms of things that are intrinsically *nonlogical* in nature.

Third and most important, however, a fundamental upshot of the self-refuting fate of scientific naturalism about logic together with the logocentric predicament is that any explanatory reduction of logic is impossible. Here is how that upshot unfolds. If logic is explanatorily reducible, then it is reducible either to the natural facts (by which, as stipulated in section 1.2, I mean the totality of basic or first-order physical facts plus the facts about sensory experience), or to some nonnatural facts. Now the reduction of logic to the natural facts self-refutingly undermines the logical strong supervenience needed to explanatorily reduce logic to the natural facts (section 1.4). But then on the other hand, if we try to reduce logic to any class of nonnatural facts, we come to realize, by way of the logocentric predicament, that logic must be presupposed and used in any explanation and justification of logic (chapter 3). So logic is not explanatorily reducible to *anything*. This in turn implies either (1) that logic is inexplicable and unjustified, hence groundless, or else (2) that we must opt for my cognitivist solution to the logocentric predicament. Since (1) is rationally intolerable, that leaves us with (2).

Or so I argued. In any case, this returns us to a point I noted briefly in section 1.2: no explanatory reduction can avoid an appeal to logical strong supervenience. But certainly there can be genuinely explanatory connections, even those involving lawlike connections, that are weaker than explanatory reduction.[11] Hence my naturalistic thesis that logic is cognitively constructed by rational animals is quite compatible with the nonreductive thesis that logic in general and logical necessity in particular are something over and above rational animals themselves. More precisely, I hold that even though logic is cognitively constructed by rational animals, it is still *objectively real*, in the twofold sense of being both (a) intersubjectively knowable and also (b) not dependent on the existence of actual individual minds. What I mean is that logic is a set of cognitively constructed abstract *linguistic structures* that have multiple actual and possible instantiations in space and time. Logical necessity, in turn, is a set of cognitively constructed abstract *linguistic substructures* within logic itself. Consequently, logical necessity is a real property or fact in a larger world that includes but is not exhausted by rational animals, whether human or nonhuman. So the nonreductive explanation of logic offered by the logical faculty thesis is ultimately dual: (i) logic is cognitively constructed by rational animals, and (ii) logic is objectively real *via language*, and consequently logical necessity is an objectively real property or fact in a world that objectively and really contains linguistic structures.

6.1 Kripke's Wittgensteinian Paradox and Logic

Wittgenstein's rule-following paradox, also known as "the rule-following considerations,"[12] is nicely illustrated by the following passage from Lewis Carroll's *Through the Looking Glass*:

Alice was puzzled. "In *our* country," she remarked, "there's only one day at a time." The Red Queen said, "That's a poor thin way of doing things. Now *here*, we mostly have days and nights two or three at a time, and sometimes in the winter we take as many as five nights together—for warmth, you know."
"Are five nights warmer than one night, then?" Alice ventured to ask.
"Five times as warm, of course."
"But they should be five times as *cold*, by the same rule—"
"Just so!" cried the Red Queen. "Five times as warm, *and* five times as cold—just as I'm five times as rich as you are, *and* five times as clever!"
Alice sighed and gave it up. "It's exactly like a riddle with no answer!" she thought.[13]

Here we vividly sense Alice's deep puzzlement about how to apply her rule for counting days, in the face of the outrageously deviant interpretation offered by the Red Queen. I do not know whether Wittgenstein actually read the Alice books, although I am willing to place a medium-sized bet that he did. In any case, seventy years after their first publication, in *Philosophical Investigations (PI)*, he accurately captured and clearly articulated the serious skeptical implications of the Red Queen's seemingly bizarre remarks.

The nub of Wittgenstein's skeptical argument is this.

(1) Assume that the meaning of any linguistic expression E is nothing but a rule for operating with that sign in a formal calculus or in some other language-system such as a natural language. Let us call this the assumption of *rule-based semantics.*

(2) It follows from rule-based semantics that understanding the meaning of any linguistic expression E consists in a speaker's being able to follow the rule for operating with E, that is, being "guided" by the rule for E. (*PI*, §§ 172–184)

(3) Every rule is expressible as a function-sign that determines a systematic mapping from inputs (arguments of the function) to outputs (values of the function). (*PI*, §§ 143–46, 151, 185)

(4) And the meaning of that function-sign (hence all its systematic mappings) is understood by grasping the rule in a "flash," hence by grasping it introspectively, privately, and instantaneously. (*PI*, §§ 186–197)

(5) But every function-sign can be multiply differently interpreted, such that although the interpretations yield the same mappings to outputs/values for all existing inputs/arguments, they diverge on some future inputs. (*PI*, § 185)

(6) And since every interpretation is in turn expressible as a higher-order function sign, each interpretation itself stands in need of further interpretation, which itself in turn can be multiply differently interpreted, ad infinitum. (*PI*, § 198)

(7) So anything the speaker does with E can, on some interpretation or another, be in accordance with the rule. (*PI*, § 201) {From (1)–(6).}

(8) Correspondingly, anything the speaker does with E can, on some interpretation or another, be also in conflict with the rule. (*PI*, § 201) {From (1)–(6).}

(9) So the speaker's actions, no matter what they are, neither accord with the rule nor conflict with the rule. (*PI*, § 201) {From (7)–(8).}

(10) Therefore it is impossible for a speaker to follow a rule. {From (9).}

(11) Therefore rule-following both actually occurs and also is impossible. Paradox! {From (1) and (10).}

There is, however, a further twist in this story. The famous Wittgensteinian paradox I have just described, namely *the rule-following paradox* (or RFP for short), is also essential to the much celebrated but also much controverted *private language argument*[14] (or PLA for short). The PLA says that semantically solipsistic languages (for example, phenomenalistic languages in which words stand for phenomenal qualia) are impossible. And the essential connection between the RFP and the PLA is that one straightforward way of preventing the paradox that arises in step (10) is simply to reject step (3):

[T]o *think* one is obeying a rule is not to obey a rule. Hence it is not possible to obey a rule 'privately': otherwise thinking one was obeying the rule would be the same thing as obeying it.[15]

The rationale here is this. If understanding the meaning of a linguistic expression is necessarily equivalent to following a rule for the use of that expression, and if any language is semantically solipsistic, then it follows that understanding the meaning of a word in that language will be the same as thinking one is following a rule. But since for Wittgenstein it is conceptually true that understanding the meaning of a linguistic expression is necessarily equivalent to following a rule for the use of that expression, and since for him it is also conceptually false that understanding the meaning of a word is the same as thinking one is following a rule, then according to him necessarily there are no private languages.

In 1982, in *Wittgenstein on Rules and Private Language*, Kripke worked out a highly influential reading of the PLA. Kripke's interpretation focuses on the RFP and its solution as developed by Wittgenstein in *Philosophical Investigations*, §§ 134–242, and argues (1) that this constitutes the essence of the PLA, which other commentators have almost always placed in *PI* §§ 243–315, and also (2) that the RFP introduces a radically new form of philosophical skepticism that should be taken every bit as seriously as Hume's skepticism about induction and necessary connection in nature, in the *Treatise of Human Nature* and *Enquiry Concerning Human Understanding*. Kripke also explicitly and relevantly compares and contrasts his version of the RFP with Quine's famous "indeterminacy of translation" and "inscrutability of reference" arguments about meaning, and with Goodman's equally famous "grue" paradox about induction.

The result is strictly speaking neither Wittgenstein's own argument nor Kripke's own, but instead a philosophical hybrid known familiarly to philosophers as "Kripkenstein's argument." Whatever its merits as a faithful interpretation of the *Investigations*, Kripkenstein's argument is a perfect example of philosophizing that actually obeys Wittgenstein's own dictum in the preface to the *Investigations*:

I should not like my writing to spare other people the trouble of thinking. But, if possible, to stimulate someone to thoughts of his own.[16]

So Kripkenstein's argument is well worth looking at both for its own sake, and also more importantly for the bearing it has on the issue of our knowledge of logic. Here is a reconstruction of the argument.

(1) Consider any meaningful use of language, but more specifically any meaningful mathematical use of language, and in particular our everyday use of the word 'plus' and the symbol '+': it is a given fact that by means of my external symbolic representation and also my internal mental representation I *grasp* the rule for addition.

(2) Although I have computed only finitely many sums in the past, the rule for addition determines my answer for indefinitely many sums that I have never considered. Indeed, the arithmetic function corresponding to the rule for addition determines a complete collection of infinitely many values/outputs for infinitely many arguments/inputs to that function.

(3) Suppose, however, that I compute '68+57' for the first time. I am confident that the correct answer is '125', and it is also true (i) that the plus function when applied to the inputs 68 and 57 yields 125 as the output, as well as (ii) that 'plus' as I intended to use it in the past denoted a function which when applied to the numbers I called '68' and '57' yields the value 125.

(4) But now a "bizarre skeptic" questions my answer, on the grounds that I might have intended (and indeed might now be intending) to use 'plus' such that the correct answer is in fact '5' and that the correct value of the function I intended is 5! That is possible because (a) in the past I computed only finitely many sums and by hypothesis had never encountered '68+57' (and let's assume for simplicity also that I had always referred to natural numbers

less than 57), and (b) it is therefore possible that the rule I followed (and am now following) corresponded in fact to the function *quus*:

If either x or y is less than 57, then x quus $y = x + y$,
but if either x or y is greater than or equal to 57, then x quus $y = 5$.

So the rule following skeptic claims that I am misinterpreting my own previous (and present) usage. More precisely, and very disturbingly, what he claims is that by 'plus' or '+' I always meant (and am now meaning) quus, *not* plus.

(5) Any adequate reply to the skeptic must satisfy two conditions: (1) it must give an account of what fact it is about my mental state that constitutes my meaning plus, not quus; and (2) it must show how I am justified in giving the answer '125' to '68+57'.

(6) But there is *no mental fact about me*, whether it is an occurrent mental representation such as a mental image or an image together with a projection that interprets it, a mental disposition, a mental state or process, or even a unique phenomenal *quale* uniformly associated with my use of 'plus' and '+', that uniquely determines what I meant (and now mean) by the use of those symbols. That is, no mental fact about me uniquely determines that I meant (or now mean) plus and not quus, precisely because the existence of each of those mental facts can be interpreted consistently with the hypothesis that I actually meant (and now mean) quus and not plus, or that (mutatis mutandis) I am "quounting" and not counting, etc. Indeed there is no mental fact about me that uniquely determines that I meant (and now mean) *any definite function whatsoever* by 'plus' or '+'. Thus I might have meant (and now mean) nothing definite at all!

(7) So I have no justification for my claim that the correct answer to '68+57' is '125' and that the corresponding value of the function is 125.

(8) Therefore, the rule-following skeptic is correct.

(9) By virtue of the RFP, Wittgenstein is committed to a generalizable and radical skepticism about the determination of future linguistic usage by the past contents of my mind. This is strongly analogous to Hume's skepticism about the determination of the future by the past (both inferentially, as skepticism about induction, and also causally, as skepticism about our knowledge of necessary connection in nature).

(10) The RFP can be resolved only by a "skeptical solution" that accepts both (i) that there is no mental fact about me that determines whether I am

following the rule for plus or the rule for quus, and also (ii) that I have no internal justification for my claim that the correct answer to '68+57' is '125', and then turns instead to look purely descriptively at the actual circumstances under which I can be correctly said to be following plus rather than quus and in which it can be asserted that the correct answer to '68+57' is '125'.

(11) If we consider a single individual in isolation, then although it is an empirical fact that the individual does confidently assert, or at least has the disposition to assert confidently, that the correct answer to '68+57' is '125', nevertheless (by (10) (ii)) there is no internal justification for this assertion.

(12) But if we take into account the fact that the individual is in a community, then the philosophical picture radically changes, and we must adopt an assertibility-conditions semantics (according to which a statement is true if and only if it is legitimately assertible) and reject truth-conditional semantics (according to which a statement is true if and only if it corresponds to the facts).

(13) The empirical fact of our successful rule-following practices (see (11)) depends essentially on the further brute empirical fact that we agree with one another in our responses to questions like 'What is 68+57?'

(14) Hence the relevant assertibility condition for the answer '125' is merely whether the individual's response agrees with everyone else's response to the same question, and this external judgment is determined just by observing the individual's behavior and surrounding circumstances. This solution to the RFP in turn is analogous to Hume's claim to have shown that the only way to make sense of a causal relation between two phenomenal events is to subsume it under a customary or habitual regularity of constant conjunctions of instances of the relevant event-types.

(15) This solution to the RFP entails that necessarily there is no private rule-following, which in turn entails the conclusion of the PLA.

For my purposes, there are three crucial points to be made about Kripkenstein's argument.

The first point is Kripkenstein's way of stating the conclusion of the RFP in step (8), which sums up steps (6) and (7): *no mental fact about me suffices either to fix the meaning or to justify my use of a mathematical rule.* Or as Kripke puts it:

An answer to the skeptic must satisfy two conditions. First, it must give an account of what fact it is (about my mental state) that constitutes my meaning plus, not quus. But further, there is a condition that any putative candidate for such a fact must satisfy. It must, in some sense, show how I am justified in giving the answer '125' to '68 + 57'.

The skeptic argues that when I answered '125' to the problem '68 + 57', my answer was an unjustified leap in the dark; my past mental history is equally compatible with the hypothesis that I meant quus, and therefore should have said '5'. We can put the problem this way: When asked for the answer to '68 + 57' I unhesitatingly and automatically produced '125', but it would seem that if I previously performed this computation explicitly I might just as well have answered '5'. Nothing justifies a brute inclination to answer one way rather than another.[17]

The second point is that there are two further serious problems standing in the way of any putative solution to the Kripkensteinian RFP that proceeds by appealing to some mental fact about me: (a) the problem of infinity and (b) the problem of normativity. These problems shape up as follows. With respect to (a), no mental fact about me, be it an occurrent mental representation such as a mental image or an image together with a projection that interprets it, a mental disposition, a mental state or process, or even a unique phenomenal *quale*, can be projected infinitely into the future in the way required by the individuation of a complete plus-function for the entire natural number system. And with respect to (b), every mental fact about me underwrites only a *descriptive* characterization of the rule I am following, not a *prescriptive* characterization. But if I am to go on following the rule for plus, as opposed to a deviant rule that is descriptively equivalent to the first one, then it must be the case that I *should be* or *ought to be* following the plus rule and not the deviant rule instead. That is, something must *obligate* me to follow the plus rule. Yet no mental fact about me has this normative force.

The third point is that the Kripkensteinian solution to the rule-following paradox is a *skeptical solution*, along the lines of Hume's famous skeptical solution to his worries about induction and causation. The Kripkensteinian–Humean skeptic thus insists that in the face of the collapse of every attempt to use some mental fact as a source of justification, nevertheless I skillfully apply the plus rule blindly, or without justification, within a long-standing public practice of such uses in my linguistic community:

199. Is what we call "obeying a rule" something it would be possible for only *one* man to do, and to do only *once* in his life? . . . It is not possible that there should have

been only one occasion on which someone obeyed a rule. It is not possible that there should have been only one occasion on which a report was made, an order given or understood, and so on.—To obey a rule, to make a report, to give an order, to play a game of chess, are *customs* (uses, institutions).

To understand a sentence means to understand a language. To understand a language means to be master of a technique. . . .

202. And hence also "obeying a rule" is a practice. And to *think* one is obeying a rule is not to obey a rule. Hence it is not possible to obey a rule "privately": otherwise thinking one was obeying a rule would be the same thing as obeying it.[18]

I want to extend Kripkenstein's argument to logic. It is clear from what Wittgenstein himself explicitly says in *Remarks on the Foundations of Mathematics* (see, for example, the second epigraph for this chapter) that an extension of the RFP is not only legitimate but also explicitly intended by him. Let us suppose, moreover, just to keep things as simple as possible for the purposes of the example, that my logic of thought contains a concept expressing classical negation.[19]

Now for the extended Kripkensteinian RFP. In addition to classical negation, there is at least one other logical concept of negation. Call it *negativity*. Negativity works exactly like classical negation for all truth-bearers that we rational animals have actually considered up to today. But after today, when negativity is applied to truth-bearers, it systematically assigns to them some nonclassical truth-value or nonclassical valuation: say, a third value, or a truth-value glut. Continuing the vaguely Sartrean spirit of my example, let us call any instance of this deviant output by the generic name "the Other." Here, then, is the hard question: Does the operator '~' or the word 'not', as I have been and currently am using them, express negation or negativity? That is, how do I know that I have not been following the rule for negativity all along? By the same reasoning used in Kripkenstein's plus–quus example, it seems obvious that both the inner and outer histories of my previous applications of the rule turn out to be fully consistent with various deviant interpretations of it. So no mental fact about me suffices either to fix the meaning or to justify my use of the rule for using '~' or 'not'.

This conclusion of course has dire implications. Suppose that I cannot now say whether it has been classical negation or negativity that I have been operating with. Then a classically false truth-bearer, sentence S, which I also believe to be classically false, shows up first thing tomorrow morning. I want to apply to *S* the operation signified by '~' or 'not'. I do not know how to deal with *S*, because I do not know whether to apply to *S* classical negation or negativity. I do not know, that is, whether the result of applying to *S* the operation signified by '~' or 'not' will be *true* in the classical sense, or *the Other*. So I do not know how to go on logicizing. Since analogues of the same worry can be retrospectively and prospectively raised about *every* use of '~' or 'not', it follows that not only all my present uses of those symbols but also all my past and future uses are undermined in the same way. I wrongly believed that I knew what I was doing and that my uses were all being *rationally guided* by the rule for classical negation. In fact, however, *I really didn't know any such thing*. And this problem is not just *my* problem, since the same problem arises for each and every rule-follower. I and everyone else who has ever used a word or other symbol for negation, that is, every speaker of a natural language, has been living all along in *logical bad faith*.

6.2 How to Follow a Logical Rule

In the previous section I spelled out Kripke's well-known Wittgensteinian argument for the RFP, or the skeptical conclusion that there is no mental fact about me that suffices either to fix the meaning or to justify my use of the rule for plus, hence my use of any mathematical rule, and then I extended this line of argument to logical rules. Nevertheless, it seems to me that Kripkenstein's RFP is a non sequitur. The invalid step in his argument derives from a crucial ambiguity in the meaning of the word 'fact' in the crucial phrase 'mental fact'. Facts can be either *empirical* (that is, sensory and contingent) or *nonempirical* (that is, underdetermined by sensory experiences and noncontingent).[20] In providing support for his skeptical conclusion, Kripkenstein considers only *empirical* mental facts. Yet the conclusion of his argument is supposed to cover *all* types of mental facts. In other words, for the purposes of his argument Kripkenstein covertly assumes the truth of empiricism, but this is highly questionable. Hence there is no valid step from the premise (which I am prepared to accept) that no empirical mental fact suffices to fix the meaning or justify my use of a mathematical or logical rule, to the conclusion that *no mental fact period* fixes the meaning or justifies my use of a mathematical or logical rule. For Kripkenstein has not considered whether a *non*empirical mental fact might instead successfully do the meaning-fixing and justificatory jobs.

In turn, this point undermines Kripkenstein's RFP. If what fixes the meaning and justifies my use of a mathematical or logical rule is a *non*empirical mental fact about me, then it is entirely unsurprising that both the inner and outer histories of my previous applications of that rule turn out to be fully consistent with deviant interpretations of it. This is because the very idea of something's being nonempirical includes its being a priori, which is to say that it is underdetermined by sensory experiences and contingent facts even though it is always actually associated with sensory experiences. Now every empirical mental fact is an a posteriori fact. So *if* there is some nonempirical mental fact about me that fixes the meaning and justifies my use of a mathematical or logical rule, then the inner and outer histories of my previous applications of the rule *must* underdetermine whatever fixes that rule's meaning and justifies my use of it.

The upshot is that in order to get a new straight solution to the logical version of Kripkenstein's RFP, and thereby also avoid Kripkenstein's Humean skeptical solution to the paradox, we can and should appeal to a nonempirical mental fact about me as the meaning-fixing and justifying ground of following a logical rule. I do not mean to say, however, that we should appeal to some sort of *supernatural* mental fact about me. I grant that nothing outside of space and time, and altogether causally irrelevant, has anything to do with following a logical rule. What we need is a mental fact about actual human animals. My proposal is that we appeal directly to a cognitive capacity for *logical intuition* in order to account for my ability to fix the meaning and justify my use of a logical rule. More precisely, what I am proposing as a solution for the logical version of the rule-following paradox is that a capacity for logical intuition is an intrinsic part of the innate logic faculty and thereby is automatically incorporated into my logic of thought. So, to return to the negation–negativity example, I am saying that on the assumption that my logic of thought contains a concept for classical negation, my innate capacity for logical intuition adequately fixes the meaning of '~' and 'not' for me as *classical negation,* and also adequately justifies my using those symbols in precisely that way by projecting that rule infinitely and by normatively supporting my use of it.

This proposal is directly linked to my earlier discussion of the logocentric predicament in chapter 3. The Kripkenstein RFP for logic is that there is no mental fact about me that suffices either to fix the meaning or to justify my use of a logical rule. And this result in turn yields Kripkenstein's Humean skeptical solution: my spade is turned, there is no justification, and I blindly do whatever is warranted by my linguistic community. Now there seem to be only two possible ways of avoiding the Kripkensteinian outcome: (1) appeal to an *inferential* justification of my use of a logical rule, or (2) appeal to a *noninferential* justification. But on the one hand, an inferential justification will obviously presuppose and use logic, and thus be subject to the logocentric predicament. And on the other hand, any noninferential justification of my use of a logical rule that we posit must also be consistent with an adequate solution to the logocentric predicament. The only adequate solution to the logocentric predicament, I argued, is via the logic faculty thesis. Therefore, any noninferential justification of my use of a logical rule must also be consistent with an adequate solution to the logocentric predicament in general, and with the logic faculty thesis in particular. This in turn rules out, for example, both communitarian and nonfactualist/noncognitivist noninferential justification strategies, simply because they fail as adequate solutions to the logocentric predicament (see section 3.5). Correspondingly, the same constraint leaves unscathed, it seems, only an *intuitional* noninferential justification strategy. Hence my proposal is that a capacity for logical intuition intrinsically belongs to the logic faculty and thereby is incorporated into my logic of thought.

6.3 Logical Intuition and Inferential Role

If the proposal about logical intuition that I offered in the last section is correct, it has some interesting and fruitful implications for another issue in the philosophy of logic, namely the acceptability of the *inferential role thesis* mentioned briefly in section 3.3. According to the inferential role thesis, the meaning of a logical constant is constituted by its inferential role, where the "inferential role" of some term T is how T functions in inferences leading to or from sentences containing T. The inferential role thesis is important, because it seems to offer a satisfactory answer to the hard question, "what is the nature of a logical constant?"

But that is not all. Logic, I have assumed all along, is the science of the necessary relation of consequence. And as Tarski and others have convincingly argued, no matter which theory of logical consequence one adopts, the concept of logical consequence is intimately bound up with the concept of a logical constant.[21] So, if the inferential role thesis gives a satisfactory explanation of the nature of a logical constant, then it will also go some not inconsiderable distance toward giving a satisfactory internal explanation of the nature of logic itself.[22]

Nevertheless, it seems to me that there are at least two basic problems with the inferential role thesis. If, as I am assuming, 'constituted by' in the thesis "the meaning of a logical constant is constituted by its inferential role," implies that the meaning of a logical constant is *exhausted* by its inferential role, then the inferential role theorist is committed to the following two-part view:

(i) that the meaning of a logical constant is exhausted by the set of all actual and possible inferences leading to or from sentences containing that constant; and

(ii) that knowledge of the meaning of a logical constant is exhausted by the knowledge of inference rules governing sentences in which that constant occurs.

But (i*) if the meaning of a logical constant is exhausted by inferences, then the meaning of a logical constant such as classical conjunction or classical negation cannot be completely or even partially determined by, say, truth tables. That seems obviously false. And (ii*) if the meaning of the logical constant is not graspable to some extent independently of one's knowledge of inference rules, then how are we to explain why the logician is justified in applying the relevant inference rule to new inferential contexts? It seems clear that the knowledge of the inference rules plus knowledge of the set of previous applications of the rule underdetermines the total set of actual and possible inference rule-applications, and thereby underdetermines knowledge of the meaning of any logical constant that is governed by those rules. Indeed, this is just another way of expressing Kripkenstein's RFP as extended to logic. So the inferential role thesis leads us back to the RFP and cannot be offered as a plausible theory independently of an adequate solution to it.

Nevertheless, it does seem to me that there is a kernel of truth in the inferential role thesis. That is, it does seem true that the meaning of a logical constant is necessarily connected to the set of all actual and possible inferences leading to or from sentences containing that constant, and also that one's knowledge of the meaning of a logical constant is necessarily connected to one's knowledge of inference rules governing sentences in which that constant occurs. What is wrong with the inferential role thesis, I think, is precisely its *reductive* component, namely that the meaning of a logical constant is *exhausted* by the actual and possible inferences into which the constant enters, and that one's knowledge of the meaning of a logical constant is *exhausted* by one's knowledge of inference rules governing sentences in which that constant occurs. So, to fix up the inferential role thesis, we need to cut out the reductive part and replace it with an appeal to some distinct and irreducible factor that has independent good grounds for its adoption.

Suppose, then, that we drop the reductive component of the inferential role thesis and supplement the inferential role of a logical constant with a direct appeal to logical intuition. This would circumvent the two difficulties. The revised inferential role thesis would say that the meaning of a logical constant (in a given logic, as constructed by rational animals possessing the logic faculty) is constituted by its inferential role *together with the capacity for logical intuition*. In this way the meaning of a logical constant would be strictly determined by all the actual and possible inferences into which the constant enters, *plus the capacity for logical intuition*, which adds supplementary factors also necessary to fixing its meaning (for example, the grasp of a corresponding truth table). Correspondingly, the knowledge of the meaning of a logical constant would be strictly determined by knowledge of inference rules governing sentences in which that constant occurs, *plus the capacity for logical intuition*, which adds supplementary factors necessary for justifying my application of logical rules governing the use of the constant (for example, the grasp of an infinite projection of a logical rule, or the grasp of the normative force of a logical rule).

In the last two sections I have used the concept of logical intuition, in conjunction with the logical faculty thesis, to offer a straight solution to Kripkenstein's RFP and thereby to block his Humean skeptical solution of the logical version of the RFP, and also to suggest how we can repair some difficulties in an inferential role approach to the nature of a logical constant. So far, however, I have not been offering anything like a *theory* of logical intuition. That is the task of sections 6.5 and 6.6. Still, even at this preliminary stage in the argument I need to say what I mean by 'intuition', and also reply to the most obvious objections to intuition. In a word, I have to show that intuition *as such* is at least minimally qualified to do the meaning-fixing and noninferential justificatory jobs required of *logical* intuition. So the next section is a prolegomenon to the theory of logical intuition I will develop in later sections.

6.4   On Intuition

Intuition in the sense I am concerned with is not a hunch or a guess. It is not divination, prognostication, or "second sight." It is not a "feeling in my heart," a "feeling in my bones," or a "feeling in my gut." And it is not an emotional attitude, a cognitive bias, or a prejudice. So what is intuition? Here is a philosophical picture of intuition that can be compared with some other philosophical conceptions of intuition, both classical and recent.[23]

First, an intuition is a *mental episode or mental act*, as opposed to being either a mental disposition or a mere mental state. Rational human animals have a mental capacity for intuition, but an intuition is not the same as that capacity, because it is what *actualizes* that capacity. Nor is an intuition the same as a mere mental state. A mere mental state is the instantiation of one or more mental properties (whether nonrelational or relational) at a particular time and place. But an intuition involves a mental process occurring over time, and it is something that a subject herself *does*.

Second, intuition is *a priori*, which is to say that it is underdetermined by inner, proprioceptive, and outer sensory experiences, even though it is always actually associated with such sensory experiences.[24]

Third, intuition is *content-comprehending*, which is to say that a subject intuits that S only if she adequately understands the semantic content (both referential and intensional) of the sentence 'S'. Misunderstood, partially grasped, or otherwise inadequately understood sentences are not targets of intuition.

Fourth, intuition is *clear and distinct*, which is to say (i) that it has a representational content, (ii) that this representational content can be directly and vividly presented to a self-conscious thinker (clarity), and (iii) that this content also displays its internal structure to that self-conscious thinker insofar as she carefully focuses her attention on it (distinctness).

Fifth, intuition is *strict-modality-attributing*, which is to say that a subject intuits that S only if she believes that necessarily S or (assuming that the belief is rational, which is equivalent) that 'S' is a necessary truth. Sentences believed to be contingent are not targets of intuition.

Sixth, intuition is *authoritative*, which is to say that if a subject intuits that S, she is thereby fully convinced that necessarily S (or again assuming that the conviction is rational, which is equivalent) that 'S' is a necessary truth. In other words, intuition is *intrinsically compelling*. An authoritative intuition that S, however, needs to be carefully distinguished from a prima facie intuition that S, which is merely an "intellectual seeming" to the effect that necessarily S.[25] For example, Chomsky's *intuitions of grammaticalness*, the data of grammatical theory in his sense, would count as prima facie intuitions in this sense.[26] Just as perceptual seemings provide prima facie evidence for perceptual beliefs, and just as grammaticality intuitions provide prima facie evidence for grammatical theories, so too prima facie modal intuitions provide prima facie evidence for strict modal beliefs. But authoritative intuition is the *self-evidence* of strict modal beliefs. That is, when a subject authoritatively intuits that S, not only does she understand the semantic content of S and find that content to be clear and distinct, she also cannot seriously entertain the possibility that it is false that necessarily S. By contrast, when a subject merely prima facie intuits that S, it thereby *intellectually seems to her* that necessarily S: but she can still seriously entertain the possibility that it is false that necessarily S.

Seventh, and crucially for my overall argument, intuition is *noninferential*, which is to say that whenever a subject intuits that S and is thereby fully convinced that necessarily S, then her belief that necessarily S is based not on any reasons or premises, but instead only on the intuitional episode itself.

Eighth, intuition is *cognitively indispensable*, which is to say that every process of reasoning or belief-justification must ultimately bottom out in an intuition of some logical principle of deductive inference that governs the relevant entailment relation between the premises and conclusion of the reasoning, or between the supporting evidence and the putatively justified belief. Otherwise there would be a vicious infinite regress of deductive inferential justificatory groundings.[27]

Ninth, intuition is *fallible*, which is to say that it is always possible for an intuition to be wrong. Neither the authoritativeness of intuition nor its cognitive indispensability implies that it cannot be mistaken.[28] Unfortunately for creatures with minds like ours, it is built into the cognitivist existential predicament (see section 5.4) that the world might be otherwise than I take it to be, no matter how intrinsically compelling the evidence for my belief is. It is plausible to hold, given the authoritativeness of intuition together with its cognitive indispensability, that an intuition that S provides *reliable evidence* for the intuiting subject's belief that necessarily S.[29] But even assuming this, an intuition that S cannot provide *an epistemic guarantee* that necessarily S.

These nine features (being a mental act, apriority, content-comprehensiveness, clarity and distinctness, strict-modality-attributivity, authoritativeness, noninferentiality, cognitive indispensability, and fallibility) make up the core of the concept of intuition as I understand it. Very shortly I will add two ancillary features (that is, the distinction between *intuition-of* and *intuition-that*, and the distinction between intuitive judgment and intuitive inference) for good measure. All eleven features, taken together as a package, yield the general concept of intuition as I am using it.

Charles Parsons draws a useful distinction between *intuition-of* and *intuition-that*.[30] Intuition-of is an intuition directed at some individual object, for example, the number 2. By contrast, intuition-that is an intuition in the form of a propositional attitude, judgment, or belief directed at a truth-bearing semantic content, for example, that 2+2=4. So I can intuit the individual number 2, or intuit the mathematical truth that 2+2=4. In the logical case, I can intuit the logical notion *classical negation*, or intuit the logical truth that if *P* and if *P* then *Q*, then *Q*. But intuiting that if *P* and if *P* then *Q*, then *Q* is not thereby to intuit an individual object, since it is merely a propositional attitude. Hence, according to Parsons, a theory of intuition need not commit itself to a theory of intuition-of. In particular, it need not commit itself to a *platonic* theory of intuition-of.

What Parsons says seems correct, as far as it goes. Yet he has overlooked the possibility that in the sentence 'I intuit that if *P* and if *P* then *Q*, then *Q*', the word 'that' can function not only as a conjunction introducing a subordinate clause, but also as a demonstrative for the sentence-type[31] which follows it. Thus when I intuit that if *P* and if *P* then *Q*, then *Q*, I also intuit *that*, i.e., 'if *P* and if *P* then *Q*, then *Q*'. More precisely, in intuiting that if *P* and if *P* then *Q*, then *Q*, I employ a mental image of the sentence-type 'if *P* and if *P* then *Q*, then *Q*' as *the mental medium or mental vehicle* of my act of intuition. Just to give it a name, I will dub this *the paratactic approach to logical intuition-that*, after Donald Davidson's paratactic approach to the analysis of belief-sentences.[32]

It needs to be emphasized that I am not offering an argument that this is the correct and unique grammatical and semantic analysis of 'intuits that' constructions. Nor do I want to defend Davidson's paratactic analysis of 'believes that' constructions. Indeed, for the purposes of this book I want to stay officially neutral on the question of the best analysis of 'believes that' constructions and propositional attitude constructions more generally. My point is simply that a paratactic *reading* of 'intuits that' constructions, and correspondingly a paratactic *approach* to logical intuition-that, are not ruled out by Parsons's useful distinction between the two types of intuition. So I am officially leaving open the possibility that both paratactic and non-paratactic readings of 'intuits that' constructions can consistently cohabit the same grammatical and semantic space: that such constructions are simply *nonproblematically ambiguous* as between the two readings, indeed, every bit as nonproblematically ambiguous as 'Flying planes can be dangerous'. Thus, in holding that logical intuition-that can be understood paratactically, I am able to hold that there is an intuition-of the necessary sentence '*S*', mediated or carried by my mental image of '*S*', that is *embedded within* the mental act of logically intuiting that *S*, where the expression 'that *S*' refers to a proposition, as in the classical Frege-style analysis of propositional attitude sentences. So although there is clearly a conceptual difference between intuition-of and intuition-that, the very same intuition can be in one respect (the paratactic respect) an intuition-of, and can also be in another respect (the classical Frege-style respect) an intuition-that. This dual aspect approach to 'intuits that' constructions and to logical intuition-that will later prove to have important philosophical payoffs.

Here is another tricky point about intuition that needs some sorting out. In *Rules for the Direction of the Mind*, Descartes distinguishes between "intuition" and "deduction."[33] What he is drawing to our attention is the difference between

(a) intuiting a single (necessary) sentence, and

(b) a sequence of such intuitions in the form of an argument, such that it terminates with an intuition of a conclusion *from* all those premises and earlier steps.

Thus there is a contrast between what can be called *intuitive judgment* on the one hand and *intuitive inference* on the other. According to Descartes, intuitive inference is supposed to be in principle *less* epistemically trustworthy than intuitive judgment. Presumably this is because intuitive inference involves holding one or more premises or inferential steps in memory and then sequentially retrieving them in the continuous advance from the premise(s) to the conclusion, and memory is, of course, notoriously untrustworthy.

This Cartesian way of looking at intuition, however, assumes that its *proper* target is a single necessary sentence. But this seems arbitrary. Surely (for example, on the temporary simplifying assumption that my logic of thought includes classical propositional logic) I can intuit the argument that is visually displayed as follows:

(1) P (premise)
(2) $P \rightarrow Q$ (premise)
(3) Q (1, 2 MPP)

every bit as easily as I can separately intuit steps (1), (2), or (3). And surely I can intuit

$$\{[P \ \& \ (P \rightarrow Q)] \rightarrow Q\},$$

that is, the corresponding conditional of the argument, just as easily as I can intuit either the whole argument or any of its steps. More generally, we can logically intuit not only single logically necessary sentences but also inference-steps and even whole arguments, since each can be treated as a direct object of logical intuition in accordance with the paratactic approach to logical intuition-that. The scope of logical intuition-that thus seems to be determined *largely by the spatial informational limitations on my ability to scan linguistic mental images.* (By 'linguistic mental image' I mean a mental image of either a natural language inscription or a bit of formal symbolism.) That is, precisely what I can or cannot logically intuit seems to be largely a function of how big and detailed my scannable linguistic mental images can be. If so, there is no reason to think that intuitive inference is inherently less trustworthy than intuitive judgment.

In light of this point, it must be particularly noted that calling an intuition "noninferential," although perfectly accurate, can be misleading. What it means is that an intuition is evidentially self-contained, or logically independent of further rational grounds or premises, as opposed to being logically dependent or based on further grounds or premises. But this is not to say that the intuition is not itself a reason for belief. On the contrary, an intuition is a mental episode or act that intrinsically carries modal and justificatory implications for belief: given the internal structure of that mental episode or act, it *metaphysically necessitates* belief, that is, it is intrinsically compelling. An intuition is something *outside* "the space of reasons," if we assume that all reasons are inferential; yet an intuition is something *inside* the space of reasons if we assume that reasons need not always be inferential. Truistically, a reason is a fact that rationally supports a human belief or intentional action, and if this support is sufficient, then it is also a justification.

This rational support is always normative and sometimes inferential, but it is not always inferential. Where noninferential, the rational support is intuitive. An intuition always stops the regress of inferential reasons precisely because it operates as a reason and thereby rationally supports a human belief, without having to be a premise for that belief. Otherwise put, intuition is reliable evidence for belief, but it *need not* be part of an inferential justification of belief precisely because its connection with belief is more cognitively basic than inferential justification. Or still otherwise put, the key Cartesian insight here is that justified belief can be *either* the rational result of an inference from reasons as premises *or* the rational result of a cognitively basic clear and distinct mental episode or act. Moreover, as Descartes also realized, the noninferentiality of an intuition does not in any way imply that an intuition cannot be applied directly either to inference-steps or to whole arguments. Intuition can also *cognitively penetrate* inferential reasoning and inferential justification more generally.

There are two obvious objections to the very idea of intuition. Wittgenstein succinctly articulates both of them:

A doubt was possible in certain circumstances. But that is not to say that I did doubt, or even could doubt. . . . So it must have been intuition that removed this doubt?—If intuition is an inner voice—how do I know *how* I am to obey it? And how do I know it doesn't mislead me? For if it can guide me right, it can also guide me wrong. ((Intuition an unnecessary shuffle.))[34]

The first objection is that because intuition presents itself as radically different from all empirical mental facts, it is therefore nothing but a *magical* empirical mental fact: an "inner voice." More precisely, it magically *causes* belief. But precisely because it causes belief, however, just like ordinary empirical mental facts, the inner voice is either perfectly consistent with deviant interpretations of the rule or else merely begs the question. In this way, intuition has no more evidential force than that of taking a drug or a receiving a blow on the head, and then immediately acquiring a belief.

The second objection is that despite an intuition's inherent claim of infallibility, it is in point of fact dubitable.

These objections can be dealt with quite easily. First, to insist that intuition must present itself as a magical *empirical-causal* mental fact, an inner voice, is just to refuse to admit the possible existence of nonempirical or a priori mental facts that are intrinsically compelling without causation. Thus the first objection simply assumes the truth of empiricism. But to the extent that empiricism is taken to be part and parcel of scientific naturalism,[35] it is self-refuting (see section 1.3), and furthermore, the poverty of the stimulus argument provides good reasons for rejecting empiricism (see section 4.1). Perhaps more importantly, however, the first objection also assumes that nothing but premises in arguments can have evidential force with respect to belief, because anything else that can bring about a belief is *merely* causal and thus nonevidential. But this overlooks the possibility that a mental episode itself can stand in an intrinsic rational and justificational noninferential noncausal relation to belief, by virtue of its internal phenomenological-*cum*-representational structure. Yet again, this is precisely what Descartes was driving at, at least implicitly, with his notion of the clarity and distinctness of a perception. The idea is that when a conscious mental episode or act in a rational animal takes on a certain *cognitively optimal* internal structure, it then necessitates belief.

Second, according to the view I am developing, it is explicitly conceded from the start that intuition is fallible or defeasible. So this heads off the worry about the implausibility of infallibilism at the pass by preemptively conceding that infallibilism about intuition is implausible.

I want to close this section with a brief excursion into the phenomenology of intuition, as a way of fleshing out its eleven features a little more fully and also cementing my straight solution to Kripkenstein's RFP. This excursion is particularly important for my view, in light of its claim that a certain class of mental episodes or acts (namely, intuitions) is intrinsically compelling and thus have noninferential, self-evident, and a priori evidential justificatory force with respect to belief. Otherwise put, I need to try to indicate more precisely just what it is about the internal structure of intuitions that makes them "cognitively optimal." As I see it then, the phenomenology of intuition has three characteristic aspects: (i) a sense of overwhelming doxic ease, (ii) "locking-onto-ness," and (iii) a sense of rational guidedness.

(i) By the notion of a sense of overwhelming doxic ease I mean the conscious experience of a maximal level of doxic first-personal self-confidence, or what is sometimes called *subjective certainty*. In having an intuition of something, or intuiting that such-and-such, I experience *no doubt or critical fussing or second-guessing whatsoever*. In intuiting something, I simply "see it," "get the point," "see how it automatically follows," or "find it obvious," and cognitively cannot help doing so. The close connection here with Quine's notion of the "obviousness" of sheer logic should be, well, obvious.

(ii) By the notion of "locking-onto-ness" I mean the momentary but intensely satisfying conscious intellectual experience of a perfect conformity between the representational content of my mental state or act, and the object I cognize. For example, as I cognize the sentence 'Annoying politicians can sometimes also be amusing', and shift spontaneously from one to another of its two syntactically and semantically distinct readings,[36] it is as if something in my mind crisply snaps one Lego block onto another. I do not have to parse this sentence laboriously and self-critically in the way I might parse a sentence in a foreign language I am trying to translate for myself: I simply lock onto the relevant chunk of semantic syntax. The syntax of the sentence which supports that particular reading thrusts itself forward as the very one my mind "wanted" all along. Similarly, on the assumption that my logic of thought contains a concept for classical negation, when I negate a false sentence I do not have to compute a value for the truth-function from some imagined truth table I have in my head, as I might if I were working with a nonstandard operator in a new formal language I am learning: I simply lock onto its being true. The output of the truth-function for classical negation thrusts itself forward as the very one my mind "wanted" all along.

(iii) Finally, by the notion of a sense of rational guidedness I mean the conscious experience of being inexorably led to a certain cognitive result. This feeling of inexorability expands to cover all my past and future applications of the same rule. That is, I do not experience an individual intuited case as an *isolated* case but rather as only one instance of an infinite series of operations of the very same kind. This is revealed, for example, in my sense that I get the result '125' for '68+57' just because it is the "right" result; and that sense of rightness stems in turn from the even deeper feeling I have that this case of addition falls smoothly into an infinite pattern that includes all possible instances of the arithmetic function '$x+y=z$'. Not only that, but for each and every one of the instances of that series that I undertake in this guided way, it is not as though I am being *forced* or *merely caused* to get this result; rather, it is as if I am obeying a command I received from someone whose authority I fully accept, or am obeying a command I freely gave to myself.

The sense of rational guidedness can be further illustrated in the following way. If, even after having carefully read my Wittgenstein and my Kripke, in the ordinary course of things one day I added up 68 and 57 and got 5 as their sum, I certainly would *not* think to myself: "Oh piffle, I wonder if I've been actually following quus all along?" Rather, I would think: "*Oh god I wonder if I'm losing my mind?*" This is not a mere calculation error. It is not as if I quickly added 68 and 57 in my head and got 115 or 135 as the answer. Given that sort of a slip, I would say to myself, "Oops," and then recalculate on paper using the familiar algorithm. But if I really did get 5 as an answer, then I would have gone completely off the computational rails. I would feel intense intellectual amazement, rising panic, and also (strange as this may sound) intense intellectual shame. This vivid experience of infinitely patterned normative cognitive inexorability also characterizes the consciousness of my intuition of classical negation, assuming for the purposes of argument, as before, that it belongs to my logic of thought. In negating a classically false truth-bearer and deriving a true truth-bearer as its result, this operation is experienced as inherently repeatable over an infinite set of cases. Against that infinite backdrop, it is also experienced as the normatively "right" answer. If one day I applied the negation sign three times successively to the symbol for a classically false truth-bearer, and successively derived falsity, a third value, and a truth-value gap, again I would not begin to wonder whether I had been using negativity all along. Rather, I would get an odd sinking feeling in the pit of my stomach, and obsessively run the truth-functional operation over and over again until I got it right. If I kept getting the deviant values or valuations, I would anxiously clutch my head and wonder whether I was beginning to go bonkers.

One last point in this connection. It is significant to note that as far as the phenomenology of intuition (its sense of overwhelming doxic ease, locking-onto-ness, and rational guidedness) is concerned, my account is perfectly in line with several often-overlooked comments about intuition made by Wittgenstein in *Remarks on the Philosophy of Psychology* and *Remarks on the Foundations of Mathematics*:

The old idea of the role of *intuition* in mathematics. Is this intuition just the seeing of complexes in different aspects?[37]

Might not one really talk of intuition in mathematics? Though it would not be a *mathematical* truth that was grasped intuitively, but a . . . psychological one. In this way I know with *great* certainty that if I multiply 25 by 25 ten times I shall get 625 every time. That is to say I know the psychological fact that this calculation will keep on seeming correct to me; as I know that if I write down the series of numbers from 1 to 20 ten times my lists will prove identical on collation.[38]

But aren't we guided by the rule? And how can it guide us, when its expression can after all be interpreted by us both thus and otherwise? I.e., when after all various regularities correspond to it. Well, we are inclined to say that an expression of the rule guides us, i.e., we are inclined to use this metaphor. . . .

Well, in our own case we surely have intuition, and people say that intuition underlies acting according to a rule.[39]

For Wittgenstein, however, the underlying ground of intuition is not in any way psychological but instead is nothing over and above a human being's embeddedness in a well-established practice:

What interests me is this immediate insight, whether it is of a truth or falsehood. I am asking: what is the characteristic demeanor of human beings who "have insight into" something "immediately," whatever the practical result of this insight is?

What interests me is not having immediate insight into a truth, but the phenomenon of immediate insight. Not indeed as a special mental phenomenon, but as one of human action.[40]

From a logical cognitivist point of view, there is a serious philosophical cost to be incurred by taking intuition to be solely a phenomenon of "human action," as if that somehow excluded its psychological dimension as a "special mental phenomenon." Above all, it overlooks the possibility that for rational human animals, some forms of human action are *also* special mental phenomena. I mean that intuition is at once and equally a phenomenon of human action *and* a special mental phenomenon. Intuition is a characteristic activity of the rational human animal.

For Wittgenstein, however, human action seems to exclude phenomenology and thus reduces intuition to nothing but a kind of socially embedded natural behavior. In turn, this reduction entails that the phenomenon of intuition is wholly empirical:

—Now is that an empirical fact? Of course—and yet it would be difficult to mention experiments that would convince me of it. Such a thing might be called an intuitively known *empirical* fact.[41]

Therefore, given another possible "form of life," I could in principle find my use of negativity to have the very same intuitional phenomenology that classical negation has for me now. Correspondingly, in the imagined new form of life, classical negation would be difficult or deviant. This seems to me incorrect. If a rational animal's logic of thought includes a concept for classical negation, then by virtue of her capacity for logical intuition, the very idea of "logicizing" as a form of life for rational animals of that sort *neces-*

sarily includes her taking classical negation to be overwhelmingly doxically easy, something she can lock onto, and something she is rationally guided by. This in turn makes the intuitability of negativity literally unthinkable for creatures like her.

Here is an argument for that claim. For the purposes of argument, suppose as before that my logic of thought includes classical negation. Now, holding that assumption fixed, try to conceive my going over to a "contrarian" world behind the looking glass, that is, a world in which my use of negativity has the same phenomenology that my use of classical negation does now, and in which classical negation is correspondingly difficult and deviant. In the contrarian world, I would spontaneously reject all classical tautologies, theorems, and valid inferences, as well as all logical laws that contain classical negation. But then I would certainly have changed the very meaning of the symbols '~' and 'not' as I currently understand them. And if I changed the meaning of such a basic logical constant, then I would have simply "changed the subject": for me it would not be logic any more; for me it would be nonlogic, a *schmogic*. Given the language of thought thesis and the logic of thought thesis (and the assumption that my logic of thought contains classical negation), it follows that both my knowledge of my own natural language and my language's total capacity for expressing my thoughts require classical negation. Combine this with the idea that I have a capacity for logical intuition, where intuition is sketched as above. Then, for me to attempt to take part in the contrarian form of life in which negativity has the phenomenology of my use of classical negation would undermine my language of thought and my logic of thought alike. That is, if it *were* possible for me to find my use of negativity to have the same phenomenology as my use of classical negation, then in that case I would not even be able to formulate the thought that it is possible, for my capacity for logical intuition (which is by hypothesis framed in terms of the concept of classical negation) would simply psychologically rule this out.

The argument I just used is intentionally similar to Quine's famous "deny the doctrine, change the subject" argument for diehard classicism. As we saw in section 2.5, Quine's argument for diehard classicism fails. My point here is that insofar as Quine's argument can be carefully restricted to the phenomenon of logical intuition, it comes out sound and undermines Wittgenstein's empiricist spin on logical intuition. I am not saying that a nonclassical logic with a negativity operator instead of a classical negation operator is impossible. I am saying that *if* for the purposes of argument we assume that classical negation belongs to my logic of thought and hence my language of thought alike, and *if* I have a capacity for logical intuition, *then* a deviant logical practice involving the negativity operator is going to be a priori unthinkable for me. Or, to use Quine's terminology, it is going to be utterly *obvious* to me that the contrarian world is impossible.

6.5   Benacerraf's Dilemma, Original and Extended

There are actually two Benacerraf worries in the philosophical literature. The first worry was described by Benacerraf in 1965 in a paper called "What Numbers Could Not Be" and is also known to Frege scholars as the *Caesar problem*.[42] It says that numbers cannot be uniquely identified with corresponding objects characterized in purely logical terms, because the very same logicized arithmetic sentences and theories can be satisfied by many distinct set-theoretic models: so what *are* the numbers? The second worry was described by Benacerraf in 1973 in a paper called "Mathematical Truth." What I am concerned with here is the second Benacerraf worry, which I will henceforth call *the Benacerraf dilemma* for convenience.

In this connection I want to argue for two claims. First, the original Benacerraf dilemma, which is about mathematical knowledge, can be smoothly extended to logical knowledge. Call this *the extended Benacerraf dilemma*. Second, a solution to the extended Benacerraf dilemma emerges if we reject two crucial steps in its explicit formulation.

The original Benacerraf dilemma is a worry about how to connect the abstract objects of true mathematical discourse with the presumed human knowability of those objects. In a nutshell, as I have already mentioned, the dilemma is that when we construe true mathematical discourse in a semantically "standard" (or Tarskian, referential) way and assume that the semantics of mathematics is homogeneous or uniform with the rest of natural language, then we are committed to the reality of humanly knowable abstract objects that are nevertheless *unknowable* according to a "reasonable epistemology," that is, our best overall theory of knowledge.

I have already taken a preliminary look at the original Benacerraf dilemma in section 1.5. But here is a more explicit formulation and rational reconstruction[43] of it.

The original Benacerraf dilemma

(1) Assuming a standard uniform semantics, mathematical truth is both objectively real and humanly knowable. (Premise.)

(2) The standard uniform semantics of mathematical truth thus implies the existence of corresponding mathematical objects not dependent on the existence of humans but still knowable by humans. (From (1) and the definition of the concept of "objective reality" as intersubjective knowability and non-dependence on actual individual minds.)

(3) According to its standard uniform semantics, mathematical truth is also necessary and a priori. (Premise.)

(4) Mathematical objects are abstract and not concrete, because the concreteness—that is, the contingency and spatiotemporality—of such objects is inconsistent with the necessity and apriority of mathematical truth. (From (2) and (3).)

(5) All and only concrete objects exist in spacetime. (Premise.)

(6) So mathematical objects do not exist in spacetime. (From (4) and (5).)

(7) All causally relevant (not to mention causally efficacious) entities exist in spacetime. (Premise.)

(8) So mathematical objects are causally inert. (From (6) and (7).)

(9) Our best overall theory of mathematical knowledge says that mathematical knowledge is some sort of intuition. (Premise.)

(10) Our best overall theory of intuition says that intuition is cognitively analogous to sense perception. (Premise.)

(11) So our best overall theory of mathematical knowledge says that mathematical intuition is cognitively analogous to sense perception. (From (9) and (10).)

(12) A reasonable epistemology is a causal theory of knowledge. (Premise.)

(13) The causal theory of sense perception is correct. (From (12).)

(14) Sense perception requires an efficacious causal link, involving direct physical contact, between the object perceived and the perceiver. (Premise.)

(15) Therefore, mathematical knowledge is impossible. (From (8), (11), (13), and (14).)

For my purposes, the original Benacerraf dilemma is important not so much because it entails skepticism about mathematical knowledge (although that is of course important enough) but rather because it easily extends to skepticism about logical knowledge. That is, we can derive a logical version of the original dilemma simply by substituting 'logical' for every occurrence of 'mathematical' in the above formulation of the problem, as below.

The Extended Benacerraf Dilemma

(1*) Assuming a standard uniform semantics, logical truth is both objectively real and humanly knowable. (Premise.)

(2*) The standard uniform semantics of logical truth thus implies the existence of corresponding logical objects not dependent on the existence of humans but still knowable by humans. (From (1*) and the definition of the concept of "objective reality" as intersubjective knowability and nondependence on actual individual minds.)

(3*) According to its standard uniform semantics, logical truth is also necessary and a priori. (Premise.)

(4*) Logical objects are abstract and not concrete, because the concreteness—that is, the contingency and spatiotemporality—of such objects is inconsistent with the necessity and apriority of logical truth. (From (2*) and (3*).)

(5*) All and only concrete objects exist in spacetime. (Premise.)

(6*) So logical objects do not exist in spacetime. (From (4*) and (5*).)

(7*) All causally relevant (not to mention causally efficacious) entities exist in spacetime. (Premise.)

(8*) So logical objects are causally inert. (From (6*) and (7*).)

(9*) Our best overall theory of logical knowledge says that logical knowledge is some sort of intuition. (Premise.)

(10*) Our best overall theory of intuition says that intuition is cognitively analogous to sense perception. (Premise.)

(11*) So our best theory of logical knowledge says that logical intuition is cognitively analogous to sense perception. (From (9*) and (10*).)

(12*) A reasonable epistemology is a causal theory of knowledge. (Premise.)

(13*) The causal theory of sense perception is correct. (From (12*).)

(14*) Sense perception requires an efficacious causal link, involving direct physical contact, between the object perceived and the perceiver. (Premise.)

(15*) Therefore, logical knowledge is impossible. (From (8*), (11*), (13*), and (14*).)

Now, instead of specifically worrying that our semantic platonism about mathematics clashes with our best overall epistemology of mathematics, we worry that our semantic platonism about logic clashes with our best overall epistemology of logic. When construed in terms of a standard uniform semantics, logical truth seems every bit as objectively real and humanly knowable as mathematical truth, and thereby commits us to the existence of abstract logical objects (for example, logical laws, logical rules, logical truths, logical proofs, logical concepts or notions) that are causally inert, and therefore unknowable according to our overall best theory of logical knowledge.

There are, as far as I can see, at least nine different strategies for solving the original Benacerraf dilemma. Furthermore, it seems obvious that since the original Benacerraf dilemma smoothly generalizes to the extended Benacerraf dilemma, we can also smoothly generalize the nine strategies for responding to the original dilemma, so that there will be nine corresponding strategies for responding to the extended dilemma. And obviously, too, the rationales behind the strategies will remain essentially the same across the generalizations. So, for my purposes, it will be useful to run briefly through each of the strategies.

The first strategy is to reject (1) and (2) while still accepting (3) and (4), and adopt *antirealism*. Mathematical truth is not objectively real, and there are no corresponding objectively real mathematical objects, because abstract mathematical objects are mind-dependent on actual humans.[44] But this human mind-dependence neatly explains how those abstract objects are humanly knowable (namely, because we mentally construct them) as well as the face-value necessity and apriority of mathematical truth (namely, because we "impose" them in the course of our constructive activity). This is the tack taken by the intuitionists.[45]

The second strategy is also to reject (1) and (2) while still accepting (3) and (4), but this time adopt *fictionalism*. Mathematical truth is not real, and there are no corresponding real mathematical objects, because abstract mathematical objects are at best logically possible and exist only in the way that Alice and the Red Queen exist, that is, by pretense, or metaphorically. But this also neatly explains how those abstract objects are humanly knowable (namely, because we freely invent them) as well as the face-value necessity and apriority of mathematical truth (namely, because what floats free of contingency and empirical facts is necessary and a priori). This is the tack taken by Hartry Field and Stephen Yablo.[46]

The third strategy is again to reject (1) and (2) while still accepting (3) and (4), but this time adopt *conventionalism*. Mathematical truth is not real, and there are no corresponding real mathematical objects, because the necessity and apriority of mathematics is nothing but the result of our individually or collectively deciding to use mathematical sentences in just this way. Therefore, both the necessity and apriority of mathematics, as well as the human knowledge of abstract mathematical objects, must be fully relativized to the language or form of linguistic practice chosen by that decision.[47] This is the tack taken by Carnap[48] and (on one reading[49]) by the later Wittgenstein.

The fourth strategy is yet again to reject (1) and (2) while still accepting (3) and (4), by adopting *nonfactualism or expressivism*. Mathematical truth is not real, and there are no corresponding real mathematical objects, because mathematical discourse expresses nothing but a set of special prescriptions (constituting a set of implicit definitions) for using mathematical language. In other words, mathematics is noncognitive. The necessary a priori truths and laws of mathematics are nothing but *ways we ought to talk when mathematizing*, and mathematical knowledge is nothing but *talking mathematically just as we ought*. This is the tack taken (on another reading[50]) by the later Wittgenstein, and by Bob Hale and Crispin Wright.[51]

The fifth strategy is to accept (1), (2), and (4), but reject (3), and adopt *radical empiricism*. Mathematical truth is real, and there are corresponding real humanly knowable mathematical objects, and these mathematical objects can even in a certain sense be legitimately regarded as abstract (namely, as *abstracted* from something else), but there is no such thing as necessity or apriority. This is because all knowledge is derived from sense experience, acquired solely by empirical methods, justified pragmatically rather than rationally, and is fallible. This is the tack taken by Hume, Mill, and Quine.

The sixth strategy is to accept (1) to (9), but reject (10) and (11), and adopt a theory of intuition *not* modeled on sense perception. I will call this strategy *antiperceptualism*. As I pointed out in section 1.5, there is no a priori reason why intuition should be cognitively analogous not to sense perception but instead to either memory, imagination, or conceptual understanding. Then, since neither memory, imagination, nor conceptual understanding requires an efficacious causal link, involving direct physical contact, between the object cognized and the cognizer, it follows that mathematical knowledge could be intuitional without any worries about the causal inertness of mathematical objects. This is the tack taken by Jerrold Katz.[52]

The seventh strategy has two steps. The first step is to accept (1) through (6), but reject (7) and (8), and adopt a theory according to which mathematical objects are in fact causally relevant, not causally inert. As I argued in section 1.5, both causal laws and functional organizations are abstract in the sense that they are not uniquely located in spacetime; yet both have fundamental causal relevance. So, if mathematical objects have the same sort of ontological status as causal laws or functional organizations, it follows that mathematical objects are causally relevant. The second step is to accept (9) through (13), but reject (14), and adopt a theory according to which intuition is modeled on the causal theory of perception but does not require that causation involves direct physical contact. As I observed in section 1.5, it is possible to adopt a counterfactual or probabilistic analysis of causation that obviates the requirement of direct physical contact between cause and effect. Putting the two steps together, if mathematical objects are causally relevant but not in direct physical contact with the cognizer, then at least in principle those objects can still be intuitively perceived by the cognizer as a result of an efficacious counterfactual or probabilistic causal relation between them. I will call this two-step strategy *noncontact causal perceptualism*.

The eighth strategy is to accept (1) through (11), but reject (12) through (14), and adopt a noncausal theory of sense perception. I will call this strategy *noncausal perceptualism*. Fred Dretske has plausibly argued (a) that there is such a thing as "nonepistemic" (that is, non-belief-based, or non-conceptual) seeing,[53] and (b) that it is possible to nonepistemically see an object O even though there is no efficacious causal link between the perceiver and O.[54] Furthermore, Quine has plausibly argued that by means of "deferred ostension"[55] I can refer directly via perception to an object O, even if by hypothesis there is no efficacious causal link between the perceiver and O. For example, I can refer directly to an actual place by means of perception simply by pointing to a spot on a map: "Here's Boulder!"[56] More to the point, I can also refer directly to abstract mathematical objects by deferred ostension, as Quine observes:

Another such example [of deferred ostension] is afforded by the Gödel numbering of expressions. Thus if 7 has been assigned as Gödel number of the letter alpha, a man conscious of the Gödel numbering would not hesitate to say "Seven" on pointing to an inscription of the Greek letter in question. This is, on the face of it, a doubly deferred ostension: one step of deferment carries us from the inscription to the letter as abstact object, and a second step carries us thence to the number.[57]

So, combining Dretske's and Quine's arguments, it follows that there is nothing in principle to prevent a nonepistemic-perception-like act of intuition from referring directly to, and thereby nothing to prevent one's knowing, causally inert mathematical objects.

The ninth and final strategy is to accept all of (1) through (14) but deny that (15) follows, by claiming that there is some way of standing in an *indirect* intuitive relation to abstract causally inert mathematical objects, by means of perceiving concrete objects. I will call this *indirect perceptualism*. For example, if mathematical objects are finitist intuitionistic constructions on symbol-types, then humans can stand in an intuitive relation to those abstract constructions by means of perceiving finite sequences of concrete symbol-*tokens*. This is the line taken by Parsons.[58] Or, if mathematical objects are abstract universals, then humans can stand in an intuitive relation to those abstract universals by means of perceiving their concrete *instances*. This is the line taken by Lawrence Bonjour.[59]

I have eight worries about these strategies, however. I certainly do not mean to imply that these worries are in any way decisive refutations of the strategies—each of which would require a chapter or even a book of its own!—but rather only that they are negative considerations that should reasonably incline us toward what I regard as the most acceptable strategy.

The first worry concerns antirealism, fictionalism, conventionalism, radical empiricism, and nonfactualism alike. Each of these theories is an explicitly skeptical or at the very least deflationary theory, an "error theory," that violates in one way or another the standard uniform semantics of mathematical or logical discourse. But surely, other things being equal, we should always prefer theories that "save face" and preserve the face-value standard uniform semantics of our discourse.

The second worry also concerns antirealism, fictionalism, conventionalism, radical empiricism, and nonfactualism. Even though the original Benacerraf dilemma smoothly extends to the extended Benacerraf dilemma, there is no clear guarantee that a skepticism or deflationism about *mathematics* will also apply to *logic*. This is bound up with the fact, captured by talking about the "topic neutrality" of logic, that logic is an essentially more comprehensive science than mathematics. It is one thing to argue that scientists and perhaps also philosophers can get along comfortably without numbers as real, necessary, or a priori knowable objects. But it is quite another to argue that scientists and philosophers can get along without the reality, necessity, or apriority of logic. Indeed, as Field has explicitly conceded, fictionalism requires the apriority of classical propositional logic in order to ground its key notion of possibility.[60] Similar points can be made about antirealism, conventionalism, radical empiricism, and nonfactualism.[61] This is not merely a point about the *indispensability* of logic for natural science and mathematics,[62] however, although it includes that point. Since each of the forms of skepticism or deflationism about logic seeks to give a *reductive explanation* of logic, it runs up against the now familiar logocentric predicament of having to use and presuppose logic in order to explain logic (see chapter 3). Furthermore, since each of them also seeks to *justify* itself, it runs up against the cognitive indispensability of intuiting principles of deductive inference.[63]

The third worry concerns antiperceptualism. I accept the basic antiperceptualist rationale: there is no a priori reason why intuition should be analogous not to sense perception but instead to some other type of cognition. Nevertheless, anti-perceptualism must accept a heavy burden of proof. To make its case, it still needs to clarify the nature of abstract mathematical or logical objects, to provide an account of the nonperception-like cognitive mechanism of mathematical or logical intuition, and to show how these are internally related to one another in mathematical or logical knowledge. Indeed it is precisely the patent lack of such a theory that is the most powerful objection to Katz's attempt to respond to Benacerraf.[64]

The fourth worry concerns noncontact causal perceptualism. It seems true that *if* mathematical objects have the same sort of ontological status as causal laws or functional organizations and thus are causally relevant, then at least in principle perceivers can stand in efficacious counterfactual or probabilistic causal relations to them. But clearly a further argument is needed to show that mathematical objects and causal laws/functional organizations really *are* ontically equivalent, and it is not obvious how such an argument would go.[65]

The fifth worry also concerns noncontact causal perceptualism. Again, I concede that if mathematical objects have the same sort of ontological status as causal laws or functional organizations and thus are causally relevant, then at least in principle perceivers can stand in efficacious counterfactual or probabilistic causal relations to them—*at least in principle*. The problem here is that there is a conceptual gap between causal *relevance* and causal *efficacy* that needs to be bridged before it can be confidently claimed that just because X is causally relevant to Y's perception, X is therefore a perceptual object that causes Y's perception of X. To be sure, nothing can be causally efficacious unless it is also causally relevant. But as Jaegwon Kim has shown in connection with nonreductive materialist approaches to the problem of mental causation, the mere causal relevance of some X to an event E does not in and of itself *entail* the causal efficacy of X in relation to E. This is because the causal relevance of X to E can be both explanatorily and ontologically trumped, and thereby exclude X from being cited in a legitimate causal explanation of E and also from being an efficacious cause of E, by the simple fact that some other distinct thing Z efficaciously causes E.[66]

The sixth worry concerns noncausal perceptualism. Suppose we grant that nonepistemic seeing is a genuine sort of sense perception that does not require an efficacious causal link between the object perceived and the perceiver. And suppose we further grant that humans can directly perceptually refer to mathematical and other abstract objects (and thereby can directly perceptually refer to causally inert objects) by deferred ostension. The problem, however, is that there still is a conceptual gap between *direct perceptual reference* and *perception*. If I directly perceptually refer to an object O via deferred ostension, does it automatically follow that I *nonepistemically perceive* O? It is at least arguable that this entailment holds when O is an actually existing concrete object.[67] But it seems to me that there is no good reason to think that it holds when O is an *abstract* object. And this of course is precisely the question at issue.

The seventh worry concerns indirect perceptualism. Doubtless it is true that if I stand in a perceptual relation to a concrete object O, and O stands in either the token-to-type or the instance-to-universal relation to X, then I stand in *some* further relation to X via my relation to O. But it simply does not follow that this relation is thereby an intuitive relation, or indeed in any other sense an epistemic relation. Just because I see O, where O = the word-token 'dog', it does not follow that I intuitively cognize, or intuitively "see," the word-*type* 'dog'. I may have no conception whatsoever of the token–type relation, or of words as abstract objects, and simply read the word. Similarly, just because I see O, where O = a pair of shoes, it does not follow that I intuitively cognize, or intuitively "see," the universal number 2. I may have no conception whatsoever of the instance–universal relation, or of numbers as abstract objects, and simply see the shoes. Perhaps the root problem here is an equivocation between *visual* seeing or concrete perception on the one hand, and *intellectual* "seeing" or abstract comprehension on the other. In any case, it seems clear that indirect perceptualism is based on a non sequitur.

The eighth and final worry concerns noncontact causal perceptualism, noncausal perceptualism, and indirect perceptualism alike. If, as I have conceded, the basic rationale behind antiperceptualism is correct and intuition is cognitively disanalogous to sense perception, then all forms of perceptualism must be questionable from the outset.

What to do now? It seems to me, all things considered, that the most promising strategy for getting around both the original and extended Benacerraf dilemmas is to affirm antiperceptualism and accept the heavy burden of proof. So in the next two sections I will sketch a solution to the extended Benacerraf dilemma that begins by rejecting (10*) and (11*), clarifies the nature of abstract logical objects, provides the beginnings of an account of the nonperception-like cognitive mechanism of logical intuition, and also shows how these are internally related to one another in logical knowledge.

6.6 Outline of an Antiperceptualist Solution to the Extended Dilemma

The heavy burden of proof for an antiperceptualist solution to the extended Benacerraf dilemma, as I have just said, is the threefold task of (i) clarifying the nature of abstract logical objects, (ii) providing an account of the nonperception-like cognitive mechanism of logical intuition, and then (iii) showing how these are internally related to one another in logical knowledge. In this section I will sketch a four-part theory of logical intuition that seems to fit the bill.

Part 1 The first part of the theory is logical structuralism. According to structuralism, abstract objects of some specific kind are not independently existing entities but instead are merely distinct roles, positions, or offices in a *structure*, that is, an abstract formal relational system consisting of a coherent set of interlinked patterns or configurations.[68] So the thesis of logical structuralism is that each logical system is an abstract formal relational totality consisting of a coherent set of logical patterns or configurations, and that logical objects are nothing but distinct roles, positions, or offices in some such system.

Both logical objects and their constitutive logical structures are abstract, so I need to say something more about the notion of abstractness I am using. On my view, something is abstract if and only if it is not uniquely located in spacetime. Concrete things, by contrast, are uniquely located in spacetime. This is a broad conception of abstractness that allows for both "platonic" abstractness and "nonplatonic" abstractness. Something is platonically abstract if and only if it has extraspatiotemporal existence: for example, *ante rem* universals, or the inhabitants of Frege's "third realm." Platonically abstract objects are intrinsically nonspatiotemporal and have no connection whatsoever to the natural causal order. So they are causally irrelevant. By contrast, something is non-platonically abstract if and only if it has an infraspatiotemporal and transspatiotemporal existence: for example, *in re* universals, rules, laws of nature, linguistic types, mental representation types (for example, concepts), multiply realizable functional organizations, sets of concrete objects, and so on.

The essential feature of nonplatonic abstractness, aside from lack of unique location in spacetime, is actual or possible repetition, or multiple occurrence, multiple instantiation, multiple realization, and so on, across spacetime. This is at least consistent with causal relevance. In this way I can assert both logical structuralism and the abstractness of logical structures while not committing myself to the highly problematic thesis that logical objects and their constitutive logical structures are platonically abstract and therefore causally irrelevant. On the contrary, if I am correct, then logical objects and their constitutive structures are nonplatonically abstract, and at least possibly causally relevant, even if not causally efficacious, precisely because they are all cognitively constructed by the logic faculty *in language* (see chapters 4–5), whether in the language of thought or in a public language. In this way the nonplatonic abstractness of logic is the abstractness of a *linguistic structure*, a formal relational system consisting of a coherent set of interlinked patterns of linguistic types.

Part 2 Assuming that logical objects and their constitutive structures are nonplatonically abstract because cognitively constructed by the logic faculty in language, I am claiming that the primary cognitive mechanism of logical intuition is *the imagination* and its capacity for phenomenal continuous isomorphism or spatial-structure-coincidence.

It seems to me, as I think it also seemed to Kant,[69] that the proper cognitive model for intuition is the faculty of imagination, and not sense perception.[70]

This is because the imagination has three basic features not shared by sense perception. First, I can imagine an object O even though O is not uniquely located in spacetime, whereas I cannot sense-perceive O unless O is uniquely located in spacetime. Second, to generate a mental image of an object O is to produce a figural or spatial image, distinct from O itself, that is directly available to introspective scanning and manipulation (for example, image rotation, zooming in, pulling back), whereas to perceive O is not *thereby*[71] to produce anything figural or spatial, distinct from O itself, that is directly available to introspective scanning and manipulation. And third, I can generate an adequate or correct image of an objectively real object O_r (for example, someone I know well) without its being the case that O_r stands either in any efficacious causal relation or in an effective "tracking" relation to my conscious image of O_r (such that I can locate O_r in an egocentric phenomenal space relative to my body and also follow O_r's movements in this "centered" space over time), whereas it is plausible to think that I cannot correctly perceive O_r without either an efficacious causal relation or an effective tracking relation between O_r and my conscious perceptual representation of O_r.[72] These three features of the imagination (that its objects can be abstract, that it generates figural or spatial images directly available to introspective scanning and manipulation, and that its adequacy or correctness conditions are not based on either efficacious causation or effective tracking) all seem to me to be deeply relevant to logical intuition.

It will be obvious, I think, that intuition is such that its objects are abstract and that its adequacy (or correctness) conditions are not based on either efficacious causation or effective tracking. That is what got us into the original and extended Benacerraf dilemmas in the first place. But the other basic feature of the imagination, its generation of figural or spatial images directly available to introspective scanning and manipulation, may not be so obviously relevant to intuition. What I want to suggest, however, is that this second of the three basic features actually clinches the case for the cognitive analogy between intuition and the imagination.

This becomes clear when we ask ourselves about the conditions under which I generate an adequate or correct mental image of an objectively real object O_r. Here I am drawing directly on a body of recent work on mental imagery in cognitive psychology by Philip Johnson-Laird, Stephen Kosslyn, and Roger Shepard.[73] According to these psychologists, the representation relation between an image (Johnson-Laird regards images as paradigm examples of mental models) and a real object is essentially *depictive or pictorial*, and not essentially *descriptive or propositional*. It should be noted here that I am taking sides in a vigorous debate in cognitive science about the nature of mental imagery, with Johnson-Laird, Kosslyn, and Shepard on the depictivist side, and Zenon Pylyshyn and others on the descriptivist or propositionalist side.[74] I am not saying that this debate has been decisively resolved, but rather only that it seems to me that the case for two irreducibly distinct types of mental representation and representational content is at this point definitely *stronger* than the case for the thesis that all mental representations and representational content are at bottom descriptive or propositional. On that assumption, then, I will plunge ahead.

A depictive or pictorial relation is based on sharing the same shape, figure, pattern, or configuration, and not based on satisfying some specific set of descriptive or propositional criteria. So an image I adequately or correctly represents its corresponding real object O_r if and only if I is continuously isomorphic or spatial-structure-coincident with O_r. When I adequately or correctly form a mental image of some object, I consciously scan and manipulate my mental image or mental model until it apparently shares the same phenomenal shape, figure, configuration, or pattern as the real object I have imaged. In other words, I *mentally simulate* the structure of the imaged object.

But here is the crucial part. Whenever in the process of mental simulation I have reached the point of what *seems to me* to be the precise or one-to-one matching of the relevant elements of the structure of my mental image or mental model with the corresponding elements of the structure of the imaged object, as I have consciously represented it (whether via memory, perception, judgment, or inference), I thereby induce in myself an unshakable belief that the imaged object really and truly is just as I have consciously represented it. That is because the criterion of adequacy or correctness for images is exact continuous isomorphism or spatial-structure-coincidence with their objects. So whenever my mental image is experienced from the inside, or phenomenally, as having the same shape, figure, configuration, or pattern as what is specified by the content of my conscious representation of the object, I am thereby fully convinced that the imaged object is just as I have represented it to be.

Of course, mistakes are always possible. The world could be otherwise than I have represented it to be. But the crucial thing is that the cognitive step—*from* the consciously experienced continuous isomorphism or spatial-structure-coincidence between my mental image and what is specified by the content of my conscious representation of the imaged object *to* an unshakable belief that the imaged object is precisely as I have represented it by means of my memory, perception, judgment, or inference—is automatic, underdetermined by sensory experiences, and self-contained. Otherwise put, the subjectively experienced "rightness of fit" between my mental image and what is specified by the content of my conscious representation of the imaged object is cognitively optimal. So I am thereby "subjectively certain" that the imaged object is precisely as I have represented it to be. And in this way the phenomenal structure-matching activity of the imagination neatly explains the apriority, clarity and distinctness, authoritativeness, and noninferentiality of intuition.

Part 3 I pointed out in section 6.0 that a commitment to the objective reality of logic and of logical necessity is forced on us by the self-refuting fate of scientific naturalism about logic, the logocentric predicament, and the nonreductive character of logical cognitivism more generally. But the objective reality of logical necessity also plays a central role in the epistemology of logical knowledge, in that the objective reality of logical necessity is the ground of logical knowledge.

What I mean is this. It seems obvious that I can have genuine knowledge that S only if it is a fact that S. Then, given the strict-modal-attributivity of intuition (that every intuition that S automatically leads to the belief that necessarily S), it follows that my logical intuition that S can be logical knowledge that S only if it is a fact that logically necessarily S. But given the fallibility of intuition, it also follows that the fact that logically necessarily S does not logically depend on me or on any other actual cognizer, human or nonhuman. No matter what I believe, or how I believe it, and no matter whether any other creature believes it, or how that creature believes it, that belief can be wrong. In other words, the fact that logically necessarily S is an *objectively real* fact.

This leads to another issue. We now know that in order for a logical intuition to be logical knowledge, logical necessity must be objectively real. But what *is* logical necessity? And for that matter, what is *necessity*? Obviously I cannot even begin to address, much less answer, such a huge question within the limited scope of this book. But since I have frequently helped myself to the concept of necessity, I should at least briefly describe the general modal framework I am using.[75]

On my view, necessity is the truth of a sentence in every member of a set of possible worlds, together with its nonfalsity in every other possible world. A possible world is nothing more and nothing less than a different total way the actual world might have been.[76] Logical possibility is the consistency of a sentence with the laws of some classical or nonclassical logic. Logically possible worlds are distinct maximal sets of mutually consistent sentences, predicates, concepts, or properties:[77] the largest distinct sets of mutually consistent sentences, predicates, concepts, or properties such that the addition of one more sentence, predicate, concept, or property to that set would yield an inconsistency. Logical necessity is the truth of a sentence by virtue of logical laws or intrinsic conceptual connections alone, and thus is the truth of a sentence in all logically possible worlds. Put in traditional terms, logical necessity is *analyticity*.[78]

Logical necessity is usually contrasted with physical or *nomological* necessity, that is, the truth of a sentence in all logically possible worlds governed by our actual laws of nature; correspondingly, physical or nomological possibility is the joint consistency of a sentence with the laws of logic and our actual laws of nature. Physical necessity is also a form of *hypothetical* or *relative* necessity. More precisely, a sentence S is hypothetically or relatively necessary if and only if it is logically necessary that $\Gamma \rightarrow S$, where Γ is some set of special axioms or postulates, for example, our laws of nature. Thus, hypothetical or relative necessity is parasitic on logical necessity or analyticity.

In addition to logical necessity and physical necessity, there is also *metaphysical* necessity. Metaphysical necessity is either (i) necessity as defined over the set of all logically possible worlds (in which case it is also logical necessity, analyticity, or *weak* metaphysical necessity), or (ii) necessity as defined over a set of possible worlds that is definitely smaller than the set of all logically possible worlds and determined by whatever it is that constitutes the underlying essence or nature of our actual world (in which case it is *strong* or *essentialist* metaphysical necessity). More precisely, a sentence S is strongly metaphysically necessary if and only if:

(i) S is true in every member of a set K of logically possible worlds, (ii) K is smaller than the set of all logically possible worlds, (iii) K is larger than the set of all physically possible worlds, (iv) K includes the class of physically possible worlds, (v) K is the class of logically possible worlds consistent with the underlying essence or nature of our actual world, and (vi) S takes no truth-value in every logical possible world not belonging to K.

Put in traditional terms, strong or essentialist metaphysical necessity is *synthetic* necessity.[79]

Needless to say, the distinction between logical or analytic necessity and synthetic necessity is highly controversial, and I include it here only to indicate (a) that I take the concept of necessity to extend beyond the concept of logical or analytic necessity, and thus my modal framework is *modally dualistic*; and (b) that the modally dualistic possible worlds framework I have adopted is nonreductive and also nonplatonic. The crucial take-away for my purposes here, in any case, is the somewhat (I hope!) less controversial thesis *that logical or analytic necessity is objectively real.*

Part 4 I have proposed that logical objects are merely distinct roles, positions, or offices in logical structures, that is, logics construed as nonplatonically abstract formal relational systems consisting of coherent sets of interlinked patterns of linguistic types; that the primary cognitive mechanism of logical intuition is the ability for consciously scanning and manipulating linguistic mental-images; and that the objective reality of logical necessity is the ground of logical knowledge. Given the philosophical picture of intuition I developed in section 6.3, a logical intuition must be a mental episode or act, a priori, content-comprehending, clear and distinct, strict-modality-attributing, authoritative, noninferential, cognitively indispensable, and fallible. And we must also distinguish between intuition-of and intuition-that on the one hand, and between intuitive judgments and intuitive inferences on the other.

In view of *all that*, my claim is that I have logical knowledge that S if and only if

(1) I logically intuit that S, and
(2) it is a real fact that logically necessarily S.

More precisely, with respect to (1), I logically intuit that S if and only if

(1.1) I intuit that S,
(1.2) I take it to be logically necessary that S, and
(1.3) I consciously scan and manipulate my linguistic mental image of the sentence 'S' to the point of phenomenal continuous isomorphism or spatial-structure-coincidence with what is specified by the semantic content of my intuitive judgment or intuitive inference that (logically necessarily) S.

So, most explicitly, my claim is that I have logical knowledge that S if and only if

(1.1) I intuit that S,

(1.2) I take it to be logically necessary that S,

(1.3) I consciously scan and manipulate my linguistic mental image of 'S' to the point of phenomenal continuous isomorphism or spatial-structure-coincidence with what is specified by the semantic content of my intuitive judgment or intuitive inference that logically necessarily S, and

(2) it is an objectively real fact that logically necessarily S.

Let me try to make this more concrete with a simplified[80] example. Consider the following:

(*) Either George W. Bush is president of the United States in 2005 or I'm the man in the moon. I'm not the man in the moon. Therefore George W. Bush is president of the United States in 2005.

Assuming my knowledge of English and of classical propositional logic, this text is read and understood by me as a disjunctive syllogism, in the form of a single sentence: 'Either George W. Bush is president of the United States in 2005 or I'm the man in the moon, and I'm not the man in the moon; therefore George W. Bush is president of the United States in 2005'. But not only do I read and understand this argument in the form of a single sentence, I also cannot help believing it to be both valid and sound. This is because insofar as I formulate (*) to myself, thereby representing a logical object (in this case an argument in the form of a single sentence), I also generate a visual mental image that looks more or less like this:

$P \vee Q, \sim Q \mid P$

Let us call this symbolic sequence '(#)'. In turn, I will label the visual mental image of the symbolic sequence (#), 'I (#)'. (#) is of course a straightforward translation of (*) into the symbolism I learned for classical propositional logic as an undergraduate. Then I (#) is used by me to intuit the argument expressed by (*) as a valid and sound argument carried out according to the rules for classical negation, disjunction, and disjunctive syllogism. This in turn happens precisely insofar as I use I (#) as a linguistic image of what is semantically represented by (*), which is a logical fact, and then consciously scan and manipulate I (#) so as to bring it into a phenomenal continuous isomorphism or spatial-structure-coincidence with that fact, which in turn is specified by the semantic content of (*). Finally, this logical intuition counts as *logical knowledge* because not only is the intuition intrinsically compelling, it is also the case that (*) semantically represents an *objectively real* logical fact, namely a genuinely valid and sound argument in classical propositional logic in the form of a single sentence.

One last point in this connection. Intuitive deduction is an interesting phenomenon in its own right. The conscious experience of intuitive deduction distinguishes it from *rote* deduction (a common occurrence in, for example, undergraduate logic courses), in which someone superficially understands an argument and self-consciously follows inference rules in order to draw the conclusion of that argument, yet fails to grasp its logical structure intuitively. And it also distinguishes it from *merely implemented* deduction, in which some physical or biological system realizes or implements the logical form or syntax of a deductive argument, but without self-consciousness (and perhaps without phenomenal consciousness or subjective experience too), which thus excludes both rote deduction and of course also intuitive deduction.[81]

This completes the outline of my antiperceptualist solution to the extended Benacerraf dilemma. I have accepted the standard uniform semantics of logical truth, and also the human knowability of objectively real abstract logical objects, construed as linguistic objects of a special kind. I have asserted the thesis of logical structuralism, and also the thesis that logical objects and their constitutive structures are nonplatonically abstract, and therefore at least possibly causally relevant. But I have denied that humans need to stand in an efficacious causal relation to objectively real logical abstract objects in order to know them, because I have denied that intuition should be modeled on sense perception. Instead I have proposed that intuition should be modeled on imagination, and that linguistic mental images (whether of ordinary natural language inscriptions or of formal-logical symbols) are the mental vehicles of logical intuition. An image need not stand in any sort of efficacious causal relation to its object in order to be adequate or correct. Instead, it need only be continuously isomorphic or spatial-structure-coincident with its object. Hence my act of logical intuition can adequately represent its logical object by virtue of the fact that its mental vehicle, a linguistic mental image, is continuously isomorphic or structure-coincident with the object of my logical judgment or logical inference. Furthermore, the imaginational cognitive mechanism of logical intuition is a process of phenomenal spatial-structure-matching between (a) the linguistic mental image of a single (perhaps fairly long and complex) sentence that I use to express my intuitive logical judgment or intuitive logical inference, and (b) what is specified by the semantic content of that judgment or inference, which in turn represents logical objects and their constitutive structures, which in turn take the form of (perhaps fairly long and complex) sentences. So the thesis that logical intuition is a type of imaginational cognition squares well with logical structuralism. Finally, I have also accounted for the fallibility of logical intuition by appealing to the objective reality of logical necessity.

Let me now add this antiperceptualist solution for the extended Benacerraf dilemma to my intuition-based solution for the extended rule-following paradox. Assuming that both solutions are philosophically acceptable, it follows that my postulation of a cognitive capacity for logical intuition, built into the innate logic faculty, explains our knowledge of logic better than the other available explanations, all of which have serious difficulties with either the extended rule-following paradox or the extended Benacerraf dilemma. So it seems to me that the logic faculty thesis is the key to the correct epistemology of logic. Moreover, given the logic-oriented approach to human rationality, it would also follow that the rational human animal, who is essentially a logical animal, is inherently capable of logical knowledge by means of logical intuition. It must be immediately added, however, that the human capacity for logical intuition is ineluctably caught up in the cognitivist existential predicament, or the consequences for rationality of our animality. Thus, even if we can get it logically right, we can get it logically wrong too: in fact, we make an awful lot of logical mistakes, as the evidence from the psychology of deductive reasoning clearly shows. In the next and last chapter I will explore the normative implications of this predicament.

7   The Ethics of Logic

When studying reasons we study normative aspects of the world. When discussing rationality we discuss our perceptions of, and responses to, reasons. Our ability to reason is central to our rationality in all its manifestations.
—Joseph Raz[1]

Logic and ethics are fundamentally the same, they are no more than duty to oneself.
—Otto Weininger[2]

Logical necessity becomes what you must think and say, and what it is necessary to think and say is what necessarily happens in the world according to logic, according to the logic of economics or the logic of race. No thought is possible outside logic. If there are such thoughts they do not exist and so may be eliminated. Each person, even in his own most private thoughts when he is alone, must think the same thought again and again, compulsively, what he must think, out of touch with all reality, ignoring the terror in the streets, the disappearances, the deaths, the concentration camps, out of touch with all others whom he assumes think like him because they must, like he must, think just like him, because he is a thing, the only thing now that he can know and they must be like him. Logic in its final perfection is insane.
—Andrea Nye[3]

7.0   Introduction

In the previous three chapters I developed a logic-oriented conception of human rationality according to which rational human animals are essentially logical animals. Logical animals, more specifically, are logico-linguistic cognizers (chapter 4), real-world logical reasoners (chapter 5), and logical knowers by means of logical intuition (chapter 6). In the first three chapters I explained the nature of logic by appealing to a rational animal's possession of a faculty for logical cognition, the logic faculty, that innately contains the protologic, which in turn is a single universal unrevisable a priori set of logical principles and concepts used for the construction, analysis, and evaluation of every logical system, be it classical or nonclassical. So logic is cognitively constructed by rational animals (centrally including of course rational *human* animals), just as natural languages are cognitively constructed by human animals (centrally including of course *rational* human animals). The conjunction of the logic faculty thesis and the logic-oriented conception of human rationality is logical cognitivism.

In this final chapter I bring together the themes of the earlier chapters under a discussion of the ethics of logic. By the notion of *the ethics of logic* I mean *the normativity of logic*. My argument has five stages. First, in section 7.1, I argue that logic is intrinsically normative, that the intrinsic normativity of logic is perfectly consistent with logic's being the science of the necessary relation of consequence, and in particular that the intrinsic normativity of logic is perfectly consistent with the existence of a single universal unrevisable a priori set of logical principles and concepts used for the construction of all logical systems, namely the protologic. Second, in section 7.2, I argue that the protologic also provides a set of unconditional prescriptive laws, or "categorical imperatives," for human reasoning.[4] Third, in section 7.3, I argue that my cognitivist account of the intrinsic categorical normativity of logic supplies a crisp and effective reply to an obvious worry: if rational human animals are essentially logical animals, how could they ever make logical mistakes? Fourth, in section 7.4, I spell out and then criticize Gilbert Harman's thesis that there is no significant connection between logic and human reasoning. Finally, in section 7.5, I address an important skeptical worry about the very idea of human reasoning, and by implication, about the very idea of human rationality: why must I be logical?

7.1 Normativity and Logic

Normativity consists in the fact that there is a set of ideals, standards, guides, recommendations, commands, rules, principles, laws, and so on (hence "norms") that govern human beliefs and intentional actions. Norms imply the existence of objective human values. Objective human values are, roughly speaking, whatever *matters* to human animals: whatever humans consciously care about, need, desire, prefer, choose, put their practical or affective or cognitive faith in, and intentionally act for the sake of. As I will construe it, the normativity of something X is expressed by saying that there is something humans *ought to (or may) believe or do because of* X. In other words, the normativity of X is the role X plays in the giving of reasons for human belief or intentional action, that is, in the justification of human belief or intentional action.[5] More precisely, then, X is normative if and only if X can be directly cited as a reason for human belief or intentional action, or at least X is intrinsic to some reason for human belief or intentional action.

In a logical context, normativity consists in a certain relation between logic and human reasoning. To say that logic is normative is to say that humans *ought* to reason soundly or validly (more generally, cogently). Otherwise put, the normativity of logic consists in the fact, if it is a fact, that the justification of human beliefs or intentional actions depends on our ability to reason cogently—*if it is a fact*. Closer inspection reveals that there are various philosophical options with respect to the normativity of logic.

The first and most basic set of options, a binary pair, concerns the fact (if it is a fact) of logic's normativity itself. The options here divide as to whether logic is either (1A) a *normative* (i.e., prescriptive or evaluative) science, or else (1B) a *nonnormative* (i.e., descriptive or factual) science.

Assuming that logic is normative, there is a further binary pair of options that divide as to whether the normativity of logic is either (2A) an *intrinsic* (i.e., a necessary, relational or nonrelational) feature of it, or else (2B) an *extrinsic* (an accidental, relational or nonrelational) feature of it.[6]

A third binary pair of options also falls under the assumption that logic is normative. These divide as to whether logic has either (3A) *categorical* (i.e., unconditional, noninstrumental) normativity, or else (3B) *hypothetical* (i.e., conditional, instrumental) normativity. This distinction can be expressed by saying that something X is categorically normative if and only if humans ought to believe or do Y because of X under all sets of circumstances and primarily because of X alone, whereas something X is hypothetically normative if and only if humans ought to believe or do Y because of X only in certain circumstances and primarily because of something else Z.

The third pair combines with the second pair to provide a fourth and seemingly exhaustive set of options, all under the assumption that logic is normative. Thus logic is either (4A) *intrinsically categorically normative*, (4B) *extrinsically categorically normative*, (4C) *intrinsically hypothetically normative*, or (4D) *extrinsically hypothetically normative*. These options in turn map fairly smoothly onto various historically real philosophical doctrines about the nature of logic.

The general idea that logic and human reasoning or human thinking are normatively connected goes at least as far back as the Cartesian logicians of Port Royal in the seventeenth century. Antoine Arnaud and Pierre Nicole construed logic as *l'art de penser*, the art of thinking.[7] According to this view, the nature of logic is that it tells us how we ought to reason or think if we want to be good metaphysicians, mathematicians, or natural scientists. Logic is strictly a systematic means to this end: in effect, it is a *technology* of thinking. So the Port Royalists held that logic is intrinsically hypothetically normative (and thus falls under 4C). Essentially the same view was later defended in the mid-nineteenth century by J. S. Mill.[8] The logic-as-technology-of-thinking view, for reasons I will mention shortly, went down in flames at the turn of the twentieth century along with logical psychologism, but it has been resuscitated in a contemporary context by logical nonfactualists or expressivists, who hold that the nature of logic is prescriptive, instrumental, and social: the laws, truths, and proofs of logic are nothing but ways we ought to talk when logicizing, and we logicize in order to serve the ends of more basic human practices, such as belief formation.[9]

The sharply different idea of logic as the science of "the laws of thought" goes back to Kant in the eighteenth century. It was later picked up by Boole in the mid-nineteenth, and again by Frege in the late nineteenth and early twentieth:

Like ethics, logic can also be called a normative science. How must I think in order to reach the goal, truth? . . . [T]he task we assign logic is only that of saying what holds with the utmost generality for all thinking, whatever its subject matter. We must assume that the rules for our thinking and for our holding something to be true are prescribed by the laws of truth.[10]

[The laws of logic] have a special title to the name 'laws of thought' only if we mean to assert that they are the most general laws which prescribe the way in which one ought to think if one is to think at all.[11]

On the Kant–Boole–Frege view, logic is the universal, topic-neutral, a priori science of the necessary laws of truth, and also a pure normative science based directly on rationality itself. Logic tells us *how we ought to reason or think in every possible set of circumstances because this is required by the nature of rationality.* So logic is intrinsically categorically normative (and thus falls under 4A): logic is a moral science. For convenience, I dub this *the moral science conception of logic*.

Ironically, despite Frege's explicit support for the moral science conception of logic, that conception did not survive the late nineteenth- and early twentieth-century critique of psychologism which Frege himself initiated. Not too surprisingly, this was mainly Husserl's doing. In the *Prolegomena to Pure Logic* Husserl vigorously rejects not only logical psychologism (see section 1.1), but also the intrinsic normativity of logic. If sound, this rejection rules out the very ideas of both the intrinsic categorical normativity of logic (i.e., 4A) and also the intrinsic hypothetical normativity of logic (i.e., 4C), thus dispensing with Kant, Boole, Frege, Arnaud and Nicole, Mill, and contemporary logical nonfactualists in one fell swoop. More positively expressed, Husserl claims that logic is an intrinsically nonnormative, descriptive, or factual science (and thus falls under 1B).

Following Husserl, most recent and contemporary writers of introductory logic textbooks start with the assumption that the intrinsic normativity of logic is as questionable as logical psychologism. So logic is an intrinsically nonnormative, descriptive, or factual science. But at the same time they also hold that logic is *extrinsically* normative, in relation to everyday human reasoning.[12] According to these writers, we rational humans *ought* to follow logical rules, either because this is required by rationality itself (and rationality is extrinsic to logic) or because it promotes other important human interests. In other words, for these writers, despite the fact that logic is an intrinsically nonnormative science, it remains either extrinsically categorically normative (and thus falls under 4B) or extrinsically hypothetically normative (and thus falls under 4D). Thus they hold that despite the fact that logic is essentially the science of the necessary relation of consequence, or the science of proof, or the science of truth, or whatever, nevertheless it is a *very good thing indeed* to study logic.

This brings us to Otto Weininger and Andrea Nye. Both Weininger, who was an early twentieth-century Viennese radical misogynist, and Nye, who is an early twenty-first century American radical feminist, fervently believe that logic is normative. Weininger, however, thinks that obeying the laws of logic is something we ought to do precisely because it belongs to the strict moral duties we have toward ourselves. He thereby holds that logic is intrinsically categorically normative (and thus falls under 4A), but for reasons quite different from those found in the moral science conception of logic.

According to the moral science conception, logic is intrinsically categorically normative because it is based on rationality itself (hence rationality is intrinsic to logic) and is also an *integral part of* human morality, namely the part that consists in justifying moral judgments and decisions, including direct moral arguments and reflective equilibrium. But Weininger takes the much stronger view that logic is *identical with* human morality. By sharp contrast to both, Nye takes the equally radical but fully skeptical view that human beings are motivated entirely by nonrational natural mechanisms and that logic is nothing but a social construct that is essentially allied with powerful cultural institutions of political tyranny, economic exploitation, sexual chauvinism, racial prejudice, and human oppression more generally. In short, we should all hate and eschew logic, because it is unhealthy, obsessive-compulsive, life-negating, discriminatory, socially destructive, and in a word, *bad*. This view constitutes a challenge to the very idea of a logic and indeed to the whole framework of options I have been using, because it denies both that normativity has any sort of rational foundation and also that there are any real distinctions to be drawn between the normative and the nonnormative, the categorical and the hypothetical, or the intrinsic and the extrinsic.

In the rest of this section and the next, I sketch and defend a cognitivist version of the moral science conception of logic. I will come back to Nye's skeptical challenge in section 7.5.

As I mentioned, Husserl's *Prolegomena* effectively prevented the transmission of Frege's moral science conception of logic into mainstream European and Anglo-American twentieth-century philosophy, by attacking the thesis that logic is intrinsically normative. Husserl has three basic objections to the intrinsic normativity thesis.[13] First, logical laws, truths, and proofs are not framed by logicians in normative terms (for example, as imperatives), but rather purely in alethic modal terms (for example, as necessary laws, truths, or proofs). Second, every normative discipline presupposes a more basic theoretical discipline that establishes the existence and nature of the facts that are taken to have the relevant normative properties. So even if there are normative disciplines connected with logic, logic itself is needed as a distinct nonnormative science in order to ground those normative disciplines. And third, the thesis that logic is normative entails logical psychologism, and logical psychologism is false: so it is also false that logic is intrinsically normative.

Husserl's first objection can be seen to fail as soon as we note that in natural deduction systems (for example, Gentzen's), logical inference rules are explicitly expressed as Rylean "inference-tickets" or generalized permissions. And even in Frege's *Begriffsschrift,* which is an axiomatic system, inference rules are explicitly not accorded the status of truth-bearers but instead are assigned a normative role.[14] So there is no inconsistency in ascribing alethic modal properties and normative properties to logical language within one and the same logical system. Indeed, since it is a general feature of logical consequence or entailment that every logical truth follows from any set of premises including the empty set of premises, then for every logical truth in every logical system there is a corresponding permission to infer that sentence from any set of premises. So for every logical truth in every logical system, a corresponding normative sentence carrying the same logical force can be formulated.

Husserl's second objection fails because it has a false assumption. He assumes that to claim that logic is intrinsically normative is thereby to make a reductive claim to the effect that logic is "nothing over and above" some set of normative facts. Assuming such a reduction, it would indeed follow that logic could be legitimately framed in alethic modal terms only if normative logical disciplines had a nonnormative logical grounding. Now, it is true that some versions of the thesis that logic is intrinsically normative are indeed reductive. Most noncognitivist theories of logic, for example, can plausibly be read as reductive. But clearly it is also possible consistently to hold (1) that logic is a nonnormative discipline in the sense that every logic explicitly contains within itself language that is intrinsically descriptive or factual; (2) that logic is *also* a normative discipline in the sense that every logic explicitly (as in natural deduction systems, and in some axiomatic systems) or implicitly (as in other axiomatic systems) contains within itself language that is intrinsically prescriptive or evaluative; and (3) that the nonnormative and normative parts of logic are complementary (so for every logical truth in every logical system, there is a corresponding normative sentence legitimating an inference to that sentence from any set of premises, and conversely) and mutually irreducible. On this nonreductive approach, logic is intrinsically descriptive *and* intrinsically prescriptive.

This leaves us with Husserl's third objection, to the effect that the intrinsic normativity of logic entails logical psychologism. And here is where the distinction between hypothetical and categorical normativity is crucially salient. Logical psychologism, we will recall from section 1.1, is a form of scientific naturalism that consists in the explanatory reduction of logic to empirical psychology. Now, Husserl makes the correct point that some forms of logical psychologism also assert the thesis that logic is intrinsically normative. Mill's theory of logic, for example, shows that it is possible to hold both that logic is intrinsically hypothetically normative and that logic is explanatorily reducible to empirical psychology. And it also seems to be true that if one wants to hold that logic is intrinsically normative *and also* that it is explanatorily reducible to empirical psychology, then one must also hold that logic is intrinsically *hypothetically* normative, because both require that logic is dependent on contingent facts of some sort, whether actual human interests or the natural facts. But the plain truth is that not every theory which takes logic to be intrinsically normative is psychologistic: at least some versions of the thesis that logic is intrinsically *categorically* normative are nonpsychologistic,[15] including Kant's and Frege's normative theories of logic.

Take, for example, Kant's normative theory of logic. Kant holds that the logical 'ought' has the same deontic force as the moral 'ought':

In logic . . . the question is not about *contingent* but about *necessary* rules; not how we think, but about how we ought to think.[16]

What I call applied logic . . . is thus a representation of the understanding and the rules of its necessary use *in concreto*, namely under the contingent conditions of the subject. . . . General and pure logic is related to [applied logic] as pure morality, which contains the necessary moral laws of a free will in general, is related to the doctrine of virtue proper, which assesses these laws under the hindrances of the feelings, inclinations, and passions to which human beings are more or less subject.[17]

Kant's ethics, in turn, explicitly states that moral prescriptions *inherently fail* to be strictly determined by either human interests or the natural facts. It is built into the very idea of pure practical reason that the categorical imperative or moral law is universally binding on rational beings, including all rational humans; that the categorical imperative is underdetermined by all human interests or natural facts; and that a rational agent is capable of acting not merely in accordance with but also *from* the categorical imperative, and (if necessary) in opposition to all human interests and all actual or possible natural facts.[18] So Kant's normative theory of logic entails that logical prescriptions are neither dependent on human interests nor logically strongly supervenient on the natural facts, that is, it entails that his logic is both noninstrumental and nonpsychologistic:

Pure logic . . . has no empirical principles, thus it draws nothing from psychology.[19]

Some logicians . . . do presuppose *psychological* principles in logic. But to bring such principles into logic is just as absurd as to derive morals from life.[20]

Kant's logic and ethics are of course controversial. The relevant point here, however, is not whether Kant's logic and ethics are *defensible,* but instead whether the thesis that logic is intrinsically categorically normative is *intelligible.* So Husserl's third objection does not apply to at least some views holding that logic is intrinsically categorically normative.

Rejecting Husserl's three objections to the intrinsic normativity of logic leaves open a window for logical cognitivism, in three ways. First, logical cognitivism asserts that despite the fact that every classical or nonclassical logical system is a nonnormative, factual, or descriptive science of the necessary relation of consequence, it *also* has a normative part corresponding directly to its nonnormative part. This can be easily seen in natural deduction systems, which explicitly contain inference rules; and it is also implicit in the very idea of a logical truth, which corresponds to a permission to deduce it from any set of premises. But even more fundamentally, this can be seen in the fact that the protologic, as a set of logical principles and concepts for constructing logical systems, is inherently normative precisely insofar as it is a set of schematic permissions to construct logical systems *in just these ways and no others.* But the protologic does not tell us how rational humans actually *do* construct logical systems under real-world conditions. Similarly, Chomsky's UG is a *prescriptive* and not a descriptive grammar: it tells us how humans are *permitted* to construct natural languages by virtue of their innate cognitive endowment for language, not how they actually *do* construct their natural languages under real-world conditions. Second, logical cognitivism asserts that the nonnormative and normative parts of logic are mutually complementary and mutually irreducible. Third, logical cognitivism asserts that the protologic is intrinsically categorically normative for human reasoning, in other words, that the protologic supplies "basic a priori rules of rationality."[21] It is the third claim that is most important and most controversial. The task of the next section is to offer an argument for it.

7.2 Logic as a System of Categorical Imperatives for Reasoning

Before getting down to the argument itself, however, I want to pursue a little further the parallel between logical cognitivism on the one hand, and

Kant's logic and ethics on the other, as a way of elaborating the thesis that logic is intrinsically categorically normative. In a seminal essay published in 1972, "Morality as a System of Hypothetical Imperatives," Philippa Foot directly and vigorously challenged all forms of Kantian ethics. Here Foot argues that Kant's categorical imperative is an overly rigid, overly abstract, and ultimately empty bit of formalism: since it provides no motivation for following it, the categorical imperative fails to yield genuinely action-guiding principles. Universal prescriptivism of any sort fails. Only a system of hypothetical imperatives, that is, a system of practical commands or prescriptive rules based explicitly on natural or empirical facts about particular human needs and aims given under particular sociohistorical conditions, can genuinely guide action.

Morality, says Foot, is more akin to rules of etiquette than it is to a system of categorical prescriptions.[22] Rules of etiquette are codes of conduct for highly localized and relatively unimportant human practices. The only difference between etiquette and morality is the extent to which members of the broader human community have achieved a certain wide solidarity or shared caring about the relevant practices. So morality is nothing but "optimally well-entrenched" or "globalized" etiquette, that is, etiquette that has established itself across all significant human practices, in the sense that all or most of the members of the human community have explicitly or tacitly adopted its otherwise merely local aims as "common aims."[23] Given these common aims, and *only* given these common aims, a set of hypothetical, conditioned, or instrumental moral prescriptions follow.

A precise analogue of Foot's complaint against Kantian ethics can be found in the conventionalist theory of logic. According to logical conventionalism, one can (whether individually or socially) create and adopt any logical system one likes by stipulating a certain set of logical axioms or postulates, and also the rules of formation, transformation, and interpretation. And it is a central feature of logical conventionalism that what motivates the formal system's creation and adoption is not itself cognitive or theoretical: indeed, it is essentially noncognitive or interest-driven, hence voluntaristic or at least pragmatic. This view is nicely encapsulated in Carnap's famous declaration in section 17 of *The Logical Syntax of Language*:

In the foregoing we have discussed several examples of negative requirements . . . by which certain common forms of language—methods of expression and of inference—would be excluded. Our attitude to requirements of this kind is given a general formulation in the *Principle of Tolerance: It is not our business to set up prohibitions, but to arrive at conventions. . . . In logic there are no morals.* Everyone is at liberty to build up his own logic, i.e., his own form of language, as he wishes.[24]

The very idea of an intrinsically categorically normative logic is undermined, Carnap thinks, by the existence of nonclassical logics. We must be radically open to the unlimited possibilities for creating alternative logics and to the diverse human motivations for modifying and challenging classical logic. In other words, for Carnap, "anything goes,"[25] at least where logic is concerned.

But there is an obvious way in which Carnap's conventionalism fails. In order to give a conventionalist definition of logical truth, conventionalism must presuppose and use preconventionalized logic. Here, of course, we are back at Quine's influential argument to the effect that conventionalism is inextricably caught up in the logocentric predicament (see section 3.2).

The crucial point, however, is that Foot's conventionalist conception of morality suffers from an analogous problem. If morality is to be interestingly different from mere etiquette, that is, if one is to crank up a local system of hypothetical imperatives into a truly global system that applies even to *prospective* and *possible* members of our community, then one must presuppose and use the very idea of a categorical imperative. For in terms of its syntactic or semantic structure, a "globalized" hypothetical imperative is nothing but a *relativized* categorical imperative. That is, a globalized hypothetical imperative to do *A* is nothing but a categorical imperative ("Everyone ought to do *A*") containing within its scope a material antecedent condition which is an arbitrarily chosen set of *human interest postulates*, call that set 'Γ':

Everyone ought (If Γ, then to do *A*).

Similarly, as we saw in section 6.6, a hypothetically or relatively logically necessary sentence *H* is nothing but a relativized logical necessity. That is, *H* is nothing but the consequent of a logically necessary material conditional containing within its scope an antecedent material condition which is an arbitrarily chosen set of theoretical postulates (for example, meaning postulates, scientific laws, etc.), call it 'Δ':

Logically necessarily ($\Delta \rightarrow H$).

Here is the moral of the story. Just as you cannot rationally escape from preconventionalized logic by making logic conventional at the level of your object language (since preconventionalized logic inevitably returns to haunt you in your metalanguage), so too you cannot rationally escape from categorical imperatives by making morality hypothetical or conventional at the level of first-order ethics (since categorical, preconventionalized morality inevitably returns to haunt you in your metaethics). And in a long footnote added to "Morality as System of Hypothetical Imperatives" in 1977, Foot as much as admitted this:

Kant's thought seems to be that universal rules are universally valid in that they are inescapable, that no one can contract out of morality, and above all that no one can say that as he does not happen to care about the ends of morality, morality does not apply to him. This thought about inescapability is very important, and we should pause to consider it. It is perhaps Kant's most compelling argument against the hypothetical imperative, and the one that may make Kantians of us all.[26]

And the same point goes, I hasten to add, for logic. To say that logic is intrinsically categorically normative is simply to say that logic is rationally humanly inescapable, or at least to say that *the protologic* is rationally humanly inescapable (see chapter 3, and the argument for the intrinsic categorical normativity of the protologic to follow shortly).

In any case, Foot seriously but instructively misinterprets Kantian ethics, as Onora O'Neill has shown.[27] It is a mistake to think of Kant's categorical imperative as a superstrong first-order principle for action (or what Kant calls a "maxim"), that is, as an all-purpose practical decision procedure or algorithm. On the contrary, the categorical imperative is a second-order procedural principle applying universally to maxims. Negatively described, the categorical imperative is a filter for screening out bad maxims; positively described, it is a constructive protocol for correctly generating maxims, given the multifarious array of concrete input-materials to practical reasoning, including beliefs, desires, habits, personal situation, sociohistorical context, and so on. Thus the categorical imperative says, roughly:

Act *only* according to those maxims that every rational human being could adopt and that remain consistent with our innate rational capacity for constructing and acting upon maxims.

This version of Kantian ethics is what O'Neill, extending and modifying the work of John Rawls, aptly calls "Kantian constructivism in ethics."[28] The crucial point is that we cannot say in advance of actual practical reasoning processes just *which* maxims will turn out to be permissible or obligatory, but we can know a priori that any maxim that will count as action-guiding *must* have a format or structure that is determined by the categorical imperative.

Where this line of argument is heading should be fairly obvious. Just as a constructivist Kantian ethics can get around both the objections and the internal problems of moral conventionalism, so too my broadly Kantian cognitivist constructivist conception of logic can get around both the objections and the internal problems of logical conventionalism. And O'Neill even supplies us with a good map of the conceptual terrain I want to occupy:

> The Categorical Imperative is the supreme principle of reasoning not because it is an algorithm either for thought or for action, but because it is an indispensable strategy for disciplining thinking or action in ways that are not contingent on specific and variable circumstances. The Categorical Imperative is a fundamental strategy, not an algorithm; it is the fundamental strategy not just of morality but *of all activity that counts as reasoned*. The supreme principle of reason is merely the principle of thinking and acting on principles that can (not "do") hold for all.[29]

O'Neill is saying that the basis for the construction of any rational scheme of principles, whether that scheme is to be thought-guiding (logic) or intentional-action-guiding (morality), is the categorical imperative. As applied to intentional action, the categorical imperative says that any first-order action-guiding principle must be universalizable, nonexploitative, and so on. But as the logical cognitivist would put it, as applied to thought, the categorical imperative says that every reasoning process must conform to the protologic:

Think *only* according to those processes of reasoning that satisfy the protologic.

Thus, according to logical cognitivism, following the principles and concepts of the protologic is not only required by human reasoning, it is also a *strict duty* for all human reasoning. And even though, as we saw in chapter 5, any such process of reasoning constructed in our logic of thought and our language of thought by means of the protologic will, by virtue of the cognitivist existential predicament, always have a more or less limited application that is determined by the inescapably contingent creature-based and world-based constraints under which that reasoning process occurs, *nevertheless*, reasoning according to the principles and concepts of the protologic remains an obligation for every rational human animal in every cognitive context whatsoever.

In this way, to say that "logic is a system of categorical imperatives for reasoning" is just to say that *the protologic is intrinsically categorically normative for human reasoning*. Here now is an explicit argument for this thesis.

(1) Something is normative if and only if it can be cited as a reason for human belief or intentional action, or is intrinsic to some reason for human belief or intentional action. (From the definition of the concept of normativity in section 7.1.)

(2) Every inferential justification of human belief or intentional action involves the logical entailment of some sentence describing a belief (call it a "belief-report") or an intentional action (call it an "action-report") by premises that describe reasons for that belief or action (call them "belief-premises" or "action-premises" respectively). (Premise.)

(3) The logical entailment of a belief-report or action-report by belief-premises or action-premises, as understood by a rational human agent, involves some concept of logical consequence in the agent's logic of thought. (From (2), and chapters 4–5.)

(4) The protologic enters intrinsically into every logical system insofar as it is both constructively and epistemically presupposed by every logical system. (From chapter 2.)

(5) So the protologic enters intrinsically into every inferential justification of human belief or intentional action. (From (2)–(4).)

(6) So the protologic is intrinsically normative. (From (1) and (5).)

(7) The intrinsic normativity of the protologic is not instrumental, or based on human interests, because the protologic is not dependent on *any* contingent facts, including human interests. (From chapters 1–2.)

(8) Therefore, the protologic is intrinsically categorically normative. (From (6) and (7).)

(9) The protologic enters intrinsically into all human reasoning, because (i) the protologic is used and presupposed in the construction, self-analysis, and self-evaluation of each rational human cognizer's own logic of thought, and (ii) each rational human cognizer's logic of thought constitutes his processes of reasoning. (From chapters 2–5.)

(10) Therefore, the protologic is intrinsically categorically normative for human reasoning. (From (8) and (9).)

As I have indicated in steps (1), (3), (4), (7), and (9), this argument relies on some claims for which I have argued elsewhere in the book, and which

I need not defend again here. The only step that needs independent support is step (2). Here the rationale is that if I am inferentially to justify a belief of mine or an intentional action of mine then obviously I must offer reasons for that belief or intentional action, and just as obviously the premises describing those reasons must *logically entail*, under some version of logical consequence or entailment, a conclusion describing my belief or my intentional action. How *else* could I inferentially justify a belief or intentional action of mine? Inferential justification, or the inferential giving of sufficient reasons for some human belief or intentional action (and this captures the very pith and marrow of a rational "because") is obviously inherently *logical* precisely because it is *inferential*.

Note that the version of logical consequence or entailment by virtue of which my belief-premises or action-premises entail my belief-report or action-report need not always be *classical* consequence or entailment (that is, the sort of consequence or entailment we find in elementary logic), but can in principle be any sort of classical or nonclassical consequence or entailment. My idea is not that any *particular* classical or nonclassical logic supplies categorical imperatives for all human reasoning. Instead, my idea is that any logical system we use in any sort of reasoning more or less implicitly, but always intrinsically, invokes the principles of the protologic as categorical imperatives for that reasoning.

7.3 Objections and Replies I: Sins and Fallacies

You have probably already thought of an obvious objection to my thesis that rational human animals are essentially logical animals in the sense that they are logical moralists. The objection is that I have falsely substituted an ideally rational, perfect, unhuman, Mr. Spock–like reasoner for real, imperfect, rational *human animals*. But we all know from first-hand experience that we constantly reason inconsistently or invalidly, shift to mere rhetoric and sophistry when moved by our desires or emotions, get confused, and so on. And the empirical research surveyed in chapter 5 shows that humans of ordinary or higher intellectual ability are generally very bad at deduction tasks. So, you will say, we cannnot possibly be essentially logical animals in the sense that we are logical moralists: how then could we ever make a logical mistake? My reply is that the obvious objection is based on an error about how the word 'logical' functions in 'logical animals'. 'Logical' as it used in this phrase is not primarily a descriptive term referring to any of the sciences of the necessary relation of consequence, that is, to any particular classical or nonclassical logical system, but instead is primarily a *prescriptive* term implicitly referring to the protologic. The protologic is intrinsically categorically normative for us, and there are good reasons to hold that we do indeed perfectly obey the protologic whenever we reason (see section 5.7); but it certainly does *not* follow that we always or even usually reason cogently under real-world conditions in relation to any particular classical or nonclassical logical system. As Boole very aptly remarks:

The . . . laws of reasoning are, properly speaking, the laws of right reasoning only, and their actual transgression is a perpetually recurring phenomenon. Error, which has no place in the material system [of physical nature], occupies a large one here. We must accept this as one of those ultimate facts.[30]

Objection that we do not actually reason thus. Reply: It is a mistake to suppose that the actual performances of our nature in any case fully answer to its faculties and capacities. We are in all things constituted with reference to an ideal standard.[31]

In other words, rational human animals are creatures with very high logical standards in one sense (as regards our conception of an ideal logical reasoner); and with very high logical success rates in another sense (as regards the protologic); but, sadly, also with very low logical success rates in still another sense (as regards any particular constructed logical system, classical or nonclassical, given the multitude of nonlogical factors affecting logical performance in the real world). The categorically normative 'ought' governing logical reasoning performance certainly does not imply a factual 'is'. The crucial general fact about our rationality with respect to logic is not our logical performance, but rather that reasoning cogently inevitably *matters* to us. Just as only an essentially moral animal would ever care about committing sins, so too only an essentially logical animal would ever care about committing fallacies. But just because it is by virtue of our being essentially moral and logical animals that we rational humans thereby care both constitutively and profoundly about morality and logic, this does not stop us from committing *lots and lots* of sins and fallacies. Alas.

This raises a further point. In his *Autobiography*, Russell tells a famous story about the early Wittgenstein:

He used to come to see me every evening at midnight, and pace up and down my room like a wild beast for three hours in agitated silence. Once I said to him: "Are you thinking about logic or about your sins?" "Both," he replied, and continued his pacing.[32]

To make a moral mistake is to sin; to make a logical mistake is to commit a fallacy. As Russell's drily told anecdote suggests, however, the early Wittgenstein's tendency to *assimilate* his worries about logic to his worries about his sins was over the top, even pathological.[33] Certainly moral wrongdoing is not necessarily or even usually connected with wrong logical reasoning; and on the other hand, wrong logical reasoning is not necessarily or even usually sinful. Wittgenstein may well have been decisively influenced by reading Weininger.[34] In any case, the point is that we need not identify logic with morality in order to hold that there is significant overlap between them, and also significant structural analogies between them. I have already suggested how there can be significant overlap between them, by way of the intrinsic role of the protologic in all inferential justification. But there are at least two important structural analogies between moral sins and logical fallacies that Wittgenstein's remarks strongly suggest, and to which, incidentally, Russell seems blithely blind.[35]

The first analogy between sins and fallacies is that moral disapproval and logical criticism both imply the freedom of the rational human animal. This of course is closely connected with the famous Kantian principle that 'ought' implies 'can'. If we are internally strictly obligated by the categorical imperative, and thus can be held responsible for violations of any first-order moral rules constructed under its higher-order constraints, then this is only because we are to some nonnegligible extent free to obey or disobey those rules of conduct. So too, if we are internally strictly obligated by the principles and concepts of the protologic, and thus can be held responsible for violations of any first-order normative logical rules constructed under its higher-order constraints, then this is only because we are to some nonnegligible extent free to obey or disobery those rules of thought. Up to a certain point, we *freely choose* to reason well or badly. Other things being equal,[36] if you are legitimately blamed for acting wrongly, then you could have done the right thing had you tried hard enough. Correspondingly, other things being equal, if you are legitimately blamed for reasoning badly, then you could have reasoned well had you tried hard enough.

The second analogy is that in logical reasoning as in morality there is an important distinction between excusable and inexcusable offenses. But this requires a brief preliminary foray into the nature of fallacies.

Fallacies are standardly divided into the formal and the informal. A formal fallacy, as it is usually construed, is a violation of one of the laws of classical logic. The class of such violations includes affirming the consequent of a conditional, denying the antecedent, illicit quantifier shift, and illicit syllogistic inference, to name a few. In each case, given the logical error, it is possible to reason from true premises to a false conclusion (invalidity), or to derive a true contradiction. Informal fallacies, by contrast, do not strictly speaking violate laws of classical logic, but do tend toward invalidity or true contradiction in particular speech contexts. Here we find such peccant argument-strategies as appeals to force, appeals to irrelevant personal or circumstantial factors, appeals to ignorance, appeals to pity, appeals to popular consensus, appeals to authority, applying a general rule in an inappropriate context, confusions about causal nexus, begging the question, complex questions (e.g., "Have you had enough to drink yet?," as asked by a policeman when he pulls you over to the side of the road), inference to an irrelevant conclusion, semantic equivocation, amphiboly or grammatical equivocation, illicit shift in emphasis, illicit composition of a whole, illicit division of a whole, black-and-white thinking, slippery slope thinking, and so on.[37]

The division between formal and informal fallacies is not an exhaustive one, since there are formal fallacies of extended or deviant logic that are strictly speaking neither formal nor informal by the previous classification: for example, confusing strict and material implication, and modal fallacies more generally. Moreover, expanding the class of formal fallacies in this way to include extended and deviant logics tends to blur the line between formal and informal fallacies by converting some of the standard informal fallacies into formal ones. Thus confusions about causal nexus go over into formal fallacies of causal logic, the fallacy of the complex question goes over into the class of formal fallacies of the logic of presupposition, compositional and divisional fallacies go over into formal fallacies of mereological logic, black-and-white thinking and slippery-slope thinking go over into formal fallacies of the logic of vagueness, and so on.

At the very outset of modern or symbolic logic in 1847, Augustus De Morgan (the discoverer of the well-known classical truth-functional equivalences) correctly observed that there is, in principle, no way of formally delimiting or determining the class of all fallacies.[38] What he meant, I think, is that the study of fallacies is not a part of logical theory per se (not part, that is, of any particular logical system) but rather belongs to the cognitive psychology of logic. As Irving Copi aptly puts it:

It is customary in the study of logic to reserve the term 'fallacy' for arguments which, although incorrect, are psychologically persuasive. We therefore define a fallacy as a form of argument that *seems* to be correct but which proves, upon, examination, not to be so.[39]

Fallacies, in other words, are the typical illusions of logical thinking. This in turn suggests an interesting analogy between the study of fallacies and the cognitive psychology of visual illusions.[40] Visual illusions, unlike casual visual mistakes, are false visual phenomena that universally persist across normal perceivers even when recognized by those perceivers as false (this of course is evidence in favor of modularity; see section 4.1). The cognitive psychology of visual illusion is a crucial part of the cognitive psychology of vision, in that the correct explanation of a given illusion partially determines the scope and limits of the general theory of vision.[41] For example, to have a correct theory of the Necker cube illusion would be to fix a specific constraint on a correct theory of visual cognition. Correspondingly, I think, the theory of wrong logical reasoning or fallacies partially determines the nature of logical cognition by fixing specific constraints on a correct psychological theory of logic.

This point is best made, as I have indicated, by developing further the structural analogy between moral wrongdoing and wrong logical reasoning. As we have seen in chapter 5, empirical studies of deductive reasoning under real-world conditions show that various nonlogical factors significantly affect logical performance, including semantic content, the sheer complexity of the logical task, and the framing of the task. In moral contexts, evaluative judgments are almost always made by way of distinguishing between the controllable and uncontrollable aspects of a given situation or action. Other things being equal, people are not blamed for the badness of what they cannot control.[42] Similarly for logical contexts, it seems to me illuminating to distinguish between controllable and uncontrollable factors affecting logical performance. As the results in the experiments using the concrete Wason selection task show, poor performance can be reversed by varying relevant features of the task. Whenever any of these situations obtains, then we have a controllable factor of fallacy, rather like a casual visual mistake. If the big black spot in the middle of your visual field is caused by a drop of ink on your glasses, then you can take your glasses off and clean them. It makes sense to criticize a reasoner for committing fallacies if she can control the factors adversely affecting her logical performance. By contrast, other things being equal, if the reasoner cannot control the factors adversely affecting her logical performance, it makes little or no sense to criticize her for failing to perform the task successfully under those conditions.

The uncontrollable factors adversely affecting logical performance are very like visual illusions, in that they cannot be made to go away but can only be noted and explained. Everyone, for example, has a blind spot or scotoma in his or her visual field corresponding to special areas on the retina where there are no rods or cones. Even when this blind spot has been demonstrated to you by making visual objects "disappear" into it, you still cannot *see* it because the mind–brain somehow fills it in. Whatever the correct psychological explanation of visual filling-in turns out to be, it will bring out a structural feature intrinsic to our capacity for vision. Similarly, whatever the correct psychological explanations for the uncontrollable factors adversely affecting logical performance turn out to be, they will establish features intrinsic to our real-world logical capacity. For example, although the logical forms of our language permit recursive constructions involving an infinitely large number of embeddings of the two modal operators 'necessarily' and 'possibly', it is a brute fact about us humans that we inevitably lose track of their meanings after a fairly small finite number of embeddings. Similarly, it is a brute fact about humans that our linguistic parsing ability is easily defeated by multiple embeddings even in short grammatically well-formed sentences, for example, the well-known "garden path sentences":

The dog the stick the fire burned beat bit the cat.[43]

Unlike our memories, which can with practice or by means of mnemonics be developed to handle staggeringly large and complex memory tasks, our logical processing capabilities apparently cannot be extended beyond a severely restricted level of cognitive construction. If we could explain this, then we would know both something about why modal fallacies in reasoning are so common and difficult to eradicate[44] and also something important about the real-world constraints on our capacity for logical intuition.

7.4 Objections and Replies II: Harman's Thesis

At this point I should respond to an objection that has been lingering, at least implicitly, since chapter 5. There I argued that the debate about human rationality in cognitive science and philosophy between rationalists and irrationalists depends heavily on four shared assumptions, and that a logical cognitivist resolution of the debate can be carried out if two of those assumptions are rejected. But I did accept the following two assumptions, along with the rationalists and the irrationalists alike:

(α) that there are normative logical principles, and

(β) that human rationality at least partially consists in reasoning in accordance with these normative logical principles.

Harman denies both (α) and (β). More positively put, he holds that there is no significant connection between reasoning and logic, and thus no significant connection between rationality and logic:

> Reasoning is conceived as reasoned revision [of belief]. Reasoning in this sense must not be confused with proof or argument, and the theory of reasoning must not be confused with logic. Psychological relations of immediate implication and immediate consistency are important in reasoning, but this is not to say that logical implication and logical inconsistency are of any special relevance.[45]

> Issues about inference and reasoning need to be distinguished from issues about implication and consistency. Inference and reasoning are psychological processes leading to possible changes in belief (theoretical reasoning) or possible changes in plans and intentions (practical reasoning). Implication is most directly a relation among propositions. Certain propositions imply another proposition when and only when, if the former proposition were true, so too is the latter proposition. It is one thing to say

> (1) A, B, C, imply D.

> It is quite another thing to say

> (2) If you believe A, B, C, you should (or may) infer D.

> Statement (1) is a remark about implication; (2) is a remark about inference. Statement (1) says nothing special about belief or any other psychological state (unless one of A, B, C has psychological content), nor does (1) say anything normative about what anyone 'should' or 'may' do. Statement (1) can be true without (2) being true.[46]

In this way, Harman's basic justification for claiming that there is no significant relation between logic and reasoning is that whereas logic is about a nonpsychological relation between propositions (logical consequence), reasoning is about a psychological phenomenon (change of belief): hence there is no automatic step from features of the former to features of the latter, or conversely. This, says Harman, is further supported by the following facts:

(a) although classical logic is closed under entailment (i.e., if P entails Q, and Q entails R, then P entails R), nevertheless someone might have a good reason to fail to believe something that is logically entailed by other things she already believes;[47]

(b) although classical logic bans the ascription of truth to logical inconsistencies, nevertheless someone might have a good reason to believe an inconsistency;[48]

(c) although classical logic places no constraints on the number of trivial logical consequences of a given set of premises, nevertheless a crucial feature of reasoning is that "one should not clutter one's mind with trivialities";[49] and

(d) some immediate implications in reasoning are not logical implications in classical logic.[50]

From the standpoint of logical cognitivism, however, there are two things wrong with Harman's argument.

The first problem is that Harman merely assumes without further argument that logic is intrinsically nonpsychological, presumably because of worries about logical psychologism.[51] But as I argued in chapter 1, not every theory that posits necessary connections between logic and psychology is psychologistic: for example, logical cognitivism does not entail logical psychologism. If logic is in some respects intrinsically psychological, however, then there is no a priori barrier to holding that principles of logic necessarily carry over to human reasoning. In particular, there is no a priori barrier to holding that the principles of the protologic necessarily carry over to human reasoning as intrinsic categorically normative principles of the latter.

The second problem is that Harman identifies logic with classical or elementary logic and thus is implicitly committed to diehard classicism. This, we will remember, is the view that there is one and only One True Logic, namely classical logic or some minor variant on classical logic. But diehard classicism is questionable (see section 2.5), and Harman has done nothing to show that some or another nonclassical logic cannot bear a significant relation to human reasoning. The same point holds equally for the protologic. None of the gaps Harman identifies between classical logic and human reasoning is a gap between the protologic and human reasoning. This is because the protologic allows for the construction of logical systems that are not closed under classical entailment, that include truth-value gluts, that include constraints on trivial classical consequences, and that include all manner of different sorts of nonclassical logical consequence or entailment. The protologic is presupposed by *every* logical system, whether classical or nonclassical. So again, there is no a priori barrier to holding that the principles of the protologic, whatever they turn out to be, are also intrinsic categorically normative principles of human reasoning.

I conclude that despite Harman's arguments for his thesis, there is no a priori barrier to holding that the protologic is intrinsically categorically normative for human reasoning, and thus no a priori barrier to holding both (α) and (β).

7.5 Objections and Replies III: Why Must I Be Logical?

So far I have tried to anticipate and take preemptive action against two possible criticisms of the thesis that the protologic is intrinsically categorically normative for human reasoning. But there is one fundamental worry I have not yet addressed. Not unexpectedly, Lewis Carroll is its wittiest and most articulate spokesman:

> Alice laughed. "There's no use trying," she said: "One *ca'n't* believe impossible things."
>
> "I daresay you haven't had much practice," said the Queen. "When I was your age, I always did it for half-an-hour a day. Why, I sometimes believed as many as six impossible things before breakfast!"[52]

Carroll, of course, is just kidding. But what about someone who takes the White Queen *very seriously*—someone who holds that it is possible sincerely and self-consciously to assert the truth of an unlimited number of contradictions? Why *couldn't* every sentence be both true and false? Let us call this strange view *white-queenism*.

It should be clear right off the bat that white-queenism has little or nothing to do with the fact that we often commit logical fallacies in our reasoning. As I pointed out in section 7.3, this fact is consistent with the thesis that we perfectly obey the protologic in our reasoning processes. A slightly more subtle point is that white-queenism should not be confused with advocating the serious study of dialetheic logics. The dialetheic logician asserts that some sentences are true contradictions or truth-value gluts, and hence are both true and false; but dialetheism does not say that *every* sentence is both true and false.[53] Even more subtly, white-queenism should not be confused with what Ruth Garrett Millikan calls "white queen psychology." Millikan's white queen psychology is part of an externalist theory of content (that is, a theory holding that the intentional content of a mental state is partially or wholly determined by causal, environmental, or social factors outside the creature possessing that mental state) which states that since it is possible for speakers sincerely and self-consciously to assert sentences that are later discovered a posteriori to be impossible, then speakers can believe some logical contradictions.[54] But again, since the principles of the protologic are perfectly consistent with dialetheism, they are also perfectly consistent with the possibility of sincere, self-conscious belief in some logical contradictions; and since the refined standard cognitivist model of the mind (see chapter 4) lying behind the logic faculty thesis is also perfectly consistent with externalism,[55] it follows that the thesis that the protologic is intrinisically categorically normative for human reasoning is perfectly consistent with white queen psychology.

By the notion of white-queenism, then, I mean the radical skeptical attempt sincerely and self-consciously to reject logic *completely*. I will consider two versions of white-queenism: (1) classical or Cartesian white-queenism, and (2) postmodern or neo-Nietzschean white-queenism. The latter has rarely been discussed by philosophers of logic.[56] But I think that it poses a serious philosophical challenge that cannot be ignored or casually dismissed.

(1) Classical or Cartesian white-queenism As every philosopher knows, in the *Discourse on the Method*, and again in the *Meditations on First Philosophy*, Descartes proposes to doubt absolutely everything, with an eye to finding out what is originally indubitable, which is thereby an Archimedean point or absolute foundation for scientific knowledge. In turn, this absolute epistemic foundation is the *cogito*: I think, therefore I exist.[57] It is a peculiar feature of Descartes's methodological skepticism, however, that although every single ordinary perceptual, scientific, and mathematical belief falls within the scope of the doubt, logic is never explicitly doubted. This is because logic is presupposed by the *cogito*. You cannot validly argue that *cogito, ergo sum* without the relation of logical consequence between premise and conclusion, which is expressed by the *ergo*. But since Descartes explicitly grants in the *Rules for the Direction of the Mind* that deductions are less certain than intuitions because of the fallibility of our memory of the premises,[58] what then rules out the possibility that my thinking process, and thus the inference itself, might go totally awry between the *cogito* and the *sum*?

For this reason, I think, Descartes quietly switches over in the *Meditations* to an intuition-based version of the *cogito*, according to which what occurs at the end of the application of his skeptical method is not a logical deduction, but instead the clear and distinct intuition of a single analytically necessary conditional truth.[59] But even this will not quite do, if one's general conception of logic is intensional, as Descartes's logic at least implicitly is, and thereby is a logic that legitimates analytic truths about existence based on the "conceptual containment" of the predicate-term in the subject-term.[60] For analytic inference requires a linear transition in thinking from a self-conscious comprehension of the subject-term to a self-conscious comprehension of the predicate term. So, in that case, one can also easily imagine a skeptic who doubts that the thinking subject's memory of the initial idea or subject-concept *cogito* or *my thinking* is epistemically or semantically sufficient to ground the analytic transition to the second idea or predicate-concept *sum* or *my existence*.

So the Cartesian skeptic I have in mind is one who goes slightly beyond Descartes's own application of his skeptical method and proposes explicitly to doubt every logical truth, law, deduction, notion, and principle whatsoever: a skeptic, in short, who proposes explicitly to doubt whatever logical apparatus is presupposed by and implemented in any and every argument or belief, including the *cogito*. This doubt would, at least implicitly, comprehend every classical or nonclassical logical system. The basic outlook of such a skeptic is nicely glossed by C. S. Peirce:

[I]t is quite possible that a person should doubt every principle of inference. He may not have studied logic, and though a logical formula may sound very obviously true to him, he may feel a little uncertain whether some subtle deception may not lurk in it.[61]

Here is a brief reply to the ultra-Cartesian or Peircean logic skeptic. Every explicit doubt about logic involves, at the very least, the conceivability of the denial of some sentence or another that is taken to be "true by virtue of logic alone" by those who believe in logic. Now, the relevant notion of conceivability entails the idea of *logical possibility*; the relevant notion of denial entails the idea of *logical negation*; and the relevant notion of "truth by virtue of logic alone" entails the idea of *logical truth*. This holds no matter what sort of classical or nonclassical logical system is supposed to be doubted by the skeptic. Thus, any explicit doubt about logic already presupposes and uses *the skeptic's own logic of thought*, at the very least. But every logic of thought is a logical system of some sort. Therefore all ultra-Cartesian or Peircean doubts about logic are self-refuting. The logocentric predicament strikes again.

(2) Postmodern or neo-Nietzschean white-queenism[62] The classical or Cartesian skeptic about logic traps himself by presupposing and using logic in order to challenge logic. Nevertheless, someone—for example, Andrea Nye—can point out that it is possible to be a skeptic about logic in a completely different and specifically non-Cartesian way. The most elegant method for doing this is, in effect, to marry Carnap to Nietzsche. This is not such an odd coupling as it may seem to be at first glance. Recall how logically promiscuous the principle of tolerance is:

It is not our business to set up prohibitions, but to arrive at conventions. . . . In logic there are no morals. Everyone is at liberty to build up his own logic, i.e., his own form of language, as he wishes.

To get to neo-Nietzschean skepticism about logic, then, we need only start with Carnap's tolerance principle and extend it a little by introducing Nietzsche's attack on what he calls the "will to truth":

The will to truth which will still tempt us to many a venture, that famous truthfulness of which all philosophers so far have spoken with respect—what questions has this will to truth not laid before us! What strange, wicked, questionable questions! . . . *Who* is it really that puts such questions to us? *What* in us really wants "truth"? . . . We asked about the *value* of this will. Suppose we want truth: *why not rather* untruth? and uncertainty? even ignorance?[63]

From here, it is only a short step indeed to the conclusion that every logic whatsoever, whether classical or nonclassical, is (a) nothing but a product of "will to power," and is (b) merely "socially constructed," not objectively real. This, in turn, is just to say that every logic is nothing but a product of natural, nonrational forces that more generally drive human action, and that the existence and specific character of any logic depends entirely on implicit or explicit communal agreements and contractual relations. Note that but for the antiscientific sounding rhetoric typically employed by Nietzscheans, this view is at bottom equivalent to that held by irrationalist cognitive scientists who adopt a "heuristics and biases" approach to human reasoning, and in particular to those who appeal to a biological, evolutionary, contractualist foundation for reasoning, for example, Leda Cosmides (see section 5.5).

In any case, the crucial point is that the Carnapian–Nietzschean or neo-Nietzschean skeptic becomes a logic skeptic not by explicitly doubting logic, but instead by simply *opting out* of the social construct that constitutes the will to truth: that is, by deciding to liberate herself from logic, and by undertaking to live a form of human life that expresses a total lack of concern for logic. It should be noted that this post-modern or neo-Nietzschean version of white-queenism is not the same as the "logical contrarianism" I briefly considered in section 6.2, which involved communities that adopt radically nonstandard logical systems based on the denial of some or another logical principle or concept assumed to belong to the logic of thought. Logical contrarianism still at least promotes the general linguistic practice of logicizing, even if the logic in question is necessarily, for those of us who by hypothesis think according to a different logic of thought, an antilogic or a "schmogic." Neo-Nietzschean logical skepticism is, by contrast, *logical nihilism*.

The reference to nihilism points up the important fact that the neo-Nietzschean logical skeptic is a mirror-reflection of the Nietzschean *antimoralist* skeptic:

This problem of pity and of the morality of pity . . . seems at first to be something merely detached, an isolated question mark; but whoever sticks with it and *learns* to ask questions here will experience what I experienced—a tremendous new prospect opens up for him, a new possibility comes over him like a vertigo, every kind of mistrust, suspicion, fear leaps up, his belief in morality, in all morality, falters—finally a new demand becomes audible. Let us articulate this *new demand*: we need a *critique* of moral values, *the value of these values themselves must be first called in question*—and for that there is needed a knowledge of the conditions and circumstances in which they grew, under which they evolved and changed . . . , a knowledge of a kind that has never yet existed or even been desired.[64]

The Nietzschean antimoralist skeptic must in turn be sharply distinguished from classical moral skeptics. Classical moral skeptics, it seems, are of three basic kinds:

(1) prototyrants like Plato's Thrasymachus in the *Republic*, who claims that might, or the advantage of the stronger over the weaker, is right;

(2) perverse immoralists like Milton's notorious "bad boy" in *Paradise Lost*, Satan, who says:

Farewell remorse! All good to me is lost;
Evil, be thou my good[65]

and

(3) egoistic hedonists like Hume's "sensible knave" in the *Treatise of Human Nature*, who claims that what is right is whatever he himself wants and can manage to get away with.

In each case, the classical moral skeptic challenges some or another classical notion of morality by proposing to establish as morally right something that is morally *wrong* according to the classical moralist. By sharp contrast, however, the Nietzschean antimoralist skeptic rejects what Bernard Williams aptly dubs the "special ethical system" or "peculiar institution" (notice the allusion to the practice of slavery) *of morality itself*,[66] by "revaluating" all practical values and thereby "going beyond good and evil."[67] Similarly, the post-modern or neo-Nietzschean logic skeptic completely rejects the special scientific system or "peculiar institution" *of logic itself*, by "revaluating" all theoretical values, and "going beyond truth and untruth."

Thus, a neo-Nietzschean logic skeptic cannot be put off merely by indicating one's deep commitment to the very idea of logic, just as a Nietzschean antimoralist skeptic cannot be put off merely by repeating one's deep commitment to the very idea of a moral 'ought'. The Nietzschean antimoralist skeptic asks: why must she be moral, if the very idea of a moral 'ought' is nothing but a product of the will to power and thus socially constructed, and if the peculiar institution of morality leads to something the Nietzschean finds deeply unpalatable, for example, the "slave morality" of Christianity? So too the neo-Nietzschean logic skeptic asks: why must she be logical, if the very idea of logical truth or consequence is nothing but a product of the will to power and thus socially constructed, and if the peculiar institution of logic (as Nye asserts in *Words of Power*) is directly aligned with powerful cultural institutions of political tyranny, economic exploitation, sexual chauvinism, racial prejudice, and human oppression more generally? Logic is therefore "insane," and to liberate ourselves from logic is to attack these forces at their very sources.

Philosophers of logic have so far either simply ignored this form of logic skepticism or dismissed it as ridiculous,[68] just as the casual reader of *Through the Looking Glass* might regard the White Queen as ridiculous. But I think that this is a serious mistake. As Foot, Raymond Geuss, and Williams have stressed,[69] the precisely analogous case of Nietzschean antimoralist skepticism is a serious worry that is surprisingly resistant to refutation and cannot be philosophically ignored. Similarly, although it may admittedly be somewhat of a stretch for Nye to attempt to link the *Begriffsschrift* with *Mein Kampf*, this is in fact no more odd *in principle* than Plato's Ring of Gyges thought experiment in the *Republic*, Descartes's Evil Demon hypothesis in the *Meditations*, or Nietzsche's myth of eternal recurrence. And in any case, the general thrust of the neo-Nietzschean worry about logic depends solely on premises that are also accepted by some highly respectable contemporary cognitive scientists.

So, assuming that we should take this extreme form of logic skepticism seriously, how can we respond? Remember that once we have granted the premises that are also accepted by some highly respectable contemporary cognitive scientists, then the neo-Nietzschean skeptic makes no explicit assertions that might be targets for the sort of self-refutation strategy I used against the ultra-Cartesian or Peircean logic skeptic, but instead *only expresses a radically strong antilogical attitude.*

What we need to do, I think, is explore the consequences of rejecting logic in this way. Let us try to imagine a community of fully logic-liberated people. Here is Frege's characteristically vigorous response to the possibility of such a community:

> [W]hat if beings were found whose laws of thought flatly contradicted ours and therefore frequently led to contrary results even in practice? . . . I should say: we have here a hitherto unknown type of madness.[70]

This, I think, is essentially the right line of response. Taking the neo-Nietzschean hypothesis seriously, it would follow that both the social-psychological and individual psychological profiles of the logical nihilists would be pathological by *any* criterion of the meaningful use of the term 'pathological'. Inconsistency and fallacy would be endemic, entrenched among them. Neither truth nor truthfulness would mean anything to them, or untruth or untruthfulness for that matter. They could not have beliefs, but instead only unreflective attitudes. They could not give reasons for anything, hence could not justify anything, hence would be without cognitive or practical norms of any kind. Without cognitive or practical norms, their emotional and volitional states would be without internal constraint or structure and utterly wanton, without any reasons for caring one way or the other about their direct or "first-order" desires or preferences.[71]

In fact, *the self-induced condition of the logical nihilist would be equivalent to removing or permanently disabling her logic faculty, which in turn would undermine her metaphysical and moral personhood.* In this way, the logical nihilist would lack any constructive or epistemic access to the protologic, and so would be without a properly functioning capacity of rational cognition or thought, and also without a properly functioning capacity for normative self-reflection, and thus would become a *postperson*. Now, to be sure, there is nothing in itself wrong with being a protorational, nonrational, or subrational animal. All of us started out as protorational human animals, and it is a sobering thought that many of us will also involuntarily become nonrational or subrational human animals before this Big Parade is over. But there is something deeply incoherent, sad, and truly awful about a rational animal's deliberately deciding—that is, rationally choosing—to become a nonrational or subrational animal. Indeed, it is the ultimate self-stultifying act. If someone we deeply respected were to suffer a catastrophic accident and be permanently reduced to the cognitive and emotional level of a happy human infant or cat, we would regard this as a great misfortune, even despite the newfound contentment. But if someone we deeply respected were to *choose* this fate, we would regard this not merely as a great misfortune but also as profoundly perverse and entirely tragic.

So it is the logic-liberated people, not the logicians, who are in this sense "insane." By endorsing the antilogical life, the postmodern or neo-Nietzschean logic skeptic intentionally commits cognitive and emotional suicide. One does not need to be a Kantian moralist to see that this is psychologically and morally self-destructive, and thus provides an effective pragmatic refutation of this extreme brand of logic skepticism.

Where does this leave us? If one were a logical cognitivist, she could point out that our being essentially logical animals is fully consistent with our self-consciously and vigorously opposing political tyranny, economic exploitation, sexual chauvinism, racial prejudice, and systematic human oppression more generally. In fact, it is hard to see how anyone could manage to pursue a progressive political agenda *without* being a logical animal. For logic, in the protean guise of the logic faculty and its innately contained protologic, is at the very basis of human rationality, hence at the very basis of all justifications of human belief and intentional action, hence at the very basis of our moral animality as well. But this is not to say that rational humans are always, or even usually, equal to the fundamental rational task of cognitively constructing decisively sufficient reasons for their beliefs and actions. The human logical and moral animal must also learn to make allowances for, and to live with, the consequences for rationality of its animality.

The specific purpose of this chapter has been to argue that logic is infused with the categorically normative character of rationality, and that the categorically normative character of human rationality is infused with logic. The overall purpose of the book has been to argue for various necessary connections between the psychological and the logical. The full recognition of the thought that categorical normativity appears intrinsically on both sides of the necessary connections—that cognitive science, just like the science of logic, is fundamentally a *moral* science and not fundamentally a *natural* science—could be expected to change both cognitive science and the science of logic quite radically. Cognitive scientists would have to study rational information-processing under two novel guiding assumptions, (i) that cognition and thought are *intrinsically* and not merely *extrinsically* linked with the affective, volitional, and decision-making capacities of sentient active organisms in the world,[72] and (ii) that these necessarily interlinked cognitive, thought-generating, affective, volitional, and decision-making capacities are inherently guided not merely by instrumental or hypothetical norms, but also and more fundamentally by *noninstrumental and categorical* norms. And logicians would have to study the necessary relation of consequence, in classical and nonclassical systems alike, as inextricably embedded within the framework of this substantially "Kantified" and "thickened" cognitive science. But in the nature of these things, such profound changes would be neither facile nor fast. So less tub-thumpingly and in the shorter term, my hope is that cognitive psychologists and logicians will find the joint project of working out a precise structural description of the protologic to be something worth doing.

Notes

Introduction

1. Aristotle, *Prior Analytics*, 24b18, in *The Collected Works of Aristotle*, trans. A. J. Jenkinson, ed. R. McKeon (New York: Random House, 1941); translation slightly modified.

2. I. Kant, "The Jäsche Logic," in I. Kant, *Lectures on Logic*, trans. J. M. Young (Cambridge: Cambridge University Press, 1992), p. 531.

3. G. Boole, "Logic and Reasoning," in his *Studies in Logic and Probability* (London: Watts, 1952), p. 212.

4. W. V. O. Quine, *Philosophy of Logic*, 2d. ed. (Cambridge, Mass.: Harvard University Press, 1986), p. 81.

5. N. Chomsky, *Language and Problems of Knowledge* (New York: Columbia University Press, 1980), p. 99.

6. J. Fodor, *The Language of Thought* (Cambridge, Mass.: Harvard University Press, 1975), p. 173.

7. This traditional gloss on the a priori–a posteriori distinction needs refinement; see sec. 6.3.

8. Sober, "Psychologism," *Journal for the Theory of Social Behavior* 8 (1978), p. 165.

9. This formulation also needs refinement; see sec. 1.2.

10. Classical or elementary logic is bivalent truth-functional first-order quantified polyadic predicate calculus with identity.

11. An extended logic is a logical system that permits the occurrence of some logical constants, axioms, inference rules, interpretation rules, valid sentences, or theorems that do not occur in elementary logic, but otherwise preserves elementary logic intact.

12. A deviant logic is a logical system that rejects some of the logical constants, axioms, inference rules, interpretation rules, valid sentences, or theorems of elementary logic.

13. A paraconsistent logic is a logical system that permits the occurrence of formally inconsistent theorems by explicitly preventing *inferential explosion*: the derivation of any sentence whatsoever from a formal contradiction.

14. A dialetheic logic is a logical system that permits the occurrence of true contradictions or truth-value "gluts." All dialetheic logics are paraconsistent but not all paraconsistent logics are dialetheic.

15. See J. Fodor, *The Mind Doesn't Work That Way* (Cambridge, Mass.: MIT Press, 2000); H. Gardner, *The Mind's New Science* (New York: Basic Books, 1985); U. Neisser, *Cognitive Psychology* (New York: Appleton-Century Crofts, 1967); U. Neisser, *Cognition and Reality* (San Francisco: W. H. Freeman, 1976); and S. Pinker, *How the Mind Works* (New York: W. W. Norton, 1997).

16. In what follows I will use 'sentence', 'predicate', and 'name' to mean *meaningful indicative sentence, meaningful predicate*, and *meaningful name*. So unless otherwise noted, 'sentence' as I use it will have the same philosophical content and force as either 'linguistically expressed proposition' or 'statement', depending on your favorite theory of truth-bearers.

17. Philosophers of logic often distinguish between *syntactic consequence* or provability, and *semantic consequence* or necessary truth-preservation. I accept that distinction, but for convenience will use 'consequence' simply to *mean* semantic consequence.

18. See J. Bennett, *Rationality* (Indianapolis: Hackett, 1989).

19. See H. Frankfurt, "Freedom of the Will and the Concept of a Person," in his *The Importance of What We Care About* (Cambridge: Cambridge University Press, 1988); and B. O'Shaughnessy, *The Will*, 2 vols. (Cambridge: Cambridge University Press, 1980).

20. See R. Audi, *The Architecture of Reason* (New York: Oxford University Press, 2001), chaps. 1–2 and 4–5; and J. Raz, *Engaging Reason* (Oxford: Oxford University Press, 1989), chaps. 1–4.

21. Sometimes truth in all logically possible worlds is called *metaphysical necessity*, as distinguished from analyticity or again from apriority. See S. Kripke, *Naming and Necessity*, 2d. ed. (Cambridge, Mass.: Harvard University Press, 1980), pp. 34–39, 158–160. This is highly contested territory, which I have attempted to map elsewhere: see R. Hanna, *Kant and the Foundations of Analytic Philosophy* (Oxford: Clarendon/Oxford University Press, 2001), chaps. 3–5; and R. Hanna, "A Kantian Critique of Scientific Essentialism," *Philosophy and Phenomenological Research* 58

(1998): 497–528. In sec. 6.5 of this book I briefly sketch a modal framework that identifies logical necessity, analyticity, and one type of metaphysical necessity ("weak" metaphysical necessity). And also, for the purposes of this book, I accept the general validity of inferences from conceivability to logical or (weak) metaphysical possibility. I fully realize that this is a highly controversial assumption— see, e.g., T. Gendler and J. Hawthorne (eds.), *Conceivability and Possibility* (Oxford: Clarendon/Oxford University Press, 2002). But the contexts in which I employ the inferential step from conceivability to possibility do not, I think, fall within the class of cases (essentially involving mathematics or natural kinds) that are normally taken to provide counterexamples to the thesis that conceivability entails possibility.

22. See, e.g., A. Mele and P. Rawling (eds.), *The Oxford Handbook of Rationality* (Oxford: Oxford University Press, 2003).

23. See G. Boolos and R. Jeffrey, *Computability and Logic*, 3d. ed. (Cambridge: Cambridge University Press, 1996). Gödel's incompleteness theorems may be taken to show that validity and soundness outrun provability and computability. The precise character of logical validity will vary according to whether the logic is classical or nonclassical, and within nonclassical logics, according to whether the logic is either merely a conservative extension of classical logic or a deviant logic. See chapter 2. What this means is that not all rational procedures are classically logically valid.

24. What I am calling "principled rationality" is closely connected with the traditional notion of *pure reason*; see, e.g., L. Bonjour, *In Defense of Pure Reason* (Cambridge: Cambridge University Press, 1998).

25. See Audi, *The Architecture of Reason*; N. Daniels, "Wide Reflective Equilibrium and Theory Acceptance in Ethics," *Journal of Philosophy* 76 (1979): 256–282; N. Daniels, "On Some Methods of Ethics and Linguistics," *Philosophical Studies* 37 (1980): 21–36; N. Daniels, "Reflective Equilibrium and Archimedean Points," *Canadian Journal of Philosophy* 10 (1980): 83–103; J. Rawls, "The Independence of Moral Theory," *Proceedings and Addresses of the American Philosophical Association* 48 (1974): 5–22; and J. Rawls, *A Theory of Justice* (Cambridge, Mass.: Harvard University Press, 1971). Reflective equilibrium is the mutual harmonization, constructive reworking, and global interrelatedness of beliefs such that they ultimately form a principled and coherent set. Narrow reflective equilibrium restricts this to a specific domain of beliefs; and wide reflective equilibrium applies it comprehensively over all domains. I am assuming that the method of reflective equilibrium can be extended from beliefs to desires, emotions, intentions, and volitions as well.

26. See M. Hollis and G. Sugden, "Rationality in Action," *Mind* 102 (1993): 1–35; D. Hubin, "The Groundless Normativity of Instrumental Rationality," *Journal of Philosophy* 98 (2001): 445–468; R. Nozick, *The Nature of Rationality* (Cambridge,

Mass.: Harvard University Press, 1993); and H. Simon, *Reason in Human Affairs* (Stanford: Stanford University Press, 1983).

27. For me, something is *multiply embodiable* just in case it is a cognitive structure that can occur in two or more distinct biological individuals or species. I apologize for the neologism. Unfortunately, however, the more familiar term 'multiple realization', deriving from Putnam's early presentation of functionalism in "The Nature of Mental States" (in his *Mind, Language, and Reality: Philosophical Papers,* vol. 2 [Cambridge: Cambridge University Press, 1975]) is now ambiguous in view of Kim's well-known but specifically *reductive* interpretation of the notion of a physical realization. See J. Kim, "Multiple Realization and the Metaphysics of Reduction," in his *Supervenience and Mind* (Cambridge: Cambridge University Press, 1993). Multiple embodiment does not entail physical realization in Kim's sense.

28. L. Wittgenstein, *Tractatus Logico-Philosophicus*, trans. C. K. Ogden (London: Routledge and Kegan Paul, 1981), prop. 5.552, p. 145.

29. L. Wittgenstein, *Philosophical Investigations*, trans. G. E. M. Anscombe (New York, Macmillan, 1953), §242, p. 88e.

30. See R. Hanna, "Kant, Wittgenstein, and the Fate of Analysis," in M. Beaney (ed.), *The Analytic Turn* (London: Routledge, forthcoming).

31. A great many recent and contemporary philosophers, especially in North America, are scientific naturalists in one way or another; some but not all of the scientific naturalists also explicitly want to naturalize mathematics; and some but not all of the naturalizers of mathematics also explicitly want to naturalize logic. Of those who explicitly want to naturalize logic, very few are also defenders of logical psychologism. Indeed, I would hazard a guess that almost every introductory logic course taught almost anywhere in the world begins with a stern warning against confusing logical facts and principles with psychological facts and principles.

32. By this I mean only that that my theory of rationality and logic shares some basic themes and theses with certain parts of Kant's Critical philosophy. But of course my theory is intended to be independently philosophically justified and motivated; and also it is not in any way intended to be an interpretation of or commentary on Kant's logical writings.

Chapter 1

1. S. Haack, *Philosophy of Logics* (Cambridge: Cambridge University Press, 1978), p. 238.

2. See A. Coffa, *The Semantic Tradition from Kant to Carnap* (Cambridge: Cambridge University Press, 1991); Hanna, *Kant and the Foundations of Analytic*

Philosophy; and P. Hylton, *Russell, Idealism, and the Emergence of Analytic Philosophy* (Oxford: Oxford University Press, 1990).

3. See F. Brentano, *Psychology from an Empirical Standpoint*, trans. A. C. Rancurello et al. (London: Routledge and Kegan Paul, 1995); and E. G. Boring, *History of Experimental Psychology* (New York: Appleton-Century-Crofts, 1950).

4. See M. Dummett, "Can Analytical Philosophy Be Systematic, and Ought It to Be?" in his *Truth and Other Enigmas* (London: Duckworth, 1978); M. Dummett, *Frege: Philosophy of Language*, 2d. ed. (Cambridge, Mass.: Harvard University Press, 1981); and M. Dummett, *Origins of Analytical Philosophy* (Cambridge, Mass.: Harvard University Press, 1993).

5. See, e.g., W. Kneale and M. Kneale, *The Development of Logic* (Oxford: Oxford University Press, 1984), chaps. 7–12; see also sec. 2.1 of this book.

6. J. Van Heijenoort, "Logic as Calculus and Logic as Language," *Synthese* 17 (1967): 324–330, p. 324.

7. G. Frege, *The Foundations of Arithmetic*, 2d. ed., trans. J. L. Austin (Evanston, Ill.: Northwestern University Press, 1953), p. vi.

8. Frege, *Foundations of Arithmetic*, p. vii.

9. G. Frege, *The Basic Laws of Arithmetic*, trans. M. Furth (Berkeley: University of California Press, 1964), pp. 12–13.

10. G. Frege, "Logic [1897]," in his *Posthumous Writings*, trans. P. Long et al. (Chicago: University of Chicago Press, 1979), pp. 127–151, here pp. 143–144.

11. Frege, "Logic [1897]," pp. 147–149.

12. G. Frege, "Thoughts," in his *Collected Papers on Mathematics, Logic, and Philosophy*, trans. M. Black et al. (Oxford: Basil Blackwell, 1984), pp. 351–372, here p. 368.

13. See J. Cohen, "Frege and Psychologism," *Philosophical Papers* 27 (1998): 45–68.

14. See M. Kusch, *Psychologism* (London: Routledge, 1995).

15. Some classes are members of themselves, and some are not. If we assume a principle of unrestricted class-formation, we are then able to form the class of all classes that are not members of themselves. Call this class K. K is a member of itself if and only if it is not a member of itself. Hence K is self-contradictory and any system that includes a principle of unrestricted class-formation is also self-contradictory. See Frege, *Basic Laws of Arithmetic*, appendix; and R. Monk, *Bertrand Russell: The Spirit of Solitude* (London: Jonathan Cape, 1996), pp. 153–154.

16. See Kusch, *Psychologism*, pp. 203–210.

17. See E. Husserl, *Prolegomena to Pure Logic*, chaps. 1–2, in his *Logical Investigations*, vol. 1, trans. J. N. Findlay (London: Routledge and Kegan Paul, 1970). Frege, by contrast, takes the traditional Kantian line that logic is unconditionally obligatory for human reasoning. See I. Kant, *Critique of Pure Reason*, trans. P. Guyer and A. Wood (Cambridge: Cambridge University Press, 1997), A53/B77; Kant, "The Jäsche Logic," p. 529; Frege, *Basic Laws of Arithmetic*, p. 12; Frege, "Logic [1897]," p. 128; and Frege, "Thoughts," pp. 368–371.

18. Cognitive relativism states that truth logically strongly supervenes on individual beliefs, socially shared beliefs, or species-specific beliefs. This must be distinguished from moral relativism, which states that moral good or right logically strongly supervenes on individual, socially-shared, or species-specific beliefs and corresponding emotional attitudes. See Husserl, *Prolegomena to Pure Logic*, chap. 7; also J. Meiland and M. Krausz (eds.), *Relativism: Cognitive and Moral* (Notre Dame: University of Notre Dame Press, 1982).

19. Husserl, *Prolegomena to Pure Logic*, p. 99.

20. Husserl, *Prolegomena to Pure Logic*, pp. 100–101, translation slightly modified. Husserl does not say precisely what he takes to be "the apodeictically self-evident, and so non-empirical and absolutely exact laws which form the core of all logic"; but read charitably, this could count as an anticipation of the notion of a protologic that I develop in sec. 2.5.

21. Husserl, *Prolegomena to Pure Logic*, p. 104.

22. Husserl, *Prolegomena to Pure Logic*, p. 98.

23. See E. Husserl, "A Reply to a Critic of My Refutation of Logical Psychologism," in P. McCormick and F. Elliston (eds.), *Husserl: Shorter Works* (Notre Dame: University of Notre Dame Press, 1982), pp. 152–158, here p. 156.

24. See D. Chalmers, *The Conscious Mind* (New York: Oxford University Press, 1996), pp. 42–51.

25. For the distinction between concepts and properties, see note 50 of this chapter.

26. See, e.g., J. Levine, "Materialism and Qualia: The Explanatory Gap," *Pacific Philosophical Quarterly* 64 (1983): 354–361.

27. See J. Fodor, "Special Sciences, or the Disunity of Science as a Working Hypothesis," *Synthese* 28 (1974): 97–115; and I. Gold and D. Stoljar, "A Neuron Doctrine in the Philosophy of Neuroscience," *Behavioral and Brain Sciences* 22 (1999): 809–830.

28. Frege, *Foundations of Arithmetic*, p. 83.

29. Frege also has worries about two metaphysical doctrines he sometimes closely associates with logical psychologism: (a) the reduction of numbers to concrete, par-

ticular entities; and (b) idealism. But these doctrines do not properly speaking belong to logical psychologism. Psychologicists need not commit themselves to any particular view of the nature of numbers or to idealism; indeed, a psychologicist could also consistently be a platonist and a scientific realist.

30. See, e.g., W. V. O. Quine, *Word and Object* (Cambridge, Mass.: MIT Press, 1960). Quine's scientific naturalism also contains a holistic component, inherited from the neo-Hegelian tradition or possibly from Pierre Duhem; but I leave this component aside because it is occasionally in direct conflict with other elements of Quine's naturalism. This is the case, e.g., in the Quine–Putnam "indispensability argument" for mathematical platonism; see S. Shapiro, *Thinking about Mathematics* (Oxford: Oxford University Press, 2000), pp. 212–220. A similar problem arises from Quine's views on the indispensability of classical logic; see R. Fogelin, "Quine's Limited Naturalism," *Journal of Philosophy* 94 (1997): 543–563.

31. See A. Danto, "Naturalism," in P. Edwards (ed.), *The Encyclopedia of Philosophy*, vol. 5 (New York: Macmillan, 1967), pp. 448–450; D. Papineau, *Philosophical Naturalism* (Oxford: Blackwell, 1993); and B. Stroud, "The Charm of Naturalism," *Proceedings and Addresses of the American Philosophical Association* 70 (1996): 43–55. Quine explicitly endorses psychologism in "Epistemology Naturalized," in his *Ontological Relativity and Other Essays* (New York: Columbia, 1969); see also H. Kornblith (ed.), *Naturalizing Epistemology* (Cambridge, Mass.: MIT Press, 1985).

32. There is of course a circularity here: physicalists define first-order physical properties by appealing to the fundamental natural sciences, and then define the fundamental natural sciences by appealing to whatever instantiates first-order physical properties. This points up an even more basic problem for physicalism. Because 'the physical' is defined either by current science, which is highly fallible, or by future science, which is highly speculative, physicalists cannot say precisely what 'the physical' is; see B. Montero, "The Body Problem," *Noûs* 33 (1999): 183–200. I think that this is an insoluble problem for physicalism. But my arguments against naturalism in secs. 1.3 and 1.4 should work even if the body problem is soluble.

33. See Chalmers, *The Conscious Mind*, pp. 32–42; T. Horgan, "From Supervenience to Superdupervenience: Meeting the Demands of a Material World," *Mind* 102 (1993): 555–586; and Kim, *Supervenience and Mind*.

34. See Kim, "Concepts of Supervenience," pp. 64–67, and "'Strong' and 'Global' Supervenience Revisited," pp. 71–82, both in his *Supervenience and Mind*.

35. See Kim, "Concepts of Supervenience," pp. 57–64.

36. The origins of the concept of supervenience go back to British emergentism, G. E. Moore, and R. M. Hare. Oddly enough, however, Kim ascribes *weak* rather

than moderate or strong supervenience to these philosophers—see "Concepts of Supervenience," pp. 57–64.

37. There is another dimension of variation within supervenience: it might be either "local" or "regional" or "global," depending respectively on whether the supervenience base (the "subvenient domain") comprises individuals, spacetime regions, or whole worlds. Local supervenience entails regional and global supervenience; regional supervenience entails global supervenience but does not entail local supervenience; and global supervenience does not entail either regional or local supervenience. See Chalmers, *The Conscious Mind*, chap. 2.

38. These nomological connections may be sufficient only, or both necessary and sufficient; they may also be causal, but they do not have to be. The crucial thing is that they somehow guarantee what Kim calls "microdeterminism" or the necessary nomological determination of the higher-level properties by lower-level microphysical properties.

39. See Horgan, "From Supervenience to Superdupervenience."

40. For more details on types of necessity, see sec. 6.5.

41. See Chalmers, *The Conscious Mind*, chap. 2.

42. Knowledge of the *A*-properties and of the causal laws can be either a priori or a posteriori. So physicalist logical supervenience is consistent with a priori physicalism and most versions of a posteriori physicalism alike.

43. See, e.g., J. McDowell, "Two Sorts of Naturalism," in R. Hursthouse et al. (eds.), *Virtues and Reasons* (Oxford: Clarendon/Oxford University Press, 1995), pp. 149–179.

44. See W. Cooper, *The Evolution of Reason* (Cambridge: Cambridge University Press, 2001); P. Maddy, "A Naturalistic Look at Logic," *Proceedings and Addresses of the American Philosophical Association* 76 (2002): 61–90; and E. Nagel, "Towards a Naturalistic Conception of Logic," in H. Kallen and S. Hook (eds.), *American Philosophy Today and Tomorrow* (New York: L. Furman, 1935), pp. 377–391.

45. Cooper and Maddy, e.g., explicitly distinguish their naturalistic approaches to logic from psychologism. See note 44 directly above.

46. G. E. Moore, *Principia Ethica* (Cambridge: Cambridge University Press, 1903), p. xx.

47. Moore, *Principia Ethica*, p. 40.

48. Moore, *Principia Ethica*, p. 58.

49. Moore, *Principia Ethica*, p. 73.

50. Moore fails to distinguish between concepts and properties, and again between properties and predicates. See G. Bealer, *Quality and Concept* (Oxford: Clarendon/Oxford University Press, 1982); A. Oliver, "The Metaphysics of Properties," *Mind* 105 (1996): 1–80; and H. Putnam, "On Properties," in his *Mathematics, Matter, and Method: Philosophical Papers,* vol. 1, 2d. ed. (Cambridge: Cambridge University Press, 1979). This is, of course, controversial territory. But for my purposes I will make the fairly standard assumptions that concepts are intersubjectively accessible psychological intensional entities whose identity criterion is definitional equivalence; that predicates are linguistic intensional entities whose identity criterion is synonymy; and that properties are nonpsychological, nonlinguistic intensional entities whose identity criterion is sharing extensions across possible worlds. Predicates express concepts as their meanings, and concepts pick out corresponding properties in the world. For convenience, however, in the following discussion of Moore's argument against naturalism I will allow the term 'property' to range over all three sorts of intensional entity. The flaws in his argument will persist no matter which sort of intensional entity is at issue.

51. Moore, *Principia Ethica*, p. 15.

52. Moore, *Principia Ethica*, p. 44.

53. Indeed Moore adopts Bishop Butler's Monty-Python-esque dictum, "everything is what it is and not another thing," as the motto of *Principia Ethica,* and also uses it repeatedly as an axiom in his arguments.

54. Moore, *Principia Ethica*, pp. 16–17.

55. It is likely that Moore inherited the phenomenal criterion of the identity of properties from his teacher James Ward, who in turn inherited it from Brentano—thus by an ironic twist returning us full circle to psychologism. See J. Ward, "Psychology," in *Encyclopedia Britannica*, 11th. ed., vol. 22 (New York: Encyclopedia Britannica, 1911), pp. 547–604.

56. Moreover, the phenomenal criterion of property identity leads directly to the paradox of analysis: If only *phenomenal* identity will suffice for property identity, and property identity is a necessary condition of a correct analysis, then every correct analysis must be epistemically trivial. See C. H. Langford, "The Notion of Analysis in Moore's Philosophy," in P. Schilpp (ed.), *The Philosophy of G. E. Moore* (New York: Tudor, 1952), pp. 321–342; and Moore's reply to Langford, "Analysis," in the same volume, pp. 660–667.

57. Moore, *Principia Ethica*, pp. 206–207.

58. G. E. Moore, "The Conception of Intrinsic Value," in his *Philosophical Studies* (New York: Harcourt Brace, 1922), pp. 253–275, here p. 261.

59. See G. E. Hughes and M. J. Cresswell, *An Introduction to Modal Logic* (London: Methuen, 1972), p. 46.

60. Frege, "Thoughts," p. 363.

61. See P. Benacerraf, "Mathematical Truth," *Journal of Philosophy* 70 (1973): 661–679.

62. I am assuming that causation is nomological singular event causation. Something is causally relevant if and only if it either provides a necessary condition for the occurrence of a causal process, or at least directly contributes some explanatorily illuminating information about that causal process. By contrast, something is causally efficacious if and only if it either is, or is an indispensable proper part of, a cause.

63. In chap. 6 below I argue that rational intuition is a function of the imagination, and more specifically that logical intuition—for rational humans—is a function of the linguistic imagination.

64. See D. Lewis, "Causation," in his *Philosophical Papers,* 2 vols. (New York: Oxford University Press, 1986), vol. 2, pp. 159–172.

65. See J. Schaffer, "Causes as Probability Raisers of Processes," *Journal of Philosophy* 98 (2001): 75–92.

66. See J. Fodor, *The Modularity of Mind* (Cambridge, Mass.: MIT Press, 1983), for the notion of a mental module. See also sec. 4.5.

67. See I. Kant, *Critique of Pure Reason*, trans. P. Guyer and A. Wood (Cambridge: Cambridge University Press, 1997); and I. Kant, *Lectures on Logic,* trans. J. M. Young (Cambridge: Cambridge University Press, 1992).

68. See G. Boole, *An Investigation of the Laws of Thought* (Cambridge: Macmillan, 1854); and G. Boole, *Studies in Logic and Probability* (London: Watts, 1952).

69. W. V. O. Quine, *Philosophy of Logic*, 2d. ed. (Cambridge, Mass.: Harvard University Press, 1986), p. 81. Although Quine never wavers on the universality and indispensability of sheer logic in the holistic web of beliefs and theories, he occasionally wobbles as to sheer logic's unrevisability. Compare, e.g., his "Two Dogmas of Empiricism," sec. 6, in his *From a Logical Point of View,* 2d. ed. (New York: Harper and Row); *Word and Object*, secs. 13–14; and *Philosophy of Logic,* chap. 6.

70. See chap. 5.

71. See sec. 2.5.

72. See S. Haack, *Deviant Logic, Fuzzy Logic* (Chicago: University of Chicago Press, 1996); and G. Priest, *An Introduction to Non-Classical Logic* (Cambridge: Cambridge University Press, 2001). The fact of the plurality of logics poses a serious philosophical problem: see chap. 2.

73. See secs. 2.6 and 4.4.

74. See N. Block (ed.), *Readings in Philosophy of Psychology*, vol. 1 (Cambridge: Cambridge University Press, 1980), part 3; and J. Fodor, *RePresentations* (Cambridge, Mass.: MIT Press, 1981). Computational or machine functionalism is the thesis that mental properties are identical to computational functional properties and strongly supervenient on first-order physical properties. Fodor carefully restricts his computational functionalism to doxic intentionality and rationality.

75. See sec. 4.3.

76. See sec. 4.5.

77. See secs. 4.2–4.8.

78. See A. Bezuidenhout, "Resisting the Step toward Naturalism," *Philosophy and Phenomenological Research* 56 (1994): 743–770; and R. Hanna, "Logical Cognition: Husserl's *Prolegomena* and the Truth in Psychologism," *Philosophy and Phenomenological Research* 53 (1993): 251–275.

Chapter 2

1. S. Shapiro, "The Status of Logic," in P. Boghossian and C. Peacocke (eds.), *New Essays on the A Priori* (Oxford: Clarendon/Oxford University Press, 2000), pp. 333–338, here p. 338.

2. So as not to multiply philosophical debates beyond necessity, I am making the distinction between formal and nonformal logic seem somewhat less controversial than it actually is. See, e.g., T. Parsons, "What Is an Argument?" *Journal of Philosophy* 93 (1996): 164–185; A. N. Prior, "What Is Logic?" in his *Papers in Logic and Ethics* (Amherst, Mass.: University of Amherst Press, 1976), pp. 122–129; S. Toulmin, *The Uses of Argument* (Cambridge: Cambridge University Press, 1958); and D. Walton, "What Is Reasoning? What Is an Argument?" *Journal of Philosophy* 87 (1990): 399–419.

3. As I understand them, the strong laws of bivalence, excluded middle, and noncontradiction are as follows:

Strong bivalence: For every sentence S, S is assigned one and no more than one of the two truth-values, true and false.
Strong excluded middle: For every sentence S, 'S or not-S' is true.
Strong noncontradiction: For every sentence S, 'S and not-S' is not true.

Aristotle does not distinguish explicitly between strong bivalence and strong excluded middle (though he is hardly alone in this). But since he explicitly accepts strong excluded middle and also worries that future-contingent propositions such as "There will be a sea-battle here tomorrow" are neither true nor false, he implicitly questions strong bivalence and therefore implicitly distinguishes between them. Hence he is sometimes said to be the first deviant logician, although it turns out that his arguments for deviance are not very good ones; see Haack, Deviant Logic, Fuzzy Logic, chap. 4. In any case, it is possible to reject strong excluded middle without challenging strong bivalence (e.g., intuitionism); to reject strong bivalence without challenging strong excluded middle (e.g., dialetheism); to reject strong noncontradiction without rejecting strong excluded middle (e.g., dialetheism again); and to reject strong bivalence without rejecting strong noncontradiction (e.g., three-valued logic).

4. See Kneale and Kneale, *The Development of Logic*, p. 511: "Frege's work . . . contains all the essentials of modern logic, and it is not unfair either to his predecessors or his successors to say that 1879 [i.e., the year the *Begriffsschrift* was published] is the most important date in the history of the subject."

5. Deep, but not *unconflicted*. As I mentioned in note 3 above, Aristotle implicitly raises a question about strong bivalence in connection with future contingent sentences—the question of a third truth-value. Frege also implicitly raises a question about strong bivalence, but only in connection with sentences containing nonreferring singular terms—the question of truth-value gaps.

6. C. I. Lewis, *Survey of Symbolic Logic* (Berkeley: University of California Press, 1918), p. 1.

7. The formalist approach to logic found in the Hilbertian tradition, and especially in R. Carnap's *Logical Syntax of Language* (trans. A. Smeaton [London: Routledge and Kegan Paul, 1937]), gives signs a sort of half-living-half-dead or "zombie" status: otherwise meaningless or dead signs are supposed to be manipulated according to arbitrary rules and *thereby* acquire a meaning. But it is now generally held that formalism fails because no amount of arbitrary rule-governed syntactical manipulation in and of itself yields a determinate logical semantics. See, e.g., Carnap, "Intellectual Autobiography," in P. Schilpp (ed.), *The Philosophy of Rudolf Carnap* (La Salle, Ill.: Open Court, 1963), pp. 60–67.

8. Wittgenstein, *Tractatus Logico-Philosophicus*, prop. 3.326, p. 57.

9. See, e.g., Carnap, *Logical Syntax of Language*, sec. 2, p. 6. He correctly observes that from a purely formal point of view, a match lying on the page would function as well as the term 'and'. But it does not follow from this that a calculus constructed in matchsticks would be a symbolic logic. In a symbolic logic, the symbolism not only expresses logical forms, logical contents, logical relationships, and logical operations but also iconically displays those very forms, contents, relationships, and operations in an optimally perspicuous format. I am not saying that a symbolic logic could *not* be constructed in matchsticks, but rather only that its sign-design must be optimally nonmessy, or perspicuous. Could a *mere heap* of matchsticks lying on the page be a symbolic logic?

10. Lewis, *Survey of Symbolic Logic*, pp. 1–2.

11. Wittgenstein, *Tractatus Logico-Philosophicus*, pp. 129 and 157, translation slightly modified.

12. A. Tarski, "The Semantic Conception of Truth and the Foundations of Semantics," *Philosophy and Phenomenological Research* 4 (1943–1944): 341–375, here p. 371, n. 12, italics added.

13. Supposedly: but not in actual fact. See Hanna, *Kant and the Foundations of Analytic Philosophy*, secs. 3.1 and 3.3.

14. See P. Benacerraf, "Frege: The Last Logicist," in P. French et al. (eds.), *The Foundations of Analytic Philosophy*, Midwest Studies in Philosophy, vol. 6 (Minneapolis: University of Minnesota Press, 1981), pp. 17–35.

15. See, e.g., B. Russell, *The Problems of Philosophy* (Indianapolis: Hackett, 1995), pp. 77, 84, and 89.

16. See Wittgenstein, *Tractatus Logico-Philosophicus*, pp. 155–163, props. 6.1–6.1224.

17. See Carnap, *Logical Syntax of Language*, pp. 39, 100–102.

18. See R. Carnap, *Meaning and Necessity*, 2d. ed. (Chicago: University of Chicago Press, 1956), pp. 8–11, 222–228.

19. See W. V. O. Quine, "Truth by Convention," in his *The Ways of Paradox*, 2d. ed. (Cambridge, Mass.: Harvard University Press, 1976), pp. 77–106.

20. See W. V. O. Quine, "Two Dogmas of Empiricism," in his *From a Logical Point of View*, 2d. ed. (New York: Harper and Row, 1961), pp. 20–46; and Quine, "Carnap and Logical Truth," in his *The Ways of Paradox*, pp. 107–132.

21. See B. Russell, *Principles of Mathematics*, 2d. ed. (New York: W. W. Norton, 1996), pp. 79–80, 101–107.

22. See B. Russell, "Mathematical Logic as Based on the Theory of Types," in his *Logic and Knowledge* (New York: G. P. Putnam's Sons, 1971), pp. 59–102.

23. B. Russell, *Introduction to Mathematical Philosophy* (London: Routledge, 1993), p. 191.

24. See M. Potter, *Reason's Nearest Kin* (Oxford: Clarendon/Oxford University Press, 2000), chap. 5.

25. See B. Russell, "On Denoting," in his *Logic and Knowledge*, pp. 41–56.

26. See, e.g., A. P. Martinich (ed.), *Philosophy of Language*, 3d. ed. (New York: Oxford University Press, 1996), parts 3 and 4; and G. Ostertag (ed.), *Definite Descriptions: A Reader* (Cambridge, Mass.: MIT Press, 1998).

27. See, e.g., M. Potter, *Sets: An Introduction* (Oxford: Oxford University Press, 1990).

28. See J. Van Heijenoort, "Logical Paradoxes," in P. Edwards (ed.), *The Encyclopedia of Philosophy*, vol. 5 (New York: Macmillan, 1967), pp. 45–51.

29. See K. Gödel, "On Formally Undecidable Propositions of *Principia Mathematica* and Related Systems," in J. Van Heijenoort (ed.), *From Frege to Gödel* (Cambridge, Mass.: Harvard University Press, 1967), pp. 596–617; Boolos and Jeffrey, *Computability and Logic*, chaps. 15–16 and 28; and G. Hunter, *Metalogic* (Berkeley: University of California Press, 1971), pp. 256–257.

30. See B. Hale, *Abstract Objects* (Oxford: Blackwell, 1987); and Wright, *Frege's Conception of Numbers as Objects*.

31. See A. Tarski, "The Concept of Truth in Formalised Languages," in his *Logic, Semantics, and Metamathematics* (Oxford: Clarendon/Oxford University Press, 1956); and Tarski, "The Semantic Conception of Truth and the Foundations of Semantics."

32. See A. Tarski, "On the Concept of Logical Consequence," in his *Logic, Semantics, and Metamathematics*, pp. 409–420.

33. See C. I. Lewis, "Alternative Systems of Logic," *Monist* 42 (1932): 481–507; and Carnap, *Logical Syntax of Language*.

34. See S. Haack, *Deviant Logic* (Cambridge: Cambridge University Press, 1974), chap. 1. See also Haack, *Deviant Logic, Fuzzy Logic*; Haack, *Philosophy of Logics*, chaps. 1 and 9; and Priest, *Introduction to Non-Classical Logic*. Interestingly, Haack includes second-order logic within classical logic, which yields a slightly stronger logic than elementary logic; see *Philosophy of Logics*, p. 4.

35. Actually there are a few kinks in Haack's formulation of the distinction. The main problem is that she spells out the distinction strictly syntactically, or wholly in terms of theorems and valid inferences. On her view, extensions preserve all the theorems and valid inferences of elementary logic, whereas deviants do not. But this won't quite work. First, since tautologousness (or sentential validity) and theoremhood are not the same notion, it is in principle possible to fiddle with one while preserving the other. So in principle a logic might preserve all the theorems and valid inferences of elementary logic, while dropping some of its tautologies (or valid sentences). This is true, e.g., of Bochvar's and Smiley's three-valued logics. See D. Bochvar, "On a Three-Valued Calculus and Its Application to the Analysis of Contradictories," *Matematceskij Sbornik* 4 (1939): 287–308; and T. Smiley, "Sense without Denotation," *Analysis* 20 (1960): 125–135. Second, even preserving all the tautologies (or valid sentences), theorems, and valid inferences of elementary logic will not guarantee that the logic is non-deviant. For a system could in principle give up the law of bivalence for atomic and contingent molecular wffs without letting that infect the other three factors. B. van

Fraassen's "supervaluationist" system, e.g., assigns truth to every truth-functional compound that would be a tautology in classical logic even when it contains truth-valueless constituents; see his "Singular Terms, Truth-Value Gaps, and Free Logic," *Journal of Philosophy* 63 (1966): 481–495; and his "Presuppositions, Supervaluations, and Free Logic," in K. Lambert (ed.), *The Logical Way of Doing Things* (New Haven, Conn.: Yale University Press, 1969), pp. 67–91. And although van Fraassen does in fact give up some valid inferences of elementary logic, there seems to be nothing preventing us from accepting as valid every inference that is valid in elementary logic even when it contains truth-valueless constituents, on the grounds that the corresponding conditional would be a tautology of elementary logic. For these reasons, my version of the extended logic vs. deviant logic distinction appeals not merely to syntactic features but also to semantic features of the systems.

36. See Lewis, *Survey of Symbolic Logic*, chap. 5.

37. See C. I. Lewis and C. H. Langford, *Symbolic Logic*, 2d. ed. (New York: Dover, 1959).

38. See S. Kripke, "Semantical Analysis of Modal Logic I, Normal Propositional Calculi," *Zeitschrift für mathematische Logik und Grundlagen der Mathematik* 9 (1963): 67–96; and S. Kripke, "Semantical Considerations on Modal Logic," *Acta Philosophica Fennica* 16 (1963): 83–94.

39. See L. E. J. Brouwer, "Historical Background, Principles, and Methods of Intuitionism," *South African Journal of Science* 49 (1952): 139–146; Haack, *Deviant Logic*, chap. 5; and Priest, *Introduction to Non-Classical Logic*, chap. 6.

40. See Priest, *Introduction to Non-Classical Logic*, chaps. 9–10.

41. See Haack, *Deviant Logic*, chap. 3; and Haack, *Deviant Logic, Fuzzy Logic*, pp. 229–258. As Haack points out (*Deviant Logic*, p. 63), some nominally many-valued systems in fact preserve an underlying adherence to bivalence and excluded middle. One way of doing this is to insist that the third value is epistemological and not logical. Another way is to reinterpret a many-valued system in such a way as to have two jointly exhaustive and mutually exclusive truth-values—i.e., classical truth and classical falsity—but allow that a wff can be true or false in a variety of ways. E.g., we can divide all the logical truths/theorems of elementary logic into the analytic ones and the synthetic ones. Such systems are not strictly speaking deviants but instead only extensions: they retain classical logical syntax and classical proof theory, yet adopt a nonclassical formal semantics.

42. See Haack, *Deviant Logic*, chap. 7.

43. See G. Priest, *In Contradiction* (Dordrecht: Martinus Nijhoff, 1987); Priest, *Introduction to Non-Classical Logic*, chaps. 7–8; Priest, "The Logic of Paradox," *Journal of Philosophical Logic* 8 (1979): 219–241; and Priest, "What Is So Bad about Contradictions?" *Journal of Philosophy* 95 (1998): 410–426.

44. There are many subtly distinct versions of the diehard classicist, diehard non-classicist, and unconstrained pluralist options. Haack, e.g., opts for an explicitly instrumentalist or pragmatic version of pluralism:

> logic is a theory, a theory on a par, except for its extreme generality, with other 'scientific' theories; and . . . choice of logic, as of other theories, is to be made on the basis of an assessment of the economy, coherence, and simplicity of the overall belief set. (*Deviant Logic*, p. 26)

Also, one might put the diehard classicist and diehard nonclassicist options together under the single rubric of *monism*: the thesis that there is One True Logic. For a good critical survey of the various versions of monism and pluralism, see M. Resnik, "Ought There to Be But One Logic?" in B. J. Copeland (ed.), *Logic and Reality: Essays on the Legacy of Arthur Prior* (Oxford: Clarendon/Oxford University Press, 1996), pp. 489–517.

45. Quine, *Philosophy of Logic*, p. 82. I agree with Quine that obviousness is the epistemic criterion of logicality. But it begs a deeper explanation. See G. Sher, "Is Logic a Theory of the Obvious?" in A. C. Varzi (ed.), *The Nature of Logic* (Stanford: CSLI Publications, 1999), pp. 207–238.

46. Quine, *Philosophy of Logic*, p. 82.

47. Quine, *Philosophy of Logic*, p. 81.

48. One way of individuating logical theories is by laying down criteria for identifying the set of logical constants recognized by a given theory. In "Logical Constants" (*Mind* 108 [1999]: 503–538), K. Warmbröd suggests that there is on the one hand "a secure, core logical theory [that] recognizes only a minimal set of constants needed for deductively systematizing scientific theories," and also on the other hand there are many "extended logical theories whose objectives are to systematize pre-theoretic, modal intuitions [and which] may recognize a variety of additional constants as needed in order to formalize a given set of intuitions" (p. 503). Warmbröd's general distinction between the core logic and extended logics is somewhat similar to my distinction between the protologic and all classical or non-classical systems. Where Warmbröd and I sharply disagree, however, is about his proposal that the core logic is classical or elementary logic minus identity, on the grounds that this is all that is required by the deductive systematization of the natural sciences (pp. 517–534). In my view there are three problems with this. First, it needlessly privileges elementary logic, which as we have seen is primarily the result of a historically contingent intellectual negotiation process between logicism and Tarski. Second, since it makes identifying the logical constants of the core logic dependent on the *deductive* systematization of the sciences, it falls into the logocentric predicament (see chap. 3): that is, in attempting to explain logic it presupposes logic. And third, since it makes the core logic dependent on the natural sciences, it is clearly a form of scientific naturalism about logic (albeit a nonpsychologistic one): so it is self-refuting by the argument laid out in sec. 1.4. To the extent that Warmbröd ties the core logic to the natural sciences, and also ties his "extended

logical theories" to pretheoretic modal intuitions, his proposal is rather like Haack's instrumentalist or pragmatic version of unconstrained pluralism; see note 44 above.

49. H. Putnam offers an interesting and plausible argument for the absolute a priori status of this logical principle in "There Is at Least One *A Priori* Truth," in his *Realism and Reason: Philosophical Papers,* vol. 3 (Cambridge: Cambridge University Press, 1983), pp. 98–114.

50. See the following works (listed in chronological order): N. Chomsky, *Syntactic Structures* (The Hague: Mouton, 1957); Chomsky, "Review of B. F. Skinner's Verbal Behavior," *Language* 35 (1959): 26–58; Chomsky, *Aspects of the Theory of Syntax* (Cambridge, Mass.: MIT Press, 1965); Chomsky, *Cartesian Linguistics* (New York: Harper and Row, 1966); Chomsky, *Language and Mind,* 2d. ed. (New York: Harcourt Brace Jovanovich, 1972); Chomsky, *Reflections on Language* (New York: Pantheon, 1975); Chomsky, *Rules and Representations* (Cambridge, Mass.: MIT Press, 1980); Chomsky, *Knowledge of Language* (Westport, Conn.: Praeger, 1986); and Chomsky, *Language and Problems of Knowledge* (New York: Columbia University Press, 1988). See also V. J. Cook and M. Newson, *Chomsky's Universal Grammar*, 2d. ed. (Oxford: Blackwell: 1996).

51. Phrase-structure grammar was an essential feature of "structuralist" or Bloomfieldian linguistics in the 1930s and '40s. But Bloomfieldian linguistics was also behaviorist. Chomsky effectively attacked and thereby sliced out the behaviorism, then embedded the basic elements of phrase-structure grammar within a rationalist (i.e., generative, innatist, mentalist) theoretical framework. See J. Lyons, *Introduction to Theoretical Linguistics* (Cambridge: Cambridge University Press, 1968); and Lyons, *Noam Chomsky*, 2d. ed. (Harmondsworth, Middlesex: Penguin, 1978).

52. In Chomsky's terminology, something is *generative* if and only if it essentially involves an explicit, formal procedure for assigning well-defined structures to something else (see *Aspects of the Theory of Syntax*, pp. 8–9). Generativity is not to be confused with the distinct but closely related notion of *productivity* or *creativity*, whereby something of infinite complexity (e.g., knowledge of every possible distinct sentence of English) can be generated by something having only finite resources (e.g., our language faculty). See chap. 4.

53. The very idea of an innate UG goes back to seventeenth-century thought, and in particular to the Port Royal Grammar. See Chomsky, *Cartesian Linguistics*. The Port Royalists also had the idea that logic is essentially mental or psychological; see A. Arnaud and P. Nicole, *Logic or the Art of Thinking*. But they did not, it seems, have the idea that there is a deep connection between the innate UG and logic. Kant, I think, was the first to have this idea; see "The Jäsche Logic," pp. 527–528. In any case, Kant was the first to have an explicitly generative-productive model of human cognition; see Hanna, *Kant and the Foundations of Analytic Philosophy*, chap. 1. And Kant was also the first to apply this model to logic; see *Critique of Pure Reason*,

A50–76/B74–101. In the nineteenth century, Von Humboldt applied the Kantian idea of cognitive generativity-productivity to natural language; see Von Humboldt, *On Language*, trans. P. Heath (Cambridge: Cambridge University Press, 1988).

54. Chomsky, *Reflections on Language*, pp. 12–13.

55. Competence and performance are mutually logically independent. It is possible to possess language-competence yet not actually be using language (e.g., when asleep); and it is possible to use language quite effectively up to a certain point without possessing the relevant competence (e.g., good actors can effectively deliver their lines in a language they do not know).

56. Chomsky, *Reflections on Language*, p. 29.

57. See N. Chomsky, "Language and Nature," *Mind* 104 (1995): 1–61. Chomsky is clearly a naturalist, but whether he is also a *scientific* naturalist is somewhat unclear. Certainly some of the things he says in "Language and Nature" suggest nonreductive naturalism; but others suggest that his ultimate scientific goal is the explanatory reduction of language to human biology.

58. Chomsky, *Reflections on Language*, p. 34. See also Chomsky, *Rules and Representations*, chap. 5.

59. Not only that, but if it were true, it would make Chomsky's theory of language psychologistic. See J. Katz, *Language and Other Abstract Objects* (Totowa, N.J.: Rowman and Littlefield, 1981), chap. 5.

60. Chomsky resists an *evolutionary* explanation of language. But some of Chomsky's students and collaborators have explicitly and vigorously taken this further step. See S. Pinker, *The Language Instinct* (New York: Harper Perennial, 1994), esp. chaps. 10–11 and 13. Indeed, while it remains somewhat unclear whether Chomsky himself is a scientific or a nonreductive naturalist, it seems quite clear that Pinker is a scientific naturalist.

61. Chomsky, *Language and Problems of Knowledge*, p. 99.

62. See N. Hornstein, *Logic as Grammar* (Cambridge, Mass.: MIT Press, 1984); and R. May, *Logical Form* (Cambridge, Mass.: MIT Press, 1985).

63. See, e.g., N. Chomsky, *The Minimalist Program* (Cambridge, Mass.: MIT Press, 1995); and Chomsky, *New Horizons in the Study of Language and Mind* (Cambridge: Cambridge University Press, 2000).

64. See, e.g., M. Sainsbury, *Logical Forms*, 2d. ed. (Oxford: Blackwell, 2001), chap. 6.

65. See G. Harman, "Deep Structure as Logical Form," in G. Harman and D. Davidson (eds.), *Semantics of Natural Language* (Dordrecht: D. Reidel, 1972), pp. 25–47.

66. See J. Macnamara, *A Border Dispute: The Place of Logic in Psychology* (Cambridge, Mass.: MIT Press, 1986), chap. 2.

67. Both the notion of an internalized logic and the term 'I-logic' are mine: but they obviously mimic Chomsky's "internalized language" or "I-language." It needs to be stressed that I-logic in my sense is not the same as what Johnson-Laird calls "mental logic, " which he criticizes and sharply opposes to "mental models"; see P. Johnson-Laird, *Mental Models* (Cambridge, Mass.: Harvard University Press, 1983), chap. 2. In secs. 5.3–5.4, I critically discuss the mental-logic and mental-models approaches to the psychology of reasoning.

68. A natural logic is the logic that expresses our actual reasoning procedures. See M. Braine, "On the Relation between the Natural Logic of Reasoning and Standard Logic," *Psychological Review* 85 (1978): 1–21; and G. Lakoff, "Linguistics and Natural Logic," in Harman and Davidson (eds.), *Semantics of Natural Language,* pp. 545–665.

69. See Resnik, "Ought There to Be But One Logic?," pp. 498–502.

Chapter 3

1. L. Carroll, "What the Tortoise Said to Achilles," *Mind* 4 (1895): 278–280, here p. 280.

2. H. M. Sheffer, "Review of *Principia Mathematica*, Volume I, second edition," *Isis* 8 (1926): 226–231, here p. 228.

3. See, e.g., J. Thomson, "What Achilles Should Have Said to the Tortoise," *Ratio* 3 (1960): 95–105.

4. L. Carroll, *Lewis Carroll's Symbolic Logic*, ed. W. W. Bartley III (New York: Clarkson Potter, 1997), p. 472.

5. See, e.g., T. Smiley, "A Tale of Two Tortoises," *Mind* 104 (1995): 725–736.

6. This point was recognized by Frege in the *Begriffsschrift*. See T. Ricketts, "Frege, the *Tractatus*, and the Logocentric Predicament," *Philosophers' Annual* 8 (1985): 247–259, p. 251. I also remind the reader here that unless otherwise indicated, I am using 'sentence' to mean the same as 'linguistically expressed proposition or statement'.

7. Smiley, "A Tale of Two Tortoises," p. 731.

8. Wittgenstein, *Tractatus Logico-Philosophicus*, pp. 79, 149, 169.

9. B. Russell, "Introduction," *Tractatus Logico-Philosophicus*, pp. 21–22. It is part of the lore of early twentieth-century philosophy that Wittgenstein was infuriated by Russell's introduction despite the fact that the *Tractatus* could not have been published in English without it.

10. Russell, "Introduction," *Tractatus Logico-Philosophicus*, p. 23.

11. W. V. O. Quine, "Truth by Convention," in his *The Ways of Paradox,* p. 104.

12. W. V. O. Quine, "Carnap and Logical Truth," in his *The Ways of Paradox,* p. 115.

13. It is somewhat unclear what Quine means by 'the most elementary part of logic' in the second text I quoted, and also somewhat unclear just which logic Quine thinks is presupposed by and employed in the attempt to constitute logic by convention. I am inclined to think (i) that 'the most elementary part of logic' picks out either classical sentential logic or monadic logic, i.e., predicate logic with quantification into one-place predicates only, and (ii) that the logic Quine thinks is presupposed by and employed in the attempt to constitute logic by convention is full-strength elementary logic.

14. See D. Johnson, "Conventionalism about Logical Truth," *Philosophical Topics* 23 (1995): 189–212.

15. I say this because Prior's official reply to criticisms of "The Runabout Inference Ticket" (*Analysis* 21 [1960]: 38–39), in "Conjunction and Contonktion Revisited" (*Analysis* 24 [1964]: 191–195), focuses on what seems to me the less important and less defensible part of his argument, namely, the rejection of an inferential role semantics for the logical constants. But this may be largely a result of the dialectical context, since his critics had concentrated on semantic issues.

16. This is slightly inaccurate. In "The Runabout Inference Ticket," p. 38, Prior officially attributes the theory he is attacking to Karl Popper and William Kneale. But as Belnap indicates in passing ("Tonk, Plonk, and Plink," *Analysis* 22 [1961]: 130–134, p. 130), Gentzen's essay is the locus classicus: "Investigations into Logical Deduction," in G. Gentzen, *The Collected Papers of Gerhard Gentzen,* trans. M. Szabo (Amsterdam: North Holland, 1969), pp. 68–131.

17. Prior, "The Runaround Inference Ticket," p. 37.

18. See Belnap, "Tonk, Plonk, and Plink." Belnap retains the inferential role thesis but avoids Prior's reductio by weakening conventionalism to the thesis that the inferential role of a logical constant is strictly determined by conventions *together with* the further constraint that every introduction of a new constant to a logical system be at most a conservative extension of the preexisting system. C. Stevenson, in "Roundabout the Runaround Inference Ticket" (*Analysis* 21 [1961]: 124–129), also avoids Prior's reductio by weakening conventionalism, but he does so by dropping the inferential role thesis and adding a supplementary noninferential specification of logical meaning via truth tables. It is also worth noting that it is possible to hold the inferential role thesis while denying that inferential roles are in any way determined by conventions. See, e.g., C. Peacocke, "Understanding Logical Constants: A Realist Account," *Proceedings of the British Academy* 73 (1987): 153–200.

19. This is not to say that there are no problems intrinsic to the inferential role thesis. See sec. 6.3.

20. A much-elaborated version of the same argument is presented by M. Dummett in *The Logical Basis of Metaphysics* (Cambridge, Mass.: Harvard University Press, 1991), chaps. 8–15. See also S. Haack, "Dummett's Justification of Deduction," *Mind* 95 (1982): 216–239.

21. M. Dummett, "The Justification of Deduction [1973]," in his *Truth and Other Enigmas* (London: Duckworth, 1978), p. 292; see also N. Goodman, "The New Riddle of Induction," in his *Fact, Fiction, and Forecast,* 4th ed. (Cambridge, Mass.: Harvard University Press, 1983), pp. 62–64. One might wonder why Dummett does not attribute the classical circularity objection to Carroll. The answer, I think, is that it is strategic for Dummett to pick a defender of the circularity objection who also explicitly defends a holistic semantics and antirealism.

22. Dummett takes it to be a necessary condition of such a semantics that the language as a whole be a conservative extension of some definite fragment of it; see "The Justification of Deduction," p. 315. For formal languages, this has the effect of ruling out the metalogical introduction of "tonk"-like logical constants by means of inference rules governing "contonktion"-like logical operations. See also Belnap, "Tonk, Plonk, and Plink."

23. Here I am rationally reconstructing Haack's argument. She officially uses a dilemma between "inductive" and "deductive," but I think that her aims are much better served by making the dilemma exhaustive from the start, and using induction as an *example* of nondeductive justification. Another distinct kind of nondeductive justification is holistic reasoning.

24. In "Dummett's Justification of Deduction," *Mind* 95 (1982): 216–239, here pp. 222–225, Haack argues that soundness proofs are necessary but not sufficient for the justification of deduction. This is because although soundness proofs show that a certain object language is deductively cogent, they do not discriminate between competing sound logics, one of which may be classical while the other is extended or even deviant. So soundness proofs will not do the philosophical job that Dummett expects of them, and they furthermore will require some distinct supplementary ground if they are to be adequately justificatory. Dummett tends to frame all discussion of classical vs. nonclassical or alternative logics in semantic terms. So I think that my vicious regress objection to Dummett's justification strategy is very much in the same spirit as Haack's objection.

25. See L. Carroll, *Symbolic Logic* (Oxford: Oxford University Press, 1986), introduction.

26. L. Wittgenstein, *Philosophical Investigations*, p. 88e. See also Ricketts, "Frege, the *Tractatus,* and the Logocentric Predicament."

27. See Resnik, "Ought There to Be But One Logic?," pp. 510–515; and C. Wright, "Inventing Logical Necessity," in J. Butterfield (ed.), *Language, Mind, and Logic* (Cambridge: Cambridge University Press, 1986), pp. 187–209.

28. See W. V. O. Quine, *Methods of Logic*, 4th ed. (Cambridge, Mass.: Harvard University Press, 1982), introduction; Quine, "Two Dogmas of Empiricism"; and Quine, *Philosophy of Logic*, chap. 7.

29. Goodman, "The New Riddle of Induction," pp. 63–64.

30. See Putnam, *Realism and Reason: Philosophical Papers*, vol. 3, chaps. 5–7; and Putnam, "Re-Thinking Mathematical Necessity," in his *Words and Life* (Cambridge, Mass.: Harvard University Press, 1994), pp. 245–263.

31. H. Putnam, "Philosophy of Logic," in his *Mathematics, Matter, and Method: Philosophical Papers,* vol. 1, pp. 323–357.

32. Haack, "The Justification of Deduction," p. 191.

33. In "The Universality of Logic" (*Mind* 110 [2001]: 335–367), S. Evnine persuasively argues that there are certain logical abilities that any rational creature must have. If he is right, then since it is obvious that rational creatures are responsible for all rational discourse and rational inquiry, it follows that logical principles and logical concepts are built into the very structure of all rational discourse and rational inquiry. Evnine's thesis is clearly quite similar to the logic faculty thesis. So I can help myself to his argument as an independent source of support for my view. There are two differences between my view and Evnine's, however. First, he accepts the inferential role thesis without qualification (pp. 339–340), whereas I am prepared to accept it only under some significant constraints (see sec. 6.3); and second, the logical abilities he ascribes to any rational creature (i.e., possession of concepts of conjunction, conditionality, exclusive disjunction, negation, quantification, identity, and possibility, pp. 359–360) are closely tied to elementary logic and its conservative extensions, which suggests that he is committed to some version of diehard nonclassicism, which in turn is questionable (see sec. 2.5).

34. Haack makes the very good point that holism is consistent with the thesis that some logical sentences are necessary in a metaphysical sense; see "Dummett's Justification of Deduction," p. 206. E.g., neo-Hegelians are frequently friendly to the idea of logical necessity. But holistic logical and metaphysical necessities cannot be guaranteed to be *really* true, that is, true by virtue of facts existing *outside* the holistic network of concepts and beliefs.

35. See N. Rescher, *The Coherence Theory of Truth* (Oxford: Oxford University Press, 1973), esp. chaps. 2, 6–8, and 13.

36. This objection (i.e., that logical semantics is inherently realistic, while logical holism is antirealistic) holds against all forms of logical holism, including justificatory

logical holism. See, e.g., J. Bickenbach, "Justifying Deduction," *Dialogue* 4 (1979): 500–516; and P. Boghossian, "Knowledge of Logic," in Boghossian and Peacocke, *New Essays on the A Priori*, pp. 229–254.

37. A third type of epistemic circularity is justificational holism. In "Knowledge of Logic," Boghossian introduces a justificationally holistic sort of epistemic circularity that he calls "rule-circularity": in order to justify the use of a rule R, you use the rule R. He argues that rule-circular justifications of logical rules are legitimate, on the assumption that an inferential-role semantics of the logical constants is correct. Since I have serious doubts about inferential-role approaches to the semantics of the logical constants (see sec. 6.3), I have corresponding doubts about the cogency of a rule-circular justification of logical rules. So I will leave that option aside here.

38. See P. F. Strawson, *Introduction to Logical Theory* (London: Methuen, 1952), pp. 174–179.

39. Presuppositional arguments are both structurally and historically related to *transcendental* arguments, which of course derive originally from philosophical argument strategies employed by Kant in the *Critique of Pure Reason*. There is an enormous literature on transcendental arguments, in part because of P. F. Strawson's qualified endorsement of transcendental arguments against skepticism in his highly influential books *Individuals* (London: Methuen, 1959) and *The Bounds of Sense* (London: Methuen, 1966). For recent discussion, see R. Stern (ed.), *Transcendental Arguments: Problems and Prospects* (Oxford: Oxford University Press, 1999); and Stern, *Transcendental Arguments and Skepticism* (Oxford: Oxford University Press, 2000). Roughly speaking, transcendental arguments in Strawson's sense are antiskeptical presuppositional arguments from some epistemic or cognitive premises shared by the skeptic, to some metaphysical, epistemic, or cognitive conclusions that refute the skeptic. But a fundamental worry about transcendental arguments is that their soundness requires semantic verificationism or some form of idealism; see, e.g., B. Stroud, "Transcendental Arguments," *Journal of Philosophy* 65 (1968): 241–256. Otherwise how could the argument ever generate a metaphysical conclusion from epistemic or cognitive premises? Now, it is true that my presuppositional argument for the logic faculty thesis involves a cognitive *conclusion*. But its *premise* is factual (i.e., it asserts the existence of a plurality of nonclassical logics), not epistemic or cognitive; hence my presuppositional argument for the logic faculty thesis is not a transcendental argument. Moreover, the presuppositional argument I have offered is *explanatory*, not antiskeptical, so it does not require verificationism or idealism in order to guarantee its soundness.

Chapter 4

1. L. Carroll, *Alice's Adventures in Wonderland* (New York: Dial, 1988), p. 102.

2. See R. Descartes, *Discourse on the Method*, pp. 139–141 (AT 56–59); and Chomsky, *Cartesian Linguistics*.

3. Of course, 'thought' and its cognates are often used very broadly and loosely to mean, roughly, *cognition* or *mental activity* or *conscious mentality*. But in this chapter and generally throughout the book I am using them more narrowly and precisely (in the manner of, e.g., Kant in the second edition of the first *Critique*) to mean *cognition that is specifically rational in character*. On this usage nonrational animals can be cognizers, but not thinkers.

4. See Hanna, *Kant and the Foundations of Analytic Philosophy*, chaps. 1–2.

5. See the references in note 15 of the introduction.

6. This is to allow for the possibility of occasional, unfortunate, and illusory conditions of cognitive triggering, e.g., the cow on a dark night that is mistakenly represented as a horse.

7. Chomsky originally said that 'I-language' stands for 'internalized language' (*Knowledge of Language*, p. 22). In more recent writings, however, he says that the 'I' in 'I-language' more broadly stands for *internal, individual, and intensional* ("Explaining Language Use," in his *New Horizons in the Study of Language and Mind* [Cambridge: Cambridge University Press, 2000], p. 26).

8. See the following works (listed in chronological order): J. Fodor, *The Language of Thought* (Cambridge, Mass.: Harvard University Press, 1975); Fodor, *RePresentations* (1981); Fodor, *The Modularity of the Mind* (Cambridge, Mass.: MIT Press, 1983); Fodor, *Psychosemantics* (Cambridge, Mass.: MIT Press, 1987); Fodor, *A Theory of Content and Other Essays* (Cambridge, Mass.: MIT Press, 1990); Fodor, *The Elm and the Expert* (Cambridge, Mass.: MIT Press, 1994); Fodor, *Concepts* (Oxford: Clarendon/Oxford University Press, 1998); and Fodor, *The Mind Doesn't Work That Way* (2000).

9. Intentionality requires attentive focusing, and attentive focusing is a spontaneous function of consciousness, so intentionality requires consciousness. See R. Hanna and E. Thompson, "Neurophenomenology and the Spontaneity of Consciousness," in E. Thompson (ed.), *The Problem of Consciousness* (Calgary: University of Alberta Press, 2005), pp. 133–162. For a somewhat similar view, see J. Searle, *The Rediscovery of the Mind* (Cambridge, Mass.: MIT Press, 1992), p. 132.

10. See Frege, "Logic [1897]"; Frege, "On Sense and Meaning," in his *Collected Papers on Mathematics, Logic, and Philosophy*, trans. M. Black et al. (Oxford: Basil Blackwell, 1984), pp. 157–177; and Frege, "Thoughts."

11. Precisely and correctly characterizing the "very close connection" between the contentfulness of intentionality and the opacity of language is not by any means easy, however; see, e.g., Fodor, *A Theory of Content*, chap. 6.

12. Just as there is a close connection between the contentfulness of intentionality and opacity in language, so too there is a close connection between the *directedness* of

intentionality and *transparency* in language. See F. Recanati, *Direct Reference* (Oxford: Blackwell, 1993).

13. J. Austen, *Northanger Abbey* (Harmondsworth, Middlesex: Penguin, 1980), p. 157.

14. See P. M. Churchland, "Eliminative Materialism and the Propositional Attitudes," *Journal of Philosophy* 78 (1981): 67–90; and Fodor, "Special Sciences, or the Disunity of Science as a Working Hypothesis."

15. See Boolos and Jeffrey, *Computability and Logic*, chaps. 1–8; H. Putnam, "Minds and Machines," in his *Mind, Language, and Reality: Philosophical Papers*, vol. 2 (Cambridge: Cambridge University Press, 1975), pp. 362–385; and A. Turing, "Computing Machinery and Intelligence," *Mind* 59 (1950): 433–460. Church's thesis says that all effective procedures or algorithms are recursive functions, and Turing has shown that every recursive function can be implemented on some Turing machine or another, so the Church–Turing thesis says that every effective procedure or algorithm can be implemented on some Turing machine or another. Turing has also shown that every Turing machine can be adequately formally modeled by the operations of a universal Turing machine or digital computer. So every effective procedure or algorithm can be adequately formally modeled by the operations of a universal Turing machine or digital computer.

16. See N. Block, "What Is Functionalism?" in N. Block (ed.), *Readings in Philosophy of Psychology*, vol. 1 (Cambridge, Mass.: Harvard University Press, 1980), pp. 171–184; and H. Putnam, "The Nature of Mental States," in his *Mind, Language, and Reality: Philosophical Papers*, vol. 2, pp. 429–440.

17. See also V. J. Cook and M. Newson, *Chomsky's Universal Grammar*, 2d. ed. (Oxford: Blackwell, 1996), pp. 81–85.

18. How is significant underdetermination established? One indicator is the determinate lack, in the external experiential stimulus, of structures present in the relevant cognitive output or manifest cognitive trait, e.g., when a child constructs a new grammatically correct sentence she has never heard before. Another indicator is the fact that the same external experiential stimulus fails to yield the same cognitive output or manifest cognitive trait when presented to other biologically and behaviorally similar animals, e.g., apes do not acquire mastery of a natural language even when presented with the same parental training. And a third indicator is the fact that the relevant cognitive or manifest cognitive trait is yielded by *K*-animals even when the external experiential stimulus varies widely in character, e.g., all medically normal and minimally well-treated human children acquire a natural language.

19. See H. Putnam, "The 'Innateness Hypothesis' and Explanatory Models in Linguistics," in N. Block (ed.), *Readings in Philosophy of Psychology*, vol. 2 (Cambridge, Mass.: Harvard University Press, 1980), p. 298.

20. Chomsky makes good critical use of this fact in "Quine's Empirical Assumptions," in D. Davidson and J. Hintikka (eds.), *Words and Objections* (Dordrecht: D. Reidel, 1969), pp. 53–68.

21. Thus the nativist vs. empiricist debate has often been mischaracterized as a debate between those who defend innatism and those who reject innatism. "Nativists" are innatists who ascribe the comparatively richest, maximal, or most fully articulated structures consistent with the empirical evidence and other theoretical constraints, to the innate factor in cognition, including innate concepts and innate beliefs, while "empiricists" are innatists who ascribe the comparatively thinnest, minimal, or least fully articulated structures consistent with the empirical evidence (and other theoretical constraints such as naturalism) to the innate factor. Indeed, the empiricists implicitly conceded very early on in the debate that they are as committed to innateness as the nativists, and ever since have almost exclusively devoted themselves to designing increasingly sophisticated models of thin or minimal innate cognitive architecture, in particular, most recently, models based on "connectionist" information-processing architecture. See, e.g., R. Cummins and D. Cummins (eds.), *Minds, Brains, and Computers* (Oxford: Blackwell, 2000), part 2; and J. L. Elman, E. A. Bates, M. H. Johnson, et al. (eds.), *Rethinking Innateness* (Cambridge, Mass.: MIT Press, 1996). On the connectionist approach to processing, information synthesis does not involve the serial transformation or construction of isolable representations by a central processing unit or "faculty of reason," but instead operates via the simultaneous lateral distribution of information over a network of mutually independent subrepresentational processing operations. More generally, connectionism suggests that cognition is essentially subrational, nonrepresentational, nongenerative (hence nonconstructive), nonmodular, and generally not language-like. See also note 38 below for some criticism of connectionism.

22. I will consider a version of radical skepticism about rational cognition in sec. 7.5.

23. These formal procedures might or might not be *rules*, if by 'rules' we mean formal procedures that necessarily have a *propositional* expression.

24. See R. Jackendoff, *Languages of the Mind* (Cambridge, Mass.: MIT Press), chap. 6; see also Kant, *Critique of Pure Reason*, A19–49/B33–73.

25. Kant, *Critique of Pure Reason*, B167.

26. See S. Mason, *A History of the Sciences*, 2d. ed. (New York: Collier, 1962), pp. 363–369.

27. This raises a slightly tricky point about the history of cognitivism and the correct formulation of the innateness thesis. Following Kant, I strongly emphasize innate mental powers (faculty innateness), and eschew innate concepts or innate beliefs (content innateness). See also note 36 below. Chomsky, however, inclines sometimes toward Cartesian rationalism and thus toward content innateness; see,

e.g., *Cartesian Linguistics*. Fodor has wobbled on this crucial point too; see, e.g., "Doing Without What's Within: Fiona Cowie's Critique of Nativism," *Mind* 110 (2001): 99–148, esp. pp. 146–147.

28. See H. Putnam, *Representation and Reality* (Cambridge, Mass.: MIT Press, 1988), pp. 15–18.

29. See Fodor, *Concepts,* chap. 6.

30. An example of experientially acquired modularity is the capacity for rectilinear depth cue recognition, as revealed, e.g., by the Ponzo and Müller–Lyer illusions, which are, it seems, to some extent culturally specific. See M. H. Segall, D. T. Campbell, and M. J. Herskovits, *The Influence of Culture on Visual Perception* (Indianapolis: Bobbs-Merrill, 1966). In her recent critique of innatism, *What's Within: Nativism Reconsidered* (Oxford: Oxford University Press, 1998), Cowie correctly notes that modularity does not entail innateness. But it seems to me that this is not a decisive point against nativism. The bare possibility that a given cognitive module is experientially acquired does not trump a well-specified and well-supported poverty-of-the-stimulus argument for its being innate. The paradigm of such an argument is Chomsky's argument for an innate language faculty. See S. Laurence and E. Margolis, "Review of Fiona Cowie, *What's Within: Nativism Reconsidered,*" *European Journal of Philosophy* 9 (2001): 242–247.

31. See D. Marr, *Vision* (San Francisco: Freeman, 1982).

32. See Pinker, *How the Mind Works*, pp. 272–274.

33. See Pinker, *The Language Instinct.*

34. See S. Dehaene, *The Number Sense* (Oxford: Oxford University Press, 1997); and K. Wynn, "Addition and Subtraction in Human Infants," *Nature* 358 (1992): 749–750.

35. This is a crucial point, because it distinguishes the standard cognitivist model of the mind from the "New Look" cognitivist thesis (see, e.g., J. Bruner, "On Perceptual Readiness," *Psychological Review* 64 [1957]: 123–152) to the effect that central or higher-level cognitive capacities or processes, including theories, beliefs, desires, and volitions, fully penetrate our peripheral or lower-level capacities or processes, including all forms of sense perception. Given encapsulation, it seems that peripheral processes are fully closed to penetration by central processes. In sec. 4.5 below, however, I will argue that peripheral processes are *narrowly open but not fully open* to penetration by central processes, via an innate, central, nonencapsulated module for logical cognition.

36. For the Kantian cognitivist, modularity is precisely what constrains innateness to mental powers, and rules out the innate ideas or beliefs of the Cartesian rationalist. In effect, the Cartesian-rationalistic postulation of innate concepts or beliefs confuses central (or higher-level) processing with peripheral (or lower-level) processing.

37. For example, in *Modularity of the Mind*, pp. 101–126, Fodor regards scientific theory construction in particular and scientific thinking more generally as essentially nonmodular. But by contrast, Chomsky postulates the existence of a "science-forming faculty" or SFF; see *New Horizons in the Study of Language and Mind*, pp. 82–83.

38. In contemporary cognitive science, the only major challenger to the mental language thesis is connectionism. But if connectionism is true, does it follow that the mental language thesis is false? No. A computer organized along connectionist lines can still compute only what is Turing-computable (i.e., recursive functions). And everything that is Turing-computable can be translated into the mental language. Thus anything that can be done by a connectionist computational system can also be done by using the mental language, although it may be fairly long-winded, slow, and complicated. Furthermore, it seems to me entirely possible that connectionism and the mental language thesis respectively characterize conceptually or explanatorily distinct "levels" of cognition: on this picture, connectionism correctly characterizes the *nonconscious neurobiological level*, and the mental language thesis correctly characterizes the *conscious intentional level*. If the latter generally requires the former, but multiple embodiability still holds so that the latter is not identical with the former, and there is no explanatory reduction of the conscious intentional level to the nonconscious neurobiological level, then the two levels are perfectly consistent with one another.

39. This is not to say, however, that Chomsky regards his I-language thesis as precisely identical to Fodor's conception of the language of thought; see *New Horizons in the Study of Language and Mind*, p. 185.

40. See A. Damasio, *The Feeling of What Happens: Body and Emotion in the Making of Consciousness* (San Diego: Harvest, 1999), pp. 109–112.

41. This view is not shared by Chomsky, who thinks that I-languages can vary across human individuals and also that I-languages can be at least syntactically identical with natural languages. In sec. 4.3 I will argue that Chomsky is right and Fodor wrong on these points. I will also argue that the presence of cognition in nonhuman animals does not entail the existence of a universal unique mental language. On the contrary, given the fact of nonhuman animal cognition together with the mental language hypothesis, it seems there must be *lots* of distinct LOTs.

42. See, e.g., F. Dretske, *Naturalizing the Mind* (Cambridge, Mass.: MIT Press, 1995), pp. 35–37.

43. See N. Block, "Troubles with Functionalism," in N. Block (ed.), *Readings in Philosophy of Psychology*, vol. 1, pp. 268–305. Block's argument in turn is an adaptation of Searle's Chinese Room argument; see sec. 4.6.

44. Previous objections to representational functionalism have appealed to externalism, meaning holism, and the normativity of meaning. See Dretske, *Naturalizing the Mind*; and Putnam, *Representation and Reality*. My argument here does not presup-

pose the cogency of any of these objections, but instead relies on the somewhat different thought that representational properties are, paradigmatically, properties of the phenomenally conscious mental states of *animals considered as basic cognitive individuals and basic sources of cognitive agency*. Of course, to the extent that the other objections to representational functionalism *are* cogent, my argument is consistent with them.

45. I have said that the mental language is not identical with any natural language. But since syntactic identity alone does not suffice for the identity of two languages (consider English and Schmenglish, where the latter has the same syntax as English but some of its words have different meanings), the mental language can still be *syntactically* identical with some natural language provided that they are not also semantically identical.

46. That is, in principle there could be aliens who are also rational animals (e.g., ET). But at the same time I think that as a matter of fact there are no actual nonhuman species capable of thought, since I also hold (i) that thought entails natural language competence, and (ii) that there are no actual nonhumans who are competent speakers of a natural language. For interesting defenses of the denial of (i) however, see, e.g., J. L. Bermúdez, *Thought without Words* (Oxford: Oxford University Press, 2003); and D. R. Griffin, *Animal Thinking* (Cambridge, Mass.: Harvard University Press, 1984).

47. See note 37 above.

48. See Chalmers, *The Conscious Mind*, chap. 3.

49. See P. Wason, "Reasoning," in B. M. Foss (ed.), *New Horizons in Psychology* (Harmondsworth, Middlesex: Penguin, 1966), pp. 135–151; P. Wason, "Reasoning about a Rule," *Quarterly Journal of Experimental Psychology* 20 (1968): 273–281; and P. Wason and P. Johnson-Laird, *The Psychology of Reasoning* (Cambridge, Mass.: Harvard University Press, 1972).

50. See R. L. Gregory, "Perceptions as Hypotheses," *Philosophical Transactions of the Royal Society of London, Series B* 290 (1980): 181–197; and I. Rock, *The Logic of Perception* (Cambridge, Mass.: MIT Press, 1983).

51. See R. Griggs and J. Cox, "The Elusive Thematic Materials Effect in Wason's Selection Task," *British Journal of Psychology* 73 (1982): 407–420.

52. See J. Searle, "Minds, Brains, and Programs," *Behavioral and Brain Sciences* 3 (1980): 417–424, pp. 417–418; and Searle, *The Rediscovery of the Mind*, pp. 201–202.

53. See Searle, "Minds, Brains, and Programs," pp. 419–421; Searle, *Minds, Brains, and Science* (Cambridge, Mass.: Harvard University Press, 1984), pp. 31–38.

54. For what it's worth, however, my own view is that Searle is correct about the entailment from intentionality to consciousness. See note 9 above, and also T. Horgan

and J. Tienson, "The Intentionality of Phenomenology and the Phenomenology of Intentionality," in D. Chalmers (ed.), *Philosophy of Mind: Classical and Modern Readings* (New York: Oxford University Press, 2002), pp. 520–533.

55. See Searle, *Minds, Brains, and Science*, chap. 1; Searle, *Rediscovery of the Mind*, chaps. 4–5. Searle does not explicitly distinguish between logical and natural strong supervenience, so in ascribing a logical strong supervenience thesis to him I am interpreting somewhat. If I am right, then Searle is a nonreductive materialist about consciousness (hence not a dualist) and an antifunctionalist about consciousness and intentionality alike.

56. Does this commit me to dualism? No. Dualism is the thesis that mental properties are real or ineliminable and modally independent of (i.e., not necessarily connected with) fundamental physical properties. But it is possible to hold that mental properties are (i) real and ineliminable, (ii) not logically strongly supervenient on fundamental physical properties, and (iii) also *necessarily interdependent* with fundamental physical properties. The basic idea is that the natural world at some basic ontological level (say, at or below the quantum level), or in some basic ontological domain (say, in the domain of sentient living organisms or animals), is "dual aspect" in that it is jointly constituted by intrinsic mental and intrinsic physical properties alike. See Chalmers, *The Conscious Mind*, chap. 8; R. Hanna and E. Thompson, "The Mind–Body–Body Problem," *Theoria et Historia Scientiarum* 7 (2003): 24–44; Hanna and Thompson, "Neurophenomenology and the Spontaneity of Consciousness"; and T. Nagel, "The Psychophysical Nexus," in Boghossian and Peacocke (eds.), *New Essays on the A Priori*, pp. 433–471.

57. Obviously here I am again adapting Searle's Chinese Room argument.

58. See note 2 above and sec. 2.7. See also D. Davidson, "Thought and Talk," in his *Inquiries into Truth and Interpretation* (Oxford: Clarendon/Oxford University Press, 1984), pp. 155–170; and Davidson, "Rational Animals," in E. Lepore and B. McLaughlin (eds.), *Actions and Events: Perspectives on the Philosophy of Donald Davidson* (Oxford: Blackwell, 1985), pp. 473–480.

59. See, e.g., J. Fodor, "Three Cheers for Propositional Attitudes," in his *RePresentations*, pp. 120–121.

60. See P. Carruthers, *Language, Thought, and Consciousness* (Cambridge: Cambridge University Press, 1998), pp. 20–22; R. Kirk, "Rationality without Language," *Mind* 76 (1967): 369–386; and R. Stalnaker, *Inquiry* (Cambridge, Mass.: MIT Press, 1984).

61. See N. Smith and I. M. Tsimpli, *The Mind of a Savant: Language Learning and Modularity* (Oxford: Blackwell, 1995); and J. Yamada, *Laura: A Case for the Modularity of Language* (Cambridge, Mass.: MIT Press, 1990).

62. See S. Baron-Cohen, *Mindblindness: An Essay on Autism and Theory of Mind* (Cambridge, Mass.: MIT Press, 1995).

63. Some autists manifest a high degree of intelligence. So I do not mean to imply that autism in and of itself implies low intelligence or even nonrationality: all I am saying is that some autists are nonrational and also linguistically competent.

Chapter 5

1. Boole, "The Claims of Science," in his *Studies in Logic and Probability* (London: Watts, 1952), pp. 187–210, here p. 208.

2. S. Stich, "Could Man Be an Irrational Animal?" *Synthese* 64 (1985): 115–135, here p. 115.

3. In contemporary scientific and philosophical psychology, the rubric of 'human reasoning' covers not only deductive cognition but also inductive judgments and hypothesis formation, probability judgments, and practical reasoning (also known as "decision making"). A great deal of empirical, formal, and philosophical work has been done in each of these areas. For the special purposes of this chapter, however, I will treat 'human reasoning' and 'human deductive reasoning' as synonymous expressions. In any case, it is tacitly or explicitly accepted by researchers in these areas that any substantive results established for the case of human deductive reasoning will constrain any theory of human reasoning more broadly construed.

4. J. Piaget, *Logic and Psychology* (New York: Basic Books, 1957), pp. 1–2.

5. M. Dummett, *Origins of Analytical Philosophy* (Cambridge, Mass.: Harvard University Press, 1993), pp. 22–27.

6. There are two widespread but importantly distinct uses of 'competence' in connection with the psychology of human reasoning, and also with the corresponding rationality debate in philosophy and cognitive science. According to the first and broader use, 'competence' means simply the ability of a subject to complete some sort of cognitive task successfully, as opposed to actual performance, which may of course (if conditions are unfavorable) deviate from the subject's ability. But according to the second and narrower use, which I have already mentioned in sec. 2.7, 'competence' means the innate ability of the ideal speaker-hearer to cognize a natural language in accordance with universal normative principles. The latter use, in turn, is closely associated with Chomsky's psycholinguistics and the standard cognitivist model of the mind. In this chapter I will use 'competence' in both senses; but it should be evident from the context which sense is at issue. But above all it is worth emphasizing that the first sense of 'competence' does *not* entail the second sense.

7. See D. Kahneman, P. Slovic, and A. Tversky (eds.), *Judgment under Uncertainty: Heuristics and Biases* (Cambridge: Cambridge University Press, 1982); and A. Tversky and D. Kahneman, "Extensional versus Intuitive Reasoning: The Conjunctive Fallacy in Probability Judgment," *Psychology Review* 90 (1983): 293–315. See also L. J. Cohen, "Are People Programmed to Commit Fallacies?

Further Thoughts about the Interpretation of Experimental Data on Probability Judgment," *Journal of the Theory of Social Behavior* (1982): 251–274.

8. Rationality, we will remember from the Introduction, is of three basic kinds (i.e., instrumental, holistic, and principled), and also has three basic dimensions running crosswise through the modal kind (i.e., logical, epistemic, and practical). This three-dimensional model can be extended to the other two types of rationality as well: instrumental and holistic rationality each have logical (inference-oriented), epistemic (belief-oriented), and practical (decision-oriented) dimensions. Corresponding to the three dimensions of rationality, moreover, are three dimensions of irrationality. Logical irrationality is the regular, systematic, and even flagrant production of errors, inconsistency, and invalidity in inference. Epistemic irrationality is the regular, systematic, and even flagrant formation of beliefs without sufficient evidence, warrant, or justification. And practical irrationality is the regular, systematic, and even flagrant pursuit of goals that do not reflect an agent's actual desires, needs, and values. It is important to note that logical irrationality does not entail irrationality in the epistemic or practical senses. Indeed, most of those (although Dostoevsky, Freud, D. H. Lawrence, and Nietzsche would be examples to the contrary) who argue that human beings are logically irrational also assume that humans are rational in the practical sense. See, e.g., J. Evans, "Bias and Rationality," in K. Manktelow and D. Over (eds.), *Rationality: Psychological and Philosophical Perspectives* (London: Routledge, 1993), pp. 6–30, here pp. 15 and 24; and S. Stich, *The Fragmentation of Reason* (Cambridge, Mass.: MIT Press, 1990).

9. L. J. Cohen, "Can Human Irrationality Be Experimentally Demonstrated?" *Behavioral and Brain Sciences* 4 (1981): 317–370.

10. See, e.g., Cohen, "Are People Programmed to Commit Fallacies?"; L. J. Cohen, "Continuing Commentary," *Behavioral and Brain Sciences* 6 (1983): 487–533; L. J. Cohen, "Reply to Stein," *Synthese* 99 (1994): 173–176; Manktelow and Over (eds.), *Rationality*; E. Stein, *Without Good Reason: The Rationality Debate in Philosophy and Cognitive Science* (Oxford: Clarendon/Oxford University Press, 1996); Stich, "Could Man Be an Irrational Animal?"; and Stich, *The Fragmentation of Reason*.

11. W. James, *Principles of Psychology*, vol. 2 (New York: Dover, 1950), pp. 346, 360, 368.

12. James, *Principles of Psychology*, vol. 2, p. 329.

13. James, *Principles of Psychology*, vol. 2, p. 329.

14. James, *Principles of Psychology*, vol. 2, p. 329.

15. James, *Principles of Psychology*, vol. 2, p. 330. Italics in the original.

16. James, *Principles of Psychology*, vol. 2, p. 341.

17. James, *Principles of Psychology*, vol. 2, p. 363.

18. James, *Principles of Psychology*, vol. 2, pp. 340–141.

19. James, *Principles of Psychology*, vol. 2, p. 342.

20. James, *Principles of Psychology*, vol. 2, pp. 343–345.

21. James, *Principles of Psychology*, vol. 2, p. 329.

22. Presumably this partially explains the fact that post-Jamesian psychologists of human reasoning have devoted a surprisingly large amount of time and energy to the study of syllogistic reasoning; see J. Evans and S. Newstead, "Creating a Psychology of Reasoning," in S. Newstead and J. Evans (eds.), *Perspectives on Thinking and Reasoning* (Hillsdale, N.J.: Lawrence Erlbaum), pp. 2–16, here pp. 2–3.

23. See B. Inhelder and J. Piaget, *The Growth of Logical Thinking from Childhood to Adolescence*, trans. A. Parsons and S. Milgram (New York: Basic Books, 1958).

24. J. Piaget, *Logic and Psychology* (New York: Basic Books, 1957), p. 18.

25. M. C. Wilkins, "The Effect of Changed Material on Ability to Do Formal Syllogistic Reasoning," *Archives of Psychology* 16 (1928): 1–83.

26. R. Woodworth and S. Sells, "An Atmosphere Effect in Formal Syllogistic Reasoning," *Journal of Experimental Psychology* 18 (1935): 451–460; and S. Sells, "The Atmosphere Effect: An Experimental Study of Reasoning," *Archives of Psychology* 29 (1936): 1–72.

27. M. Henle, "On the Relation between Logic and Thinking," *Psychological Review* 69 (1962): 366–378.

28. It bears repeating that this is irrationality only in the logical sense, which many irrationalists take to be consistent with rationality in the practical sense.

29. See chap. 4, note 49, for references.

30. L. J. Rips, "Deduction," in R. Sternberg and E. Smith (eds.), *Psychology of Human Thought* (Cambridge: Cambridge University Press, 1988), pp. 142–143.

31. R. Griggs and J. Cox, "The Elusive Thematic Materials Effect in Wason's Selection Task," *British Journal of Psychology* 73 (1982): 407–420.

32. K. Manktelow and J. Evans, "Facilitation of Reasoning by Realism: Effect or Non-Effect?" *British Journal of Psychology* 70 (1979): 477–488.

33. This phrase is a semitechnical term; see sec. 5.5.

34. Cohen's approach was anticipated, in slightly different ways, by Goodman, Quine, Davidson, and Dennett. See Goodman, "The New Riddle of Induction," pp. 63–64; Quine, *Word and Object*, pp. 58–59; Davidson, "On the Very Idea of a

Conceptual Scheme," in his *Inquiries into Truth and Interpretation* (Oxford: Clarendon/Oxford University Press, 1984), pp. 183–198; Davidson, "Radical Interpretation," in the same volume, pp. 125–139; D. Dennett, "Intentional Systems," *Journal of Philosophy* 68 (1971): 87–106; Dennett, "Making Sense of Ourselves," in his *The Intentional Stance* (Cambridge, Mass.: MIT Press, 1987), pp. 83–101; and Dennett, "Three Kinds of Intentional Psychology," in the same volume, pp. 43–68.

35. See the introduction, note 25.

36. See Stich, "Could Man Be an Irrational Animal?"; Stich, *The Fragmentation of Reason*, chaps. 2 and 4; and Stein, *Without Good Reason*, chaps. 2, 3, 5, and 7.

37. One exception is the version of the mental logic theory developed by J. Macnamara. See Macnamara, *A Border Dispute*, pp. 18, 42–44. Macnamara distinguishes between Mental Logic and mental logic per se, the latter of which is *not* justified by appealing to the former. Indeed, to the extent that it applies a Chomskyan psycholinguistic model to the cognition of logic, Macnamara's mental logic mental theory has several interesting similarities with logical cognitivism. But it is unclear to me how Macnamara manages to avoid the difficulties of logical platonism; see sec. 1.5.

38. See Fodor, "Three Cheers for Propositional Attitudes," pp. 120–121.

39. See Braine, "On the Relation between the Natural Logic of Reasoning and Standard Logic."

40. See Braine, "The 'Natural Logic' Approach to Reasoning"; and M. Braine and D. O'Brien (eds.), *Mental Logic* (Mahwah, N.J.: Lawrence Erlbaum, 1998).

41. See Macnamara, *A Border Dispute*.

42. See W. Overton, "Competence and Procedures: Constraints on the Development of Logical Reasoning," in W. Overton (ed.), *Reasoning, Necessity, and Logic: Developmental Perspectives* (Hillsdale, N.J.: Lawrence Erlbaum, 1990), pp. 1–32.

43. See L. J. Rips, "Cognitive Processes in Propositional Reasoning," *Psychological Review* 90 (1983): 38–71; Rips, "Deduction"; and Rips, *The Psychology of Proof: Deductive Reasoning in Human Thinking* (Cambridge, Mass.: MIT Press, 1994).

44. See N. Wetherick, "Human Rationality," in Manktelow and Over, *Rationality*, pp. 83–109; and Wetherick, "Psychology and Syllogistic Reasoning," *Philosophical Psychology* 2 (1989): 111–124.

45. Another problem that arises for the mental logic theory in this connection is the fact that applying inference rules without further constraints will generate an infinite number of conclusions from a given set of premises: hence the mental logic theory seemingly cannot explain why the reasoner reaches only a *single* conclusion in most cases. See Johnson-Laird, *Mental Models*, p. 34. This problem does not seem insurmountable for the mental logic theory, however, if one has a suitably refined account

of the logic module, according to which extra pragmatic factors will introduce the needed constraints ; see Braine and O'Brien (eds.), *Mental Logic*.

46. Strictly speaking, however, this is not necessitated since it is at least conceptually consistent to hold (i) that the Mental Logic is cognitively basic and takes the form of some particular classical or nonclassical system, and (ii) that diehard classicism and diehard nonclassicism are both false (say, because logical pluralism is true). But although their union is possible, the mental logic theory and logical pluralism would obviously make fairly strange bedfellows, because the former entails rationalism, while the latter pulls strongly toward irrationalism.

47. See sec. 6.3.

48. On the other hand, however, not all rationalists are mental logic theorists. Dennett, e.g., is a rationalist but also an antirealist pragmatist about intentionality and rationality, and hence not a mental logic theorist. See the references in note 34 above.

49. See K. Craik, *The Nature of Explanation* (Cambridge: Cambridge University Press, 1943).

50. See Johnson-Laird, *Mental Models*; Johnson-Laird, "Mental Models and Deduction," *Trends in Cognitive Science* 5 (2001): 434–442; Johnson-Laird, "Inference and Mental Models," in Newstead and Evans, *Perspectives on Thinking and Reasoning,* pp. 115–146; P. Johnson-Laird and R. Byrne, *Deduction* (Hillsdale, N.J.: Lawrence Erlbaum, 1991); and Johnson-Laird and Byrne, "Models and Deductive Rationality," in Manktelow and Over, *Rationality*, pp. 177–210.

51. According to the mental models theory, reasoners typically represent complete arguments by means of mental models. They search for a model that forces truth on the conclusion, given the truth of the premises. By contrast, the mental logic theory says that reasoners typically represent premises only, then produce conclusions as outputs by applying the rules to them. This leads to the problem that since an infinite number of conclusions can be generated from a given finite set of premises, the mental logic theory cannot account for the drawing of a single conclusion relative to that set of premises. So in this regard the mental models theory has at least a prima facie advantage over the mental logic theory. But see also note 45 above.

52. See C. S. Peirce, *Collected Papers of Charles Sanders Peirce*, vol. 4, book 2 (Cambridge, Mass.: Belknap/Harvard University Press, 1961).

53. See R. Smullyan, *First-Order Logic* (Berlin: Springer-Verlag, 1968).

54. See E. J. Lowe, "Rationality, Deduction, and Mental Models," in Manktelow and Over, *Rationality,* pp. 220–228.

55. See P. Carruthers and J. Boucher (eds.), *Language and Thought* (Cambridge: Cambridge University Press, 1998).

56. In sec. 6.6 I will argue that logical intuition depends on our capacity for processing linguistic mental imagery, i.e., mental models of either natural language inscriptions or formal symbolism.

57. Interestingly, however, content-free deontic analogues of the abstract selection task (i.e., analogues of the abstract selection task in which the abstract rule is formulated in terms of permission or obligation) *are* associated with high success rates. See sec. 5.5.

58. The paradigm here is of course Tarski; see his "The Concept of Truth in Formalised Languages" and "The Establishment of Scientific Semantics."

59. Johnson-Laird and Byrne, "Models and Deductive Rationality," p. 179.

60. Johnson-Laird and Byrne, "Models and Deductive Rationality," p. 194.

61. Johnson-Laird and Byrne, "Models and Deductive Rationality," p. 194.

62. See note 7 above.

63. See, e.g., Griggs and Cox, "The Elusive Thematic Materials Effect in Wason's Selection Task"; Manktelow and Evans, "Facilitation of Reasoning by Realism: Effect or Non-Effect?"; and P. Pollard, "Human Reasoning: Some Possible Effects of Availability," *Cognition* 10 (1982): pp. 65–96.

64. See A. Tversky and D. Kahneman, "Availability: A Heuristic for Judging Frequency and Probability," *Cognitive Psychology* 5 (1973): 207–232.

65. See P. Cheng and K. Holyoak, "Pragmatic Reasoning Schemas," *Cognitive Psychology* 17 (1985): 391–416, p. 394.

66. See Cheng and Holyoak, "Pragmatic Reasoning Schemas"; P. Cheng et al., "Pragmatic versus Syntactic Approaches to Training Deductive Reasoning," *Cognitive Psychology* 18 (1986): 293–328; and Holyoak and Cheng, "Pragmatic Reasoning about Human Voluntary Action: Evidence from Wason's Selection Task," in Newstead and Evans, *Perspectives on Thinking and Reasoning*, pp. 67–89.

67. Cheng and Holyoak, "Pragmatic Reasoning Schemas," p. 414.

68. Cheng and Holyoak, "Pragmatic Reasoning Schemas," p. 396.

69. Holyoak and Cheng, "Pragmatic Reasoning about Human Voluntary Action," p. 76.

70. Holyoak and Cheng, "Pragmatic Reasoning about Human Voluntary Action," p. 75.

71. I am rationally reconstructing a little here, since Cheng and Holyoak are not explicit about their commitment to the modularity thesis. In any case, it is important to remember that mental modules need not be innate; see sec. 4.5.

72. See L. Cosmides, "The Logic of Social Exchange: Has Natural Selection Shaped How Humans Reason? Studies with the Wason Selection Task," *Cognition* 31 (1989): 187–276.

73. See Cosmides, "The Logic of Social Exchange," pp. 260–261.

74. Cosmides, "The Logic of Social Exchange," pp. 201–260.

75. See Braine and O'Brien, *Mental Logic,* chaps. 7–8; O'Brien, "Finding Logic in Human Reasoning Requires Looking in the Right Places," in Newstead and Evans, *Perspectives on Thinking and Reasoning,* pp. 189–216, here pp. 205–210; and Rips, *The Psychology of Proof,* chap. 5.

76. O'Brien, "Finding Logic in Human Reasoning," p. 205.

77. See, e.g., the logical system described by D. Kaplan in secs. 18–19 of his influential "Demonstratives: An Essay on the Semantics, Logic, and Epistemology of Demonstratives and Other Indexicals," in J. Almog et al. (eds.), *Themes from Kaplan* (New York: Oxford University Press, 1989), pp. 481–614. Holyoak and Cheng deny that their mental deontology can be adequately represented by a mental deontic *logic*, on the grounds that their mental deontology is intrinsically context-sensitive; see "Pragmatic Reasoning about Human Voluntary Action," pp. 84–85, and p. 87 n. 2. But this overlooks logics specifically designed to incorporate context-sensitivity.

78. That the theory of social contract schemas can be construed as a form of rationalism squares well with the fact that the theory of natural selection is consistent with rationalism: see, e.g., W. Cooper, *The Evolution of Reason* (Cambridge: Cambridge University Press, 2001); and E. Sober, "The Evolution of Rationality," *Synthese* 46 (1981): 95–120. But it does not *entail* rationalism. See Stich, *The Fragmentation of Reason,* chap. 3; and Stein, *Without Good Reason,* chap. 6.

79. Interestingly, my reduction of irrationalism to rationalism converges, by a quite different route, with Cohen's basic conclusion in "Can Human Irrationality Be Experimentally Demonstrated?": that psychological theories of human deductive reasoning are perforce rationalistic.

80. This is slightly overstated for convenience's sake. Intentionality, nativism, modularity, and LOT are all more or less controversial. But these are domestic debates within cognitivism, not fundamental challenges to the standard cognitivist model of the mind. To reject the standard cognitivist model of the mind fundamentally, one would have to go back in some basic way to empiricism and behaviorism and give up the highly plausible and almost universally accepted idea that cognition is mainly (or at least nontrivially) what the conscious organism, from its own cognitive endowment, logico-constructively contributes to informational inputs from its environment.

81. See C. Cherniak, *Minimal Rationality* (Cambridge, Mass.: MIT Press, 1986).

82. Cherniak, *Minimal Rationality,* pp. 24–25.

83. Cherniak, *Minimal Rationality*, pp. 81–82.

84. Cherniak, *Minimal Rationality*, p. 11.

85. Cherniak, *Minimal Rationality*, chap. 2.

86. Cherniak, *Minimal Rationality*, chap. 4. I take it, though, that according to the minimal rationality theory the logic ascribed to the creature by the interpreter will mirror *some* extended or deviant logical system. That is, I am assuming that for the minimal rationality theory the creature's logic *won't* be a protologic in my sense of that notion.

87. Cherniak, *Minimal Rationality*, pp. 107–109. Cherniak's conception of rationality is thus a critical extension of Davidson's conception, with what Evnine aptly calls Davidson's "rationalist idealist" tendencies having been replaced by pragmatic naturalism. See S. Evnine, *Donald Davidson* (Stanford: Stanford University Press, 1991), p. 148.

88. Indeed, Cherniak is principally concerned to criticize and weaken the ideal rationality assumptions built into classical decision theory. But the instrumental conception of rationality does not exhaust human rationality. On the contrary, it seems to me likely that a fully worked-out conception of even instrumental rationality would require that rational animals have a much richer set of cognitive abilities than the minimal rationality theory requires: in particular, abilities involving self-consciousness and high-level reflection either over a holistic network of beliefs or intentions or over strict modal concepts. In this respect, Davidson's own conception of rationality seems closer to the truth than Cherniak's. See D. Davidson, "Psychology as Philosophy," in his *Essays on Actions and Events* (Oxford: Clarendon/Oxford University Press, 1980), pp. 229–244; and Davidson, "Rational Animals."

89. This is not to say that animals that are strictly speaking not reasoners (e.g., cats, dogs, horses, etc.) cannot temporarily be treated, by a logical fiction or from within the "intentional stance," *as if* they were reasoners, for some special purpose or another. It is one thing to have one's actions treated *as if* they fell under normative logical principles, however, and quite another to be actually carrying out an inference.

90. See R. Griggs, "The Effects of Rule Clarification, Decision Justification, and Selection Instruction on Wason's Abstract Selection Task," in Newstead and Evans, *Perspectives on Thinking and Reasoning*, pp. 17–39.

91. What would a proper reasoning test look like? At the very least, it would have to involve a genuinely representative range of human subjects, and not just mature, healthy humans of normal or higher intelligence, i.e., rational humans. And the point of such tests would be to determine whether and under what conditions humans are capable of understanding what a logical task is, and what generally counts as success or failure in performing logical tasks, and *not* whether and under what conditions humans of normal or higher intelligence are able successfully to perform more or less abstract deduction tasks in classical propositional logic.

92. It is in fact possible to deny both (α) and (β), thereby decoupling logic and reasoning; see G. Harman, *Change in View: Principles of Reasoning* (Cambridge, Mass.: MIT Press, 1986), chaps. 1–2, and appendix A; and Harman, "Rationality," in his *Reasoning, Meaning, and Mind* (Oxford: Clarendon/Oxford University Press, 1999), pp. 9–45. For a sketch and critique of Harman's argument, see sec. 7.5.

Chapter 6

1. Wittgenstein, *Philosophical Investigations*, § 201, p. 81e.

2. Wittgenstein, *Remarks on the Foundations of Mathematics*, 2d. ed., trans. G. E. M. Anscombe (Cambridge, Mass.: MIT Press, 1983), p. 38.

3. Benacerraf, "Mathematical Truth," pp. 672–673.

4. R. Shepard and L. Cooper, *Mental Images and Their Transformations* (Cambridge, Mass.: MIT Press, 1982), p. 5.

5. For an outline of the five-pronged cumulative argument, see sec. 2.7.

6. This seems to be largely a consequence of early Wittgenstein's influential misinterpretation of Frege's and Russell's doctrines of self-evidence. See Jeshion, "Frege's Notions of Self-Evidence," *Mind* 110 (2001): 937–976, esp. pp. 972–973.

7. Unfortunately there is no generally accepted account of what it is for something to be abstract. As I am using the notion, an object is abstract if and only if it is not uniquely located in spacetime. Concrete objects, by contrast, are uniquely located in spacetime. I also distinguish between *platonic* abstractness and *nonplatonic* abstractness. See sec. 6.6.

8. Structuralism says that abstract objects of some specific kind are not independently existing entities but instead merely distinct roles, positions, or offices in an abstract formal relational system consisting of a coherent set of interlinked patterns or configurations. See sec. 6.6.

9. Apart from its connections to Lewis and early Wittgenstein, my theory also has some significant parallels with Husserl's theory of "categorial intuition" in the sixth of his *Logical Investigations*.

10. So Wittgenstein's misinterpretation of Frege and Russell on self-evidence (see note 6 above) in fact pays theoretical dividends.

11. See Fodor, "Special Sciences, or the Disunity of Science as a Working Hypothesis."

12. C. Wright, *Wittgenstein on the Foundations of Mathematics* (Cambridge, Mass.: Harvard University Press, 1980), chap. 2. See also G. Baker and P. Hacker, *An Analytical Commentary on the Philosophical Investigations*, vol. 2: *Wittgenstein, Rules, Grammar, and Necessity* (Oxford: Blackwell, 1980); P. Boghossian, "The Rule-Following

Considerations," *Mind* 98 (1989): 507–549; and J. Katz, *The Metaphysics of Meaning* (Cambridge, Mass.: MIT Press, 1990), chap. 3.

13. L. Carroll, *Through the Looking Glass* (New York: Dial, 1988), pp. 178–179.

14. The private language argument is usually held to be found in the sections immediately following §243 of *Philosophical Investigations*. See, e.g., D. Pears, *The False Prison: A Study of the Development of Wittgenstein's Philosophy*, vol. 2 (Oxford: Oxford University Press, 1987), chaps. 13–15.

15. Wittgenstein, *Philosophical Investigations*, §202, p. 81e.

16. Wittgenstein, *Philosophical Investigations*, p. xe.

17. S. Kripke, *Wittgenstein on Rules and Private Language* (Cambridge, Mass.: Harvard University Press, 1982), pp. 11, 15.

18. Wittgenstein, *Philosophical Investigations*, pp. 80e–81e.

19. I leave it open whether our ordinary use of 'not' implies that normal speakers of English possess the concept of classical negation. But it is at least possible. And for an argument that it does, see H. P. Grice, "Logic and Conversation [1967, 1987]," in his *Studies in the Way of Words* (Cambridge, Mass.: Harvard University Press, 1989), pp. 1–143.

20. The empirical–nonempirical distinction is not exactly the same as the a posteriori–a priori distinction. For the latter, see note 24 below. If the scientific essentialists are correct then there are necessary a posteriori truths. For my own part, I think that doubts can be raised about the very idea of the necessary a posteriori: see R. Hanna, "A Kantian Critique of Scientific Essentialism," *Philosophy and Phenomenological Research* 58 (1998): 497–528; and also P. Tichy, "Kripke on Necessity A Posteriori," *Philosophical Studies* 43 (1983): 225–241. In any case, since everything empirical is contingent, nothing can be both necessary *and* empirical.

21. See Tarski, "On the Concept of Logical Consequence"; and Tarski, "What Are Logical Notions?" *History and Philosophy of Logic* 7 (1986): 143–154. See also W. Hanson, "The Concept of Logical Consequence," *Philosophical Review* 106 (1997): 365–409; and Warmbröd, "Logical Constants."

22. See, e.g., I. Hacking, "What Is Logic?" *Journal of Philosophy* 86 (1979): 285–319.

23. The locus classicus is of course Descartes, *Rules for the Direction of the Mind*, in *Philosophical Writings of Descartes,* vol. 1, trans. J. Cottingham et al. (Cambridge: Cambridge University Press, 1984); see also Gaukroger, *Descartes: An Intellectual Biography*, pp. 115–124. For a survey of recent work, see M. DePaul and W. Ramsey (eds.), *Rethinking Intuition* (Lanham, Md.: Rowman and Littlefield, 1998).

24. As I see it, there are three distinct versions of the a priori–a posteriori distinction: cognitive, semantic, and epistemic. *The cognitive version*: a cognition is a posteriori if and only if it is strictly determined by either inner, proprioceptive, or outer sensory experiences, whereas a cognition is a priori if and only if it is underdetermined by either inner, proprioceptive, or outer sensory experiences even though it is always actually accompanied by such sensory experiences. *The semantic version*: a sentence (statement, proposition, etc.) is a posteriori if and only if its truth conditions are strictly determined by its verification conditions, whereas a sentence (statement, proposition, etc.) is a priori if and only if its truth conditions are underdetermined by its verification conditions. *The epistemic version*: a belief is a posteriori if and only if its justification is strictly determined by sensory evidence, whereas a belief is a priori if and only if its justification is underdetermined by sensory evidence. See also R. Hanna, "How Do We Know Necessary Truths? Kant's Answer," *European Journal of Philosophy* 6 (1998): 115–145; and Hanna, *Kant and the Foundations of Analytic Philosophy*, pp. 245–246. For other conceptions of apriority, see Boghossian and Peacocke, *New Essays on the A Priori*; P. Hanson and B. Hunter (eds.), *Return of the A Priori* (Calgary: University of Alberta Press, 1992); and P. Moser (ed.), *A Priori Knowledge* (Oxford: Oxford University Press, 1987).

25. For a theory of intuition based on prima facie intuitions, see G. Bealer, "The Incoherence of Empiricism," *Proceedings of the Aristotelian Society* 66 (1992): 99–138; G. Bealer, "Intuition and the Autonomy of Philosophy," in DePaul and Ramsay, *Rethinking Intuition*, pp. 201–239; and G. Bealer, "A Theory of the A Priori," *Philosophical Quarterly* 81 (2000): 1–30.

26. See Chomsky, *Aspects of the Theory of Syntax*.

27. See Bonjour, *In Defense of Pure Reason*, chaps. 1 and 4.

28. See R. Jeshion, "On the Obvious," *Philosophy and Phenomenological Research* 60 (2000): 333–355. This is a crucial point. Once one has distinguished carefully between intuition on the one hand and hunches, guesses, etc., on the other, the philosophical ill repute of intuition stems largely from the implausibility of Cartesian infallibilism.

29. That is, it is plausible to think that the reliability of intuition is entailed by its intrinsic features and does not require a metajustification; see Bonjour, *In Defense of Pure Reason*, pp. 142–147.

30. See C. Parsons, "Mathematical Intuition," *Proceedings of the Aristotelian Society* 80 (1979–1980): 145–168.

31. As opposed to the particular sentence-token, I mean. It is also possible to hold that '[propositional attitude verb] that [sentence]' constructions are demonstratives picking out sentence-tokens of the sentence-type following the demonstrative expression 'that'.

32. See D. Davidson, "On Saying That," in his *Inquiries into Truth and Interpretation,* pp. 93–108.

33. See Descartes, *Rules for the Direction of the Mind,* rule 3, pp. 14–15.

34. Wittgenstein, *Philosophical Investigations,* p. 84e, §213.

35. I am not saying that empiricists must also be scientific naturalists, but rather only that scientific naturalists must also be empiricists.

36. See Pinker, *The Language Instinct,* pp. 207–219.

37. Wittgenstein, *Remarks on the Philosophy of Psychology,* trans. G. E. M. Anscombe, vol. 1 (Chicago: University of Chicago Press, 1980), p. 176e (translation slightly modified).

38. Wittgenstein, *Remarks on the Foundations of Mathematics,* p. 247.

39. Wittgenstein, *Remarks on the Foundations of Mathematics,* p. 347.

40. Wittgenstein, *Remarks on the Foundations of Mathematics,* p. 241.

41. Wittgenstein, *Remarks on the Foundations of Mathematics,* p. 247.

42. See Frege, *Foundations of Arithmetic,* p. 68.

43. Unfortunately Benacerraf does not spell out all the basic steps of his argument, nor does he unpack any of the steps in detail; so my reconstruction goes significantly beyond what he explicitly says.

44. Something is anthropically mind-dependent if and only if its existence is logically dependent on the existence of human minds. All idealists are antirealists, but antirealists can resist an explicit commitment to idealist metaphysics. See, e.g., Dummett, *The Logical Basis of Metaphysics.*

45. See, e.g., Brouwer, "Historical Background, Principles, and Methods of Intuitionism." For the extension of antirealism to logic, see N. Tennant, *Anti-Realism and Logic* (Oxford: Oxford University Press, 1987).

46. See H. Field, *Science without Numbers* (Princeton: Princeton University Press, 1980); and S. Yablo, "Apriority and Existence," in Boghossian and Peacocke, *New Essays on the A Priori,* pp. 197–228.

47. Conventionalism is of course consistent with antirealism. See M. Friedman, *Reconsidering Logical Positivism* (Cambridge: Cambridge University Press, 1999); and A. Richardson, *Carnap's Construction of the World* (Cambridge: Cambridge University Press, 1998).

48. Carnap explicitly applies this strategy to logic in *The Logical Syntax of Language.*

49. See M. Dummett, "Wittgenstein's Philosophy of Mathematics," *Philosophical Review* 74 (1965): 504–518.

50. See Wright, *Wittgenstein on the Foundations of Mathematics*, chap. 21; and for the extension of nonfactualism to Wittgenstein's theory of logic, see P. Railton, "A Priori Rules: Wittgenstein on the Normativity of Logic," in Boghossian and Peacocke, *New Essays on the A Priori*, pp. 170–196.

51. See B. Hale and C. Wright, "Benacerraf's Dilemma Revisited," *European Journal of Philosophy* 10 (2002): 101–129; and B. Hale and C. Wright, "Implicit Definition and the A Priori," in Boghossian and Peacocke, *New Essays on the A Priori*, pp. 286–319. For the extension of nonfactualism to logic, see C. Wright, "Inventing Logical Necessity," in Butterfield, *Language, Mind, and Logic*, pp. 187–209.

52. See J. Katz, "What Mathematical Knowledge Could Be," *Mind* 104 (1995): 491–522.

53. F. Dretske, *Seeing and Knowing* (Chicago: University of Chicago Press, 1969), chap. 1.

54. See Dretske, *Seeing and Knowing*, p. 50. What is essential for Dretske is that the perceiver be able to visually discriminate the object from its local environment; but this environment-perceiver relation does *not* imply that the object causes the perception.

55. See Quine, "Ontological Relativity," in his *Ontological Relativity and Other Essays*, pp. 26–68, here pp. 40–41.

56. See R. Hanna, "Direct Reference, Direct Perception, and the Cognitive Theory of Demonstratives," *Pacific Philosophical Quarterly* 74 (1993): 96–117.

57. Quine, "Ontological Relativity," p. 40.

58. See C. Parsons, "Finitism and Intuitive Knowledge," in M. Schirn (ed.), *The Philosophy of Mathematics Today* (Oxford: Oxford University Press, 1998), pp. 249–270; Parsons, "Intuition in Constructive Mathematics," in Butterfield, *Language, Mind, and Logic*, pp. 211–229; and Parsons, "Mathematical Intuition." Parsons's approach has been doubly influenced by Husserl's notion of categorial intuition and Hilbert's finitist intuitionism. The finitistic element of Parsons's theory seems problematic, however; see J. Page, "Parsons on Mathematical Intuition," *Mind* 102 (1993): 223–232.

59. See Bonjour, *In Defense of Pure Reason*, pp. 156–161.

60. See H. Field, "The Aprioricity of Logic," *Proceedings of the Aristotelian Society* 96 (1996): 359–379.

61. See Hanna, *Kant and the Foundations of Analytic Philosophy*, pp. 281–185; and H. Putnam, "Analyticity and Apriority: Beyond Wittgenstein and Quine," in his *Realism and Reason: Philosophical Papers*, vol. 3 (Cambridge: Cambridge University Press, 1983), pp. 115–138.

62. See Fogelin, "Quine's Limited Naturalism," pp. 550–551.

63. See note 27 above.

64. See J. Katz, *Realistic Rationalism* (Cambridge, Mass.: MIT Press, 1998). Katz's argument strategy is to criticize challenges to realism. He explicitly admits that how we intuit abstract objects is a "mystery" (pp. 32–34). Interestingly, however, in his later paper "Mathematics and Metaphilosophy" (*Journal of Philosophy* 99 [2002]: 362–390), pp. 382–383, he sketches an approach to intuition that is quite similar to the one I develop in secs. 6.4 and 6.6.

65. A first step in this direction is to note that functional organizations are all structural and then adopt mathematical structuralism. So I will shortly suggest that structuralism for mathematics and logic alike is the right way to go. Not every structural system is a functional organization, but structuralism is a necessary condition of functional organizations.

66. See Kim, *Supervenience and Mind*, part 2.

67. See Hanna, "Direct Reference, Direct Perception, and the Cognitive Theory of Demonstratives."

68. See, e.g., S. Shapiro, *Philosophy of Mathematics: Structure and Ontology* (New York: Oxford University Press, 1997), chaps. 3–5. For an extension of structuralism to logic, see, e.g., A. Koslow, *A Structuralist Theory of Logic* (Cambridge: Cambridge University Press, 1992).

69. See Hanna, "How Do We Know Necessary Truths? Kant's Answer."

70. See also Parsons, "Mathematical Intuition," p. 200.

71. Of course in perceiving an object we often generate an image of it too. But this is not, I think, absolutely necessary. Otherwise it would have to be the case that I can in principle remember absolutely everything I perceive. But surely there is some sort of representational paring-down that occurs in the transition from perceptual content to memory content.

72. I agree with Dretske that the existence of an efficacious causal relation between the object perceived and perception is not a necessary condition of perceiving (see note 54 above). In place of a causal requirement, Dretske proposes that for seeing at least, the perceiver must be able to visually discriminate the object from its local environment (although, with engaging frankness, he then also notes some counterexamples in which the perceived object merges with its local environment). My own view is that effective tracking not only handles all the cases mentioned by Dretske, but also generalizes to the other kinds of sense perception. The crucial point, however, is that although there certainly can be an efficacious causal relation between the object perceived and the perception, this is not necessary; some other noncausal sort of perceptual contact relation is what is essential. But for a causalist approach to perception,

see Grice, "The Causal Theory of Perception," in R. Swartz (ed.), *Perceiving, Sensing, and Knowing* (Berkeley: University of California Press, 1965), pp. 438–472.

73. See Johnson-Laird, *Mental Models*; S. Kosslyn, *Image and Brain* (Cambridge, Mass.: MIT Press, 1994); Kosslyn, *Image and Mind* (Cambridge, Mass.: Harvard University Press, 1980); R. Shepard, "The Mental Image," *American Psychologist* 33 (1978): 125–137; R. Shepard and S. Chipman, "Second Order Isomorphisms of Internal Representations: Shapes of States," *Cognitive Psychology* 1 (1970): 1–17; Shepard and Cooper, *Mental Images and Their Transformations*; and R. Shepard and J. Metzler, "Mental Rotation of Three-Dimensional Objects," *Science* 171 (1971): 701–703.

74. See, e.g., N. Block (ed.), *Imagery* (Cambridge, Mass.: MIT Press, 1981); Block, *Readings in the Philosophy of Psychology*, vol. 2, part 2; and Block, "The Photographic Fallacy in the Debate about Mental Imagery," *Noûs* 17 (1983): 651–661.

75. This modal framework is basically the same (with a few important differences, such as the general gloss on the notion of necessity, and the positive inclusion of synthetic necessity or "strong" metaphysical necessity) as that used by Chalmers in *The Conscious Mind*, pp. 52–71 and 136–138. See also S. Kripke, "Semantical Considerations on Modal Logic," *Acta Philosophica Fennica* 16 (1963): 83–94; R. Montague, "Logical Necessity, Physical Necessity, Ethics, and Quantifiers," in his *Formal Philosophy* (New Haven: Yale University Press, 1974); and T. Smiley, "Relative Necessity," *Journal of Symbolic Logic* 28 (1963): 113–134. For a closely related historical discussion of the analytic–synthetic distinction, see Hanna, *Kant and the Foundations of Analytic Philosophy*, chaps. 3–5.

76. See S. Kripke, *Naming and Necessity*, pp. 15–20.

77. Recall that I am using 'sentence' to mean a linguistically expressed truth-bearer, whether a proposition or statement or other sort of truth-bearer. Moreover, I realize that there are important differences between predicates, concepts, and properties, but for my limited purposes here it does not really matter which is used.

78. Chalmers's conception of logical or weak metaphysical necessity is also "two-dimensional," a conception based mainly on earlier work by Saul Kripke, David Kaplan, Robert Stalnaker, Gareth Evans, Martin Davies, and Lloyd Humberstone. See D. Chalmers, "The Foundations of Two-Dimensional Semantics," in M. García-Carpintero and J. Macia (eds.), *Two-Dimensionalism: Foundations and Applications* (Oxford: Oxford University Press, 2004), pp. 55–140. The basic idea behind two-dimensionalism is that there are two distinct types of semantic functions from worlds to extensions, depending on the type of concept or intension one uses: (1) the "primary" intension (a function from subject-centered worlds considered as actual, to extensions) and (2) the "secondary" intension (a function from worlds considered as counterfactual variants on the indexically fixed actual world, to extensions). To each

function or intension corresponds a different type of logical necessity. Analytic necessity corresponds to the primary intension; and a posteriori necessity corresponds to the secondary intension. For the notion of a posteriori necessity, see Kripke, *Naming and Necessity*. Of course, two-dimensional modal semantics is controversial. The crucial point for my purposes is that logical or analytic necessity in my sense will, in Chalmers's framework, count as logical necessity according to the primary intension.

79. Chalmers objects to strong or essentialist metaphysical necessity on the following three grounds: (a) that it is an ad hoc addition to the roster of modalities; (b) that it is brute and inexplicable; and (c) that the defenders of strong metaphysical necessity fail to provide an account of how humans get epistemic access to this modality. All of these objections may apply to conceptions of strong metaphysical necessity that take it to be a form of a posteriori necessity, and in particular identify it with physical necessity. But none of them applies to, e.g., Kant's conception of strong metaphysical necessity as synthetic a priori necessity; see Hanna, *Kant and the Foundations of Analytic Philosophy*, chap. 5. Leaving aside other worries one may have about Kant's metaphysics, the crucial point here is simply that Chalmers's objections do not generalize. Indeed, it is even arguable that strong or essentialist metaphysical necessity is more basic than logical necessity, since in the modal framework I have sketched there are going to be logical possibilities that are not *real* possibilities. For a similar idea, see S. Shalkowski, "Logic and Absolute Necessity," *Journal of Philosophy* 101 (2004): 55–82.

80. The simplification consists in separating the linguistic mental image I use in my intuition (in the example, *I* (#)) from the linguistic text (in the example, (*)) I use to represent the logical object. In most cases, the shape of the linguistic image and the shape of the linguistic text used to represent the logical object would be the same. Nevertheless the simplification is justified by psychological research strongly indicating that linguistic mental imagery is processed separately from either syntax or semantic content. See D. Schacter, "Perceptual Representation Systems and Implicit Memory: Toward a Resolution of the Multiple Memory Systems Debate," *Annals of the New York Academy of Science* 608 (1990): 543–571.

81. Not only computers, but also viruses, swarms of ants or bees, squirrels, cats, and rats can merely implement deductions. They are not reasoners in my mentalistic sense of the concept of rationality, however. Since they are incapable of normative-reflective cognition, they are incapable of representing themselves *as* reasoners, which is something that a reasoner must be able to do. Nevertheless, computers, viruses, etc., can indeed have "procedural rationality" (see the introduction). And, to the extent that computers, viruses, etc., actually do implement logical structures, they can for certain purposes be treated by reasoners *as if* they were reasoners. That is, for certain purposes we can take (to borrow Dennett's term) "the intentional stance" toward them. Thanks to William Lyons for pressing me on this point.

Chapter 7

1. J. Raz, *Engaging Reason* (Oxford: Oxford University Press, 1999), p. 75.

2. O. Weininger, *Sex and Character* (London: Heinemann, 1906), p. 159.

3. A. Nye, *Words of Power* (New York: Routledge, 1990), pp. 170–171.

4. In chap. 5, for convenience, I used 'reasoning' primarily to mean *deductive* reasoning. In this chapter I use it primarily in the broader sense that includes inductive judgments, hypothesis formation, probability judgments, and practical reasoning, in addition to deduction.

5. See Raz, *Engaging Reason*, chaps. 4–5. 'May' expresses permission as opposed to obligation. A further question, falling beyond the scope of this book, concerns the *ground* or *source* of normativity. For the record, however, I am inclined to take the Kantian line and ground normativity in the rational human animal's innate capacity for self-legislation, self-governance, or autonomy. See C. Korsgaard, *The Sources of Normativity* (Cambridge: Cambridge University Press, 1996), chaps. 1–4 and 9.

6. For convenience, I focus on options that assume the normativity of logic. But strictly speaking, since something can be both intrinsically *non-F* and also extrinsically *F* (i.e., in relation to something else), this opens up the possibility that logic is both intrinsically nonnormative and also extrinsically normative. In fact, as we shall see, most recent and contemporary writers of introductory logic textbooks hold that logic is intrinsically nonnormative and also extrinsically normative.

7. See A. Arnauld and P. Nicole, *Logic or the Art of Thinking*, trans. J. V. Buroker (Cambridge: Cambridge University Press, 1996); S. Gaukroger, *Cartesian Logic* (Oxford: Oxford University Press, 1989); and Kneale and Kneale, *The Development of Logic*, pp. 315–320.

8. See J. S. Mill, *System of Logic* (London: Longmans, Green, 1879); and Husserl, *Prolegomena to Pure Logic*, p. 76.

9. See, e.g., M. Resnik, "Logic: Normative or Descriptive? The Ethics of Belief or a Branch of Psychology?" *Philosophy of Science* 52 (1985): 221–238; and Resnik, "Ought There to Be But One Logic?," pp. 510–514. Resnik holds that logic is intrinsically prescriptive but also has some descriptive applications.

10. See Frege, "Logic [1897]," p. 128.

11. Frege, *Basic Laws of Arithmetic*, p. 12.

12. See, e.g., S. Barker, *Elements of Logic*, 3d. ed. (New York: McGraw Hill, 1980), p. 4; I. Copi, *Symbolic Logic*, 4th ed. (New York: Macmillan, 1973), p. 2; and W. Salmon, *Logic* (Englewood Cliffs, N.J.: Prentice Hall, 1963), p. 8.

13. See Husserl, *Prolegomena to Pure Logic*, chaps. 1–3.

14. See Ricketts, "Frege, the *Tractatus*, and the Logocentric Predicament."

15. It is also true that some versions of the thesis that logic is intrinsically hypothetically normative are nonpsychologistic. See, e.g., Resnik, "Logic: Normative or Descriptive? The Ethics of Belief or a Branch of Psychology?" Still, any psychologistic theory of logic that holds that logic is intrinsically normative must also hold that logic is hypothetically normative.

16. Kant, "The Jäsche Logic," p. 529.

17. Kant, *Critique of Pure Reason*, A54–55/B78–79.

18. See I. Kant, *Grounding for the Metaphysics of Morals*, trans. J. Ellington (Indianapolis: Hackett, 1981).

19. Kant, *Critique of Pure Reason*, A54/B78.

20. Kant, "The Jäsche Logic," p. 529.

21. See R. Wedgwood, "The A Priori Rules of Rationality," *Philosophy and Phenomenological Research* 59 (1999): 113–131, esp. p. 128: "a rule counts as a basic a priori rule if, and only if, it is necessary that any thinker who has [capacities for forming judgments or decisions on the basis of reasons] will be immediately inclined to follow the rule, when she is in the input conditions of the rule, exercising those capacities, and considering whether to make the change of attitude that is the output of the rule."

22. P. Foot, "Morality as a System of Hypothetical Imperatives," in her *Virtues and Vices* (Berkeley: University of California Press, 1978), p. 164.

23. Foot, "Morality as System of Hypothetical Imperatives," p. 170.

24. Carnap, *Logical Syntax of Language*, pp. 51–52, italics in the original.

25. The *Logische Syntax der Sprache* was published in late 1934; Cole Porter's *Anything Goes* first opened at the Alvin Theatre in New York on November 21, 1934. Chalk one up for the Zeitgeist.

26. Foot, "Morality as a System of Hypothetical Imperatives," p. 171.

27. See O. O'Neill, *Acting on Principle* (New York: Columbia University Press, 1975); and O'Neill, *Constructions of Reason* (Cambridge: Cambridge University Press, 1989).

28. See O'Neill, *Constructions of Reason*, chap. 11. See also T. Hill, "Kantian Constructivism in Ethics," *Ethics* 99 (1989): 752–770; J. Rawls, "Kantian Constructivism in Moral Theory," *Journal of Philosophy* 77 (1980): 515–572; and J. Rawls, "Themes in Kant's Moral Philosophy," in E. Förster (ed.), *Kant's Transcendental Deductions* (Stanford: Stanford University Press, 1989), pp. 81–113.

29. O'Neill, *Constructions of Reason*, pp. 58–59, italics added.

30. Boole, *An Investigation of the Laws of Thought*, p. 408.

31. G. Boole, "Extracts from a Paper Entitled 'On the Mathematical Theory of Logic and on the Philosophical Interpretations of Its Methods and Processes' (1855 or 1856)," in his *Studies in Logic and Probability*, pp. 230–246, here p. 246.

32. B. Russell, *Autobiography* (London: Unwin, 1975), p. 330.

33. Russell's next sentence is: "I did not like to suggest that it was time for bed, as it seemed probable both to him and me that on leaving me he would commit suicide" (*Autobiography*, p. 330).

34. See R. Monk, *Ludwig Wittgenstein: The Duty of Genius* (London: Jonathan Cape, 1990), pp. 19–25.

35. Russell's view of the nature of logic shifted radically over time from platonism (in *Problems of Philosophy*, chaps. 7–11) to psychologism (in *An Inquiry into Meaning and Truth* [London: George Allen and Unwin, 1940], pp. 194–203): but he always held fixed the idea that logic is intrinsically nonnormative.

36. Sometimes other things are not equal. In some cases there are acts, or causal consequences of acts, for which we must take moral responsibility *despite* our not being able to control them in the sense of having been able to will or do otherwise. See Frankfurt, "Alternate Possibilities and Moral Responsibility" and "Freedom of the Will and the Concept of a Person," both in his *The Importance of What We Care About*; and T. Nagel, "Moral Luck," in his *Mortal Questions* (Cambridge: Cambridge University Press, 1979), pp. 24–38. Presumably the same holds for some cases of human reasoning.

37. With the exception of the last two items, this list is taken over with some minor changes from I. Copi, *Introduction to Logic*, 2d. ed. (New York: Macmillan, 1961), chap. 3. Begging the question will never lead to invalidity, but it might lead someone to accept a contradiction as true.

38. See A. De Morgan, *Formal Logic* (London: Taylor, 1847), chap. 13.

39. Copi, *Introduction to Logic*, p. 52.

40. See also Wason and Johnson-Laird, *Psychology of Reasoning*, p. 6.

41. See, e.g., R. L. Gregory, *The Intelligent Eye* (New York: McGraw-Hill, 1970).

42. See note 36 above.

43. See Pinker, *The Language Instinct*, pp. 201–207.

44. See Wason and Johnson-Laird, *Psychology of Reasoning*, chaps. 7–8. Many modal fallacies turn on confusions between classical logical consequence and various kinds of nonclassical consequence.

45. Harman, *Change in View*, p. 115.

46. Harman, "Rationality," p. 18.

47. Harman, *Change in View*, pp. 11–12, 13–14.

48. Harman, *Change in View*, pp. 11–12, 15–17.

49. Harman, *Change in View*, p. 12.

50. Harman, *Change in View*, pp. 117–127.

51. Harman endorses antipsychologism for deductive logic in *Thought* (Princeton: Princeton University Press, 1973), pp. 15–19.

52. Carroll, *Through the Looking Glass*, pp. 91–92.

53. In sec. 2.5 I tentatively proposed that there is a weak or minimal version of a basic principle of classical logic—the weak principle of noncontradiction—which states that not every sentence is both true and false, and that this metalogical principle is part of the protologic. If this proposal is correct, it would explain why dialetheism has to restrict itself to the thesis that only *some* sentences are true contradictions.

54. See R. Millikan, "White Queen Psychology; Or, the Last Myth of the Given," in her *White Queen Psychology and Other Essays for Alice* (Cambridge, Mass.: MIT Press, 1993), pp. 279–363. See also Kripke, *Naming and Necessity*, and S. Kripke, "A Puzzle about Belief," in A. Margalit (ed.), *Meaning and Use* (Dordrecht: D. Reidel, 1979), pp. 239–283.

55. This is not to say that the refined standard cognitivist model of the mind is consistent with *every* variety of externalism; for a survey of the varieties, see C. McGinn, *Mental Content* (Oxford: Blackwell, 1989), chap. 1. But Fodor, e.g., has proposed a theory that allows for both nonexternalist ("narrow") mental content and externalist ("wide") content; see Fodor, *The Elm and the Expert*; *Psychosemantics*; and *A Theory of Content and Other Essays*.

56. Indeed, the only discussion I am aware of is Haack's one-paragraph treatment in the 1996 introduction to *Deviant Logic, Fuzzy Logic*, p. xv.

57. In *Casablanca*, Humphrey Bogart never actually says "Play it again, Sam." Similarly, Descartes never actually writes "*Cogito, ergo sum.*" In the *Discourse*, he writes in French that "this truth 'I am thinking, therefore I exist' was so firm and sure that all the extravagant suppositions of the skeptics were incapable of shaking it" (*Discourse on the Method*, in *Philosophical Writings of Descartes*, vol. 1, p. 127); and in the *Meditations* he writes in Latin that "this proposition *I am, I exist*, is necessarily true whenever it is put forward by me or conceived in my mind" (*Meditations on First Philosophy*, in *Philosophical Writings of Descartes*, vol. 2, p. 17).

58. Descartes, *Rules for the Direction of the Mind*, pp. 14–15.

59. See note 57 above. In the "Objections and Replies" (*Philosophical Writings of Descartes*, vol. 2), p. 100, Descartes rather cagily splits the difference between the formulation in the *Discourse* and the formulation in the *Meditations* by writing in Latin that "when someone says 'I am thinking, therefore I am, or I exist,' he does not deduce existence from thought by means of a syllogism, but recognizes it as something self-evident by a simple intuition of the mind."

60. See J. Katz, *Cogitations* (New York: Oxford University Press, 1986), esp. chaps. 7–12.

61. C. S. Peirce, "Grounds of the Validity of the Laws of Logic," in his *Collected Papers,* 5.318, p. 190.

62. In addition to Nye's *Words of Power*, see also R. J. Falmagne and M. Hass (eds.), *Representing Reason: Feminist Theory and Formal Logic* (Lanham, Md.: Rowman and Littlefield, 2002).

63. F. Nietzsche, *Beyond Good and Evil*, trans. W. Kaufmann (New York: Vintage, 1966), p. 9.

64. F. Nietzsche, *On the Genealogy of Morals*, in *On the Genealogy of Morals and Ecce Homo*, trans. W. Kaufmann and R. Hollingdale (New York: Vintage, 1967), pp. 13–163, here p. 20.

65. J. Milton, *Paradise Lost,* bk. 4, l. 108, in *The Poems of John Milton*, 2d. ed. (New York: Ronald, 1953).

66. B. Williams, *Ethics and the Limits of Philosophy* (London: Fontana Collins, 1985), chaps. 1–2 and 10.

67. See P. Foot, "Nietzsche: The Revaluation of Values," in her *Virtues and Vices*, pp. 81–95; and R. Geuss, "Nietzsche and Morality," *European Journal of Philosophy* 5 (1997): 1–20.

68. Haack writes: "Needless to say, my reaction to Nye's conclusion—'Logic in its final perfection is insane'—is . . . , well, needless to say"; see *Deviant Logic, Fuzzy Logic*, p. xv. I take it that this means that Nye's conclusion is philosophically ludicrous. I will argue later that Nye's conclusion is in almost a literal sense "insane," but far from being philosophically ludicrous.

69. See notes 66–67 above.

70. Frege, *Basic Laws of Arithmetic*, p. 14.

71. See Frankfurt, "Freedom of the Will and the Concept of a Person."

72. For some recent attempts to connect consciousness and perception intrinsically with embodiment, affect, and intentional action, see A. Clark, *Being There: Putting Brain, Body, and World Together Again* (Cambridge, Mass.: MIT Press, 1997); A. Damasio, *The Feeling of What Happens*; S. Hurley, *Consciousness in Action*

(Cambridge, Mass.: Harvard University Press, 1998); W. T. Rockwell, *Neither Brain nor Ghost: A Nondualist Alternative to the Mind–Brain Identity Theory* (Cambridge, Mass.: MIT Press, 2005); and F. Varela, E. Thompson, and E. Rosch, *The Embodied Mind* (Cambridge, Mass.: MIT Press, 1991). What I am proposing, then, is that this recent "bottom-up" or "existential" revolution in cognitive science focusing on embodied consciousness and perception should be correspondingly combined with a "top-down" or "Kantian" revolution focusing on categorically normative rationality and thought. Just to give it a handy label, one might call such a radical dual approach *embodied rationalism.*

Bibliography

Aristotle. *Prior Analytics*. Trans. A. J. Jenkinson. In R. McKeon (ed.), *The Collected Works of Aristotle*, 62–107. New York: Random House, 1941.

Arnaud, A., and P. Nicole. *Logic or the Art of Thinking*. Trans. J. V. Buroker. Cambridge: Cambridge University Press, 1996.

Audi, R. *The Architecture of Reason*. New York: Oxford University Press, 2001.

Austen, J. *Northanger Abbey*. Harmondsworth, Middlesex: Penguin, 1972.

Ayer, A. J., and R. Rhees. "Can There Be a Private Language?" *Proceedings of the Aristotelian Society* 28 (1954): 63–94.

Baker, G., and P. Hacker. *An Analytical Commentary on the Philosophical Investigations*, vol. 2: *Wittgenstein, Rules, Grammar, and Necessity*. Oxford: Blackwell, 1980.

Barker, S. *Elements of Logic*, 3d. ed. New York: McGraw-Hill, 1980.

Baron-Cohen, S. *Mindblindness: An Essay on Autism and Theory of Mind*. Cambridge, Mass.: MIT Press, 1995.

Bealer, G. "The Incoherence of Empiricism." *Proceedings of the Aristotelian Society* 66 (1992): 99–138.

———. "Intuition and the Autonomy of Philosophy." In M. DePaul and W. Ramsey (eds.), *Rethinking Intuition*, 201–239. Lanham, Md.: Rowman and Littlefield, 1998.

———. *Quality and Concept*. Oxford: Clarendon/Oxford University Press, 1982.

———. "A Theory of the A Priori." *Philosophical Quarterly* 81 (2000): 1–30.

Belnap, N. D. "Tonk, Plonk, and Plink." *Analysis* 22 (1961): 130–134.

Benacerraf, P. "Frege: The Last Logicist." In P. French et al. (eds.), *The Foundations of Analytic Philosophy*, Midwest Studies in Philosophy, vol. 6, 17–35. Minneapolis: University of Minnesota Press, 1981.

———. "Mathematical Truth." *Journal of Philosophy* 70 (1973): 661–679.

———. "What Numbers Could Not Be." *Philosophical Review* 74 (1965): 47–73.

Bennett, J. *Rationality*. Indianapolis: Hackett, 1989.

Bermúdez, J. L. *Thought without Words*. Oxford: Oxford University Press, 2003.

Bezuidenhout, A. "Resisting the Step toward Naturalism." *Philosophy and Phenomenological Research* 56 (1994): 743–770.

Bickenbach, J. "Justifying Deduction." *Dialogue* 4 (1979): 500–516.

Block, N. (ed.). *Imagery*. Cambridge, Mass.: MIT Press, 1981.

———. "The Photographic Fallacy in the Debate about Mental Imagery." *Noûs* 17 (1983): 651–661.

——— (ed.). *Readings in Philosophy of Psychology*, 2 vols. Cambridge, Mass.: Harvard University Press, 1980.

———. "Troubles with Functionalism." In Block, *Readings in Philosophy of Psychology*, vol. 1, 268–305.

———. "What Is Functionalism?" In Block, *Readings in Philosophy of Psychology*, vol. 1, 171–184.

Bochvar, D. "On a Three-Valued Calculus and Its Application to the Analysis of Contradictories." *Matematceskij Sbornik* 4 (1939): 287–308; 5 (1940): 119.

Boghossian, P. "Knowledge of Logic." In Boghossian and Peacocke, *New Essays on the A Priori*, 229–254.

———. "The Rule-Following Considerations." *Mind* 98 (1989): 507–549.

Boghossian, P., and C. Peacocke (eds.). *New Essays on the A Priori*. Oxford: Clarendon/Oxford University Press, 2000.

Bonjour, L. *In Defense of Pure Reason*. Cambridge: Cambridge University Press, 1998.

Boole, G. "The Claims of Science." In Boole, *Studies in Logic and Probability*, 187–210.

———. "Extracts from a Paper Entitled 'On the Mathematical Theory of Logic and on the Philosophical Interpretations of Its Methods and Processes' (1855 or 1856)." In Boole, *Studies in Logic and Probability*, 230–246.

———. *An Investigation of the Laws of Thought*. Cambridge: Macmillan, 1854.

———. "Logic and Reasoning." In Boole, *Studies in Logic and Probability*, 211–229.

———. *Studies in Logic and Probability*. London: Watts, 1952.

Boolos, G., and R. Jeffrey. *Computability and Logic*, 3d. ed. Cambridge: Cambridge University Press, 1996.

Boring, E. G. *History of Experimental Psychology*. New York: Appleton-Century-Crofts, 1950.

Braine, M. "On the Relation between the Natural Logic of Reasoning and Standard Logic." *Psychological Review* 85 (1978): 1–21.

————. "The 'Natural Logic' Approach to Reasoning." In W. Overton (ed.), *Reasoning, Necessity, and Logic: Developmental Perspectives*, 133–157. Hillsdale, N.J.: Lawrence Erlbaum, 1990.

Braine, M., and D. O'Brien (eds.). *Mental Logic*. Mahwah, N.J.: Lawrence Erlbaum, 1998.

Brentano, F. *Psychology from an Empirical Standpoint*. Trans. A. C. Rancurello et al. London: Routledge and Kegan Paul, 1995.

Brouwer, L. E. J. "Historical Background, Principles, and Methods of Intuitionism." *South African Journal of Science* 49 (1952): 139–146.

Bruner, J. "On Perceptual Readiness." *Psychological Review* 64 (1957): 123–152.

Butterfield, J. (ed.). *Language, Mind, and Logic*. Cambridge: Cambridge University Press, 1986.

Carnap, R. "Intellectual Autobiography." In P. Schilpp (ed.), *The Philosophy of Rudolf Carnap*, 3–84. La Salle, Ill.: Open Court, 1963.

————. *The Logical Structure of the World*. Trans. R. George. Berkeley: University of California Press, 1967.

————. *The Logical Syntax of Language*. Trans. A. Smeaton. London: Routledge and Kegan Paul, 1937.

————. *Meaning and Necessity*, 2d. ed. Chicago: University of Chicago Press, 1956.

Carroll, L. *Alice's Adventures in Wonderland*. New York: Dial, 1988.

————. *Lewis Carroll's Symbolic Logic*. Ed. W. W. Bartley III. New York: Clarkson Potter, 1997.

————. *Symbolic Logic*. Oxford: Oxford University Press, 1896.

————. *Through the Looking Glass*. New York: Dial, 1988.

————. "What the Tortoise Said to Achilles." *Mind* 4 (1895): 278–280.

Carruthers, P. *Language, Thought, and Consciousness*. Cambridge: Cambridge University Press, 1996.

Carruthers, P., and J. Boucher (eds.). *Language and Thought*. Cambridge: Cambridge University Press, 1998.

Chalmers, D. *The Conscious Mind*. New York: Oxford University Press, 1996.

————. "The Foundations of Two-Dimensional Semantics." In M. García-Carpintero and J. Macia (eds.), *Two-Dimensionalism: Foundations and Applications*, 55–140. Oxford: Oxford University Press, 2004.

———— (ed.). *Philosophy of Mind: Classical and Contemporary Readings.* New York: Oxford University Press, 2002.

Cheng, P., and K. Holyoak. "Pragmatic Reasoning Schemas." *Cognitive Psychology* 17 (1985): 391–416.

Cheng, P., K. Holyoak, R. Nisbett, and R. Oliver. "Pragmatic versus Syntactic Approaches to Training Deductive Reasoning." *Cognitive Psychology* 18 (1986): 293–328.

Cherniak, C. *Minimal Rationality.* Cambridge, Mass.: MIT Press, 1986.

Chomsky, N. *Aspects of the Theory of Syntax.* Cambridge, Mass.: MIT Press, 1965.

————. *Cartesian Linguistics.* New York: Harper and Row, 1966.

————. "Explaining Language Use." In Chomsky, *New Horizons in the Study of Language and Mind*, 19–45.

————. *Knowledge of Language.* Westport, Conn.: Praeger, 1986.

————. *Language and Mind*, 2d. ed. New York: Harcourt Brace Jovanovich, 1972.

————. "Language and Nature." *Mind* 104 (1995): 1–61.

————. *Language and Problems of Knowledge.* New York: Columbia University Press, 1988.

————. *The Minimalist Program.* Cambridge, Mass.: MIT Press, 1995.

————. *New Horizons in the Study of Language and Mind.* Cambridge: Cambridge University Press, 2000.

————. "Quine's Empirical Assumptions." In D. Davidson and J. Hintikka (eds.), *Words and Objections*, 53–68. Dordrecht: D. Reidel, 1969.

————. *Reflections on Language.* New York: Pantheon, 1975.

————. "Review of B. F. Skinner's *Verbal Behavior*." *Language* 35 (1959): 26–58.

————. *Rules and Representations.* Cambridge, Mass.: MIT Press, 1980.

————. *Syntactic Structures.* The Hague: Mouton, 1957.

Churchland, P. M. "Eliminative Materialism and the Propositional Attitudes." *Journal of Philosophy* 78 (1981): 67–90.

Clark, A. *Being There: Putting Brain, Body, and World Together Again.* Cambridge, Mass.: MIT Press, 1997.

Coffa, A. *The Semantic Tradition from Kant to Carnap*. Cambridge: Cambridge University Press, 1991.

Cohen, J. "Frege and Psychologism." *Philosophical Papers* 27 (1998): 45–68.

Cohen, L. J. "Are People Programmed to Commit Fallacies? Further Thoughts about the Interpretation of Experimental Data on Probability Judgment." *Journal of the Theory of Social Behavior* 12 (1982): 251–274.

———. "Can Human Irrationality Be Experimentally Demonstrated?" *Behavioral and Brain Sciences* 4 (1981): 317–370.

———. "Continuing Commentary." *Behavioral and Brain Sciences* 6 (1983): 487–533.

———. "Reply to Stein." *Synthese* 99 (1994): 173–176.

Cook, V. J., and M. Newson. *Chomsky's Universal Grammar*, 2d. ed. Oxford: Blackwell, 1996.

Cooper, W. *The Evolution of Reason*. Cambridge: Cambridge University Press, 2001.

Copi, I. *Introduction to Logic*, 2d. ed. New York: Macmillan, 1961.

———. *Symbolic Logic*, 4th ed. New York: Macmillan, 1973.

Cosmides, L. "The Logic of Social Exchange: Has Natural Selection Shaped How Humans Reason? Studies with the Wason Selection Task." *Cognition* 31 (1989): 187–276.

Cowie, F. *What's Within: Nativism Reconsidered*. Oxford: Oxford University Press, 1998.

Craik, K. *The Nature of Explanation*. Cambridge: Cambridge University Press, 1943.

Cummins, R., and D. Cummins (eds.). *Minds, Brains, and Computers*. Oxford: Blackwell, 2000.

Damasio, A. *The Feeling of What Happens: Body and Emotion in the Making of Consciousness*. San Diego: Harcourt, 1999.

Daniels, N. "On Some Methods of Ethics and Linguistics." *Philosophical Studies* 37 (1980): 21–36.

———. "Reflective Equilibrium and Archimedean Points." *Canadian Journal of Philosophy* 10 (1980): 83–103.

———. "Wide Reflective Equilibrium and Theory Acceptance in Ethics." *Journal of Philosophy* 76 (1979): 256–282.

Danto, A. "Naturalism." In Edwards, *Encyclopedia of Philosophy*, vol. 5, 448–450.

Darwall, S., A. Gibbard, and P. Railton. "Towards *fin de siècle* Ethics: Some Trends." *Philosophical Review* 101 (1992): 115–189.

Davidson, D. *Inquiries into Truth and Interpretation*. Oxford: Clarendon/Oxford University Press, 1984.

———. "On Saying That." In Davidson, *Inquiries into Truth and Interpretation*, 93–108.

———. "On the Very Idea of a Conceptual Scheme." In Davidson, *Inquiries into Truth and Interpretation*, 183–198.

———. "Psychology as Philosophy." In D. Davidson, *Essays on Actions and Events*, 229–244. Oxford: Clarendon/Oxford University Press, 1980.

———. "Radical Interpretation." In Davidson, *Inquiries into Truth and Interpretation*, 125–139.

———. "Rational Animals." In E. Lepore and B. McLaughlin (eds.), *Actions and Events: Perspectives on the Philosophy of Donald Davidson*, 473–480. Oxford: Blackwell, 1985.

———. "Thought and Talk." In Davidson, *Inquiries into Truth and Interpretation*, 155–170.

Dehaene, S. *The Number Sense*. Oxford: Oxford University Press, 1997.

De Morgan, A. *Formal Logic*. London: Taylor, 1847.

Dennett, D. *The Intentional Stance*. Cambridge, Mass.: MIT Press, 1987.

———. "Intentional Systems." *Journal of Philosophy* 68 (1971): 87–106.

———. "Making Sense of Ourselves." In Dennett, *The Intentional Stance*, 83–101.

———. "Three Kinds of Intentional Psychology." In Dennett, *The Intentional Stance*, 43–68.

DePaul, M., and W. Ramsey (eds.). *Rethinking Intuition*. Lanham, Md.: Rowman and Littlefield, 1998.

Descartes, R. *Discourse on the Method*. In Descartes, *Philosophical Writings of Descartes*, vol. 1, 109–151.

———. *Meditations on First Philosophy*. In Descartes, *Philosophical Writings of Descartes*, vol. 2, 1–65.

———. "Objections and Replies." In Descartes, *Philosophical Writings of Descartes*, vol. 2, 63–383.

———. *Philosophical Writings of Descartes*, 3 vols. Trans. J. Cottingham et al. Cambridge: Cambridge University Press, 1984.

————. *Rules for the Direction of the Mind.* In Descartes, *Philosophical Writings of Descartes*, vol. 1, 9–78.

Donagan, A. "Wittgenstein on Sensation." In A. Donagan, *Philosophical Papers of Alan Donagan,* vol. 1, 235–256. Chicago: University of Chicago Press, 1994.

Dretske, F. *Naturalizing the Mind.* Cambridge, Mass.: MIT Press, 1995.

————. *Seeing and Knowing.* Chicago: University of Chicago Press, 1969.

Dummett, M. "Can Analytical Philosophy Be Systematic, and Ought It to Be?" In Dummett, *Truth and Other Enigmas*, 437–458.

————. *Frege: Philosophy of Language,* 2d. ed. Cambridge, Mass.: Harvard University Press, 1981.

————. "The Justification of Deduction [1973]." In Dummett, *Truth and Other Enigmas*, 290–318.

————. *The Logical Basis of Metaphysics.* Cambridge, Mass.: Harvard University Press, 1991.

————. *Origins of Analytical Philosophy.* Cambridge, Mass.: Harvard University Press, 1993.

————. *Truth and other Enigmas.* London: Duckworth, 1978.

————. "Wittgenstein's Philosophy of Mathematics." *Philosophical Review* 74 (1965): 504–518.

Edwards, P. (ed.). *The Encyclopedia of Philosophy,* 7 vols. New York: Macmillan, 1967.

Elman, J. L., E. A. Bates, M. H. Johnson, et al. (eds.). *Rethinking Innateness.* Cambridge, Mass.: MIT Press, 1996.

Evans, J. "Bias and Rationality." In Manktelow and Over, *Rationality: Psychological and Philosophical Perspectives*, 6–30.

Evans, J., and S. Newstead. "Creating a Psychology of Reasoning." In Newstead and Evans, *Perspectives on Thinking and Reasoning*, 2–16.

Evnine, S. *Donald Davidson.* Stanford: Stanford University Press, 1991.

————. "The Universality of Logic." *Mind* 110 (2001): 335–367.

Falmagne, R. J., and M. Hass (eds.). *Representing Reason: Feminist Theory and Formal Logic.* Lanham, Md.: Rowman and Littlefield, 2002.

Field, H. "The Aprioricity of Logic." *Proceedings of the Aristotelian Society* 96 (1996): 359–379.

————. *Science without Numbers*. Princeton: Princeton University Press, 1980.

Fodor, J. *Concepts*. Oxford: Clarendon/Oxford University Press, 1998.

————. "Doing Without What's Within: Fiona Cowie's Critique of Nativism." *Mind* 110 (2001): 99–148.

————. *The Elm and the Expert*. Cambridge, Mass.: MIT Press, 1994.

————. *The Language of Thought*. Cambridge, Mass.: Harvard University Press, 1975.

————. "Methodological Solipsism Considered as a Research Strategy in Cognitive Psychology." In Fodor, *RePresentations*, 225–253.

————. *The Mind Doesn't Work That Way*. Cambridge, Mass.: MIT Press, 2000.

————. *The Modularity of Mind*. Cambridge, Mass.: MIT Press, 1983.

————. *Psychosemantics*. Cambridge, Mass.: MIT Press, 1987.

————. *RePresentations: Philosophical Essays on the Foundations of Cognitive Science*. Cambridge, Mass.: MIT Press, 1981.

————. "Special Sciences, or the Disunity of Science as a Working Hypothesis." *Synthese* 28 (1974): 97–115.

————. *A Theory of Content and Other Essays*. Cambridge, Mass.: MIT Press, 1990.

————. "Three Cheers for Propositional Attitudes." In Fodor, *RePresentations*, 100–123.

Fogelin, R. "Quine's Limited Naturalism." *Journal of Philosophy* 94 (1997): 543–563.

Foot, P. "Morality as a System of Hypothetical Imperatives." In Foot, *Virtues and Vices*, 157–173.

————. "Nietzsche: The Revaluation of Values." In Foot, *Virtues and Vices*, 81–95.

————. *Virtues and Vices*. Berkeley: University of California Press, 1978.

Frankfurt, H. "Alternate Possibilities and Moral Responsibility." In Frankfurt, *The Importance of What We Care About*, 1–10.

————. "Freedom of the Will and the Concept of a Person." In Frankfurt, *The Importance of What We Care About*, 11–25.

————. *The Importance of What We Care About*. Cambridge: Cambridge University Press, 1988.

Frege, G. *The Basic Laws of Arithmetic*. Trans. M. Furth. Berkeley: University of California Press, 1964.

———. *Collected Papers on Mathematics, Logic, and Philosophy*. Trans. M. Black et al. Oxford: Basil Blackwell, 1984.

———. *The Foundations of Arithmetic*, 2d. ed. Trans. J. L. Austin. Evanston, Ill.: Northwestern University Press, 1953.

———. "Logic [1897]." In G. Frege, *Posthumous Writings*, trans. P. Long et al., 127–151. Chicago: University of Chicago Press, 1979.

———. "On Sense and Meaning." In Frege, *Collected Papers on Mathematics, Logic, and Philosophy*, 157–177.

———. "Review of E. G. Husserl, *Philosophie der Arithmetik I* [1894]." In Frege, *Collected Papers on Mathematics, Logic, and Philosophy*, 195–209.

———. "Thoughts." In Frege, *Collected Papers on Mathematics, Logic, and Philosophy*, 351–372.

Friedman, M. *Reconsidering Logical Positivism*. Cambridge: Cambridge University Press, 1999.

Fumerton, R. *Reason and Morality*. Ithaca, N.Y.: Cornell University Press, 1990.

Gardner, H. *The Mind's New Science*. New York: Basic Books, 1985.

Gaukroger, S. *Cartesian Logic*. Oxford: Oxford University Press, 1989.

———. *Descartes: An Intellectual Biography*. Oxford: Clarendon/Oxford University Press, 1995.

Gendler, T., and J. Hawthorne (eds.). *Conceivability and Possibility*. Oxford: Clarendon/Oxford University Press, 2002.

Gentzen, G. "Investigations into Logical Deduction." In G. Gentzen, *The Collected Papers of Gerhard Gentzen*, trans. M. Szabo, 68–131. Amsterdam: North Holland, 1969.

Geuss, R. "Nietzsche and Morality." *European Journal of Philosophy* 5 (1997): 1–20.

Gödel, K. "On Formally Undecidable Propositions of *Principia Mathematica* and Related Systems." In Van Heijenoort, *From Frege to Gödel*, 596–617.

Gold, I., and D. Stoljar. "A Neuron Doctrine in the Philosophy of Neuroscience." *Behavioral and Brain Sciences* 22 (1999): 809–830.

Goodman, N. "The New Riddle of Induction." In N. Goodman, *Fact, Fiction, and Forecast*, 4th ed., 59–83. Cambridge, Mass.: Harvard University Press, 1983.

Gregory, R. L. *The Intelligent Eye*. New York: McGraw-Hill, 1970.

————. "Perceptions as Hypotheses." *Philosophical Transactions of the Royal Society of London, Series B* 290 (1980): 181–197.

Grice, H. P. "The Causal Theory of Perception." In R. Swartz (ed.), *Perceiving, Sensing, and Knowing*, 438–472. Berkeley: University of California Press, 1965.

————. "Logic and Conversation [1967, 1987]." In H. P. Grice, *Studies in the Way of Words*, 1–143. Cambridge, Mass.: Harvard University Press, 1989.

Griffin, D. R. *Animal Thinking*. Cambridge, Mass.: Harvard University Press, 1984.

Griggs, R. "The Effects of Rule Clarification, Decision Justification, and Selection Instruction on Wason's Abstract Selection Task." In Newstead and Evans, *Perspectives on Thinking and Reasoning*, 17–39.

Griggs, R., and J. Cox. "The Elusive Thematic Materials Effect in Wason's Selection Task." *British Journal of Psychology* 73 (1982): 407–420.

Haack, S. *Deviant Logic*. Cambridge: Cambridge University Press, 1974.

————. *Deviant Logic, Fuzzy Logic*. Chicago: University of Chicago Press, 1996.

————. "Dummett's Justification of Deduction." *Mind* 95 (1982): 216–239.

————. "The Justification of Deduction." *Mind* 85 (1976): 112–119.

————. *Philosophy of Logics*. Cambridge: Cambridge University Press, 1978.

Hacking, I. "What Is Logic?" *Journal of Philosophy* 86 (1979): 285–319.

Hale, B. *Abstract Objects*. Oxford: Blackwell, 1987.

Hale, B., and C. Wright. "Benacerraf's Dilemma Revisited." *European Journal of Philosophy* 10 (2002): 101–129.

————. "Implicit Definition and the A Priori." In Boghossian and Peacocke, *New Essays on the A Priori*, 286–319.

Hanna, R. "Direct Reference, Direct Perception, and the Cognitive Theory of Demonstratives." *Pacific Philosophical Quarterly* 74 (1993): 96–117.

————. "How Do We Know Necessary Truths? Kant's Answer." *European Journal of Philosophy* 6 (1998): 115–145.

————. "Kant, Wittgenstein, and the Fate of Analysis." In M. Beaney (ed.), *The Analytic Turn*. London: Routledge, forthcoming.

————. *Kant and the Foundations of Analytic Philosophy*. Oxford: Clarendon/ Oxford University Press, 2001.

————. "A Kantian Critique of Scientific Essentialism." *Philosophy and Phenomenological Research* 58 (1998): 497–528.

———. "Logical Cognition: Husserl's *Prolegomena* and the Truth in Psychologism." *Philosophy and Phenomenological Research* 53 (1993): 251–275.

Hanna, R., and E. Thompson. "The Mind–Body–Body Problem." *Theoria et Historia Scientiarum* 7 (2003): 24–44.

———. "Neurophenomenology and the Spontaneity of Consciousness." In E. Thompson (ed.), *The Problem of Consciousness,* pp. 133–162. Calgary: University of Alberta Press, 2005.

Hanson, P., and B. Hunter (eds.). *Return of the A Priori.* Calgary: University of Alberta Press, 1992.

Hanson, W. "The Concept of Logical Consequence." *Philosophical Review* 106 (1997): 365–409.

Harman, G. *Change in View: Principles of Reasoning.* Cambridge, Mass.: MIT Press, 1986.

———. "Deep Structure as Logical Form." In Harman and Davidson, *Semantics of Natural Language,* 25–47.

———. "Rationality." In G. Harman, *Reasoning, Meaning, and Mind,* 9–45. Oxford: Clarendon/Oxford University Press, 1999.

———. *Thought.* Princeton: Princeton University Press, 1973.

Harman, G., and D. Davidson (eds.). *Semantics of Natural Language.* Dordrecht: D. Reidel, 1972.

Henle, M. "On the Relation between Logic and Thinking." *Psychological Review* 69 (1962): 366–378.

Hill, T. "Kantian Constructivism in Ethics." *Ethics* 99 (1989): 752–770.

Hollis, M., and G. Sugden. "Rationality in Action." *Mind* 102 (1993): 1–35.

Horgan, T. "From Supervenience to Superdupervenience: Meeting the Demands of a Material World." *Mind* 102 (1993): 555–586.

Holyoak, K., and P. Cheng. "Pragmatic Reasoning about Human Voluntary Action: Evidence from Wason's Selection Task." In Newstead and Evans, *Perspectives on Thinking and Reasoning,* 67–89.

Horgan, T., and J. Tienson. "The Intentionality of Phenomenology and the Phenomenology of Intentionality." In Chalmers, *Philosophy of Mind: Classical and Modern Readings,* 520–533.

Hornstein, N. *Logic as Grammar.* Cambridge, Mass.: MIT Press, 1984.

Hubin, D. "The Groundless Normativity of Instrumental Rationality." *Journal of Philosophy* 98 (2001): 445–468.

Hughes, G. E., and M. J. Cresswell. *An Introduction to Modal Logic*. London: Methuen, 1972.

Hunter, G. *Metalogic*. Berkeley: University of California Press, 1971.

Hurley, S. *Consciousness in Action*. Cambridge, Mass.: Harvard University Press, 1998.

Husserl, E. *Logical Investigations*, 2 vols. Trans. J. N. Findlay. London: Routledge and Kegan Paul, 1970.

————. *Prolegomena to Pure Logic*. In Husserl, *Logical Investigations*, vol. 1, 51–247.

————. "A Reply to a Critic of My Refutation of Logical Psychologism." In P. McCormick and F. Elliston (eds.), *Husserl: Shorter Works*, 152–158. Notre Dame: University of Notre Dame Press, 1982.

Hylton, P. *Russell, Idealism, and the Emergence of Analytic Philosophy*. Oxford: Oxford University Press, 1990.

Inhelder, B., and J. Piaget. *The Growth of Logical Thinking from Childhood to Adolescence*. Trans. A. Parsons and S. Milgram. New York: Basic Books, 1958.

Jackendoff, R. *Languages of the Mind*. Cambridge, Mass.: MIT Press, 1992.

James, W. *Principles of Psychology*, 2 vols. New York: Dover, 1950.

Jeshion, R. "Frege's Notions of Self-Evidence." *Mind* 110 (2001): 937–976.

————. "On the Obvious." *Philosophy and Phenomenological Research* 60 (2000): 333–355.

Johnson, D. "Conventionalism about Logical Truth." *Philosophical Topics* 23 (1995): 189–212.

Johnson-Laird, P. "Inference and Mental Models." In Newstead and Evans, *Perspectives on Thinking and Reasoning*, 115–146.

————. *Mental Models*. Cambridge, Mass.: Harvard University Press, 1983.

————. "Mental Models and Deduction." *Trends in Cognitive Science* 5 (2001): 434–442.

Johnson-Laird, P., and R. Byrne. *Deduction*. Hillsdale, N.J.: Lawrence Erlbaum, 1991.

————. "Models and Deductive Rationality." In Manktelow and Over, *Rationality: Psychological and Philosophical Perspectives*, 177–210.

Kahneman, D., P. Slovic, and A. Tversky (eds.). *Judgment under Uncertainty: Heuristics and Biases*. Cambridge: Cambridge University Press, 1982.

Kant, I. *Critique of Pure Reason*. Trans. P. Guyer and A. Wood. Cambridge: Cambridge University Press, 1997.

————. *Grounding for the Metaphysics of Morals*. Trans. J. Ellington. Indianapolis: Hackett, 1981.

————. "The Jäsche Logic." In Kant, *Lectures on Logic*, 521–640.

————. *Lectures on Logic*. Trans. J. M. Young. Cambridge: Cambridge University Press, 1992.

Kaplan, D. "Demonstratives: An Essay on the Semantics, Logic, and Epistemology of Demonstratives and other Indexicals." In J. Almog et al. (eds.), *Themes from Kaplan*, 481–614. New York: Oxford University Press, 1989.

Katz, J. *Cogitations*. New York: Oxford University Press, 1986.

————. *Language and Other Abstract Objects*. Totowa, N.J.: Rowman and Littlefield, 1981.

————. "Mathematics and Metaphilosophy." *Journal of Philosophy* 99 (2002): 362–390.

————. *The Metaphysics of Meaning*. Cambridge, Mass.: MIT Press, 1990.

————. *Realistic Rationalism*. Cambridge, Mass.: MIT Press, 1998.

————. "What Mathematical Knowledge Could Be." *Mind* 104 (1995): 491–522.

Kim, J. "Concepts of Supervenience." In Kim, *Supervenience and Mind*, 53–78.

————. "Multiple Realization and the Metaphysics of Reduction." In Kim, *Supervenience and Mind*, 309–335.

————. "'Strong' and 'Global' Supervenience Revisited." In Kim, *Supervenience and Mind*, 79–91.

————. *Supervenience and Mind*. Cambridge: Cambridge University Press, 1993.

Kirk, R. "Rationality without Language." *Mind* 76 (1967): 369–386.

Kitcher, P. "The Naturalists Return." *Philosophical Review* 101 (1992): 53–114.

Kneale, W., and M. Kneale. *The Development of Logic*. Oxford: Oxford University Press, 1984.

Kornblith, H. (ed.). *Naturalizing Epistemology*. Cambridge, Mass.: MIT Press, 1985.

Korsgaard, C. *The Sources of Normativity*. Cambridge: Cambridge University Press, 1996.

Koslow, A. *A Structuralist Theory of Logic*. Cambridge: Cambridge University Press, 1992.

Kosslyn, S. *Image and Brain*. Cambridge, Mass.: MIT Press, 1994.

———. *Image and Mind*. Cambridge, Mass.: Harvard University Press, 1980.

Kripke, S. "Identity and Necessity." In A. W. Moore (ed.), *Meaning and Reference*, 162–191. Oxford: Oxford University Press, 1993.

———. *Naming and Necessity*, 2d. ed. Cambridge, Mass.: Harvard University Press, 1980.

———. "A Puzzle about Belief." In A. Margalit (ed.), *Meaning and Use*, 239–283. Dordrecht: D. Reidel, 1979.

———. "Semantical Analysis of Modal Logic I, Normal Propositional Calculi." *Zeitschrift für mathematische Logik and Grundlagen der Mathematik* 9 (1963): 67–96.

———. "Semantical Considerations on Modal Logic." *Acta Philosophica Fennica* 16 (1963): 83–94.

———. *Wittgenstein on Rules and Private Language*. Cambridge, Mass.: Harvard University Press, 1982.

Kusch, M. *Psychologism*. London: Routledge, 1995.

Lakoff, G. "Linguistics and Natural Logic." In Harman and Davidson, *Semantics of Natural Language*, 545–665.

Langford, C. H. "The Notion of Analysis in Moore's Philosophy." In Schilpp, *The Philosophy of G. E. Moore*, 321–342.

Laurence, S., and E. Margolis. "Review of Fiona Cowie, *What's Within: Nativism Reconsidered*." *European Journal of Philosophy* 9 (2001): 242–247.

Levine, J. "Materialism and Qualia: The Explanatory Gap." *Pacific Philosophical Quarterly* 64 (1983): 354–361.

Lewis, C. I. "Alternative Systems of Logic." *Monist* 42 (1932): 481–507.

———. *Survey of Symbolic Logic*. Berkeley: University of California Press, 1918.

Lewis, C. I., and C. H. Langford. *Symbolic Logic*, 2d. ed. New York: Dover, 1959.

Lewis, D. "Causation." In D. Lewis, *Philosophical Papers*, vol. 2, 159–172. New York: Oxford University Press, 1986.

Lowe, E. J. "Rationality, Deduction, and Mental Models." In Manktelow and Over, *Rationality: Psychological and Philosophical Perspectives*, 211–230.

Lyons, J. *Introduction to Theoretical Linguistics*. Cambridge: Cambridge University Press, 1968.

———. *Noam Chomsky,* 2d. ed. Harmondsworth, Middlesex: Penguin, 1978.

Macnamara, J. *A Border Dispute: The Place of Logic in Psychology.* Cambridge, Mass.: MIT Press, 1986.

Maddy, P. "A Naturalistic Look at Logic." *Proceedings and Addresses of the American Philosophical Association* 76 (2002): 61–90.

Manktelow, K., and J. Evans. "Facilitation of Reasoning by Realism: Effect or Non-Effect?" *British Journal of Psychology* 70 (1979): 477–488.

Manktelow, K., and D. Over (eds.). *Rationality: Psychological and Philosophical Perspectives* London: Routledge, 1993.

Marr, D. *Vision.* San Francisco: Freeman, 1982.

Martinich, A. P. (ed.). *Philosophy of Language,* 3d. ed. New York: Oxford University Press, 1996.

Mason, S. *A History of the Sciences,* 2d. ed. New York: Collier, 1962.

May, R. *Logical Form.* Cambridge, Mass.: MIT Press, 1985.

McDowell, J. "Two Sorts of Naturalism." In R. Hursthouse et al. (eds.), *Virtues and Reasons,* 149–179. Oxford: Clarendon/Oxford University Press, 1995.

McGinn, C. *Mental Content.* Oxford: Blackwell, 1989.

Meiland, J., and M. Krausz (eds.). *Relativism: Cognitive and Moral.* Notre Dame: University of Notre Dame Press, 1982.

Mele, A., and P. Rawling (eds.). *The Oxford Handbook of Rationality.* Oxford: Oxford University Press, 2003.

Mill, J. S. *A System of Logic.* London: Longmans, Green, 1879.

Millikan, R. "White Queen Psychology; Or, the Last Myth of the Given." In R. Millikan, *White Queen Psychology and Other Essays for Alice,* 279–363. Cambridge, Mass.: MIT Press, 1993.

Milton, J. *Paradise Lost.* In *The Poems of John Milton,* 2d. ed., 204–487. New York: Ronald, 1953.

Monk, R. *Bertrand Russell: The Spirit of Solitude.* London: Jonathan Cape, 1996.

———. *Ludwig Wittgenstein: The Duty of Genius.* London: Jonathan Cape, 1990.

Montague, R. "Logical Necessity, Physical Necessity, Ethics, and Quantifiers." In R. Montague, *Formal Philosophy,* 71–83. New Haven: Yale University Press, 1974.

Montero, B. "The Body Problem." *Noûs* 33 (1999): 183–200.

Moore, G. E. "Analysis." In Schilpp, *The Philosophy of G. E. Moore*, 660–667.

———. "The Conception of Intrinsic Value." In G. E. Moore, *Philosophical Studies*, 253–275. New York: Harcourt, Brace, 1922.

———. *Principia Ethica*. Cambridge: Cambridge University Press, 1903.

Moser, P. (ed.). *A Priori Knowledge*. Oxford: Oxford University Press, 1987.

Nagel, E. "Towards a Naturalistic Conception of Logic." In H. Kallen and S. Hook (eds.), *American Philosophy Today and Tomorrow*, 377–391. New York: L. Furman, 1935.

Nagel, T. "Moral Luck." In T. Nagel, *Mortal Questions*, 24–38. Cambridge: Cambridge University Press, 1979.

———. "The Psychophysical Nexus." In Boghossian and Peacocke, *New Essays on the A Priori*, 433–471.

Neisser, U. *Cognition and Reality*. San Francisco: W. H. Freeman, 1976.

———. *Cognitive Psychology*. New York: Appleton-Century-Crofts, 1967.

Nietzsche, F. *Beyond Good and Evil*. Trans. W. Kaufmann. New York: Vintage, 1966.

———. *On the Genealogy of Morals*. In F. Nietzsche, *On the Genealogy of Morals and Ecce Homo*, trans. W. Kaufmann and R. Hollingdale, 13–163. New York: Vintage, 1967.

Newstead, S., and J. Evans (eds.). *Perspectives on Thinking and Reasoning*. Hillsdale, N.J.: Lawrence Erlbaum, 1995.

Nozick, R. *The Nature of Rationality*. Cambridge, Mass.: Harvard University Press, 1993.

Nye, A. *Words of Power*. New York: Routledge, 1990.

O'Brien, D. "Finding Logic in Human Reasoning Requires Looking in the Right Places." In Newstead and Evans, *Perspectives on Thinking and Reasoning*, 189–216.

Oliver, A. "The Metaphysics of Properties." *Mind* 105 (1996): 1–80.

O'Neill (Nell), O. *Acting on Principle*. New York: Columbia University Press, 1975.

———. *Constructions of Reason*. Cambridge: Cambridge University Press, 1989.

O'Shaughnessy, B. *The Will*, 2 vols. Cambridge: Cambridge University Press, 1980.

Ostertag, G. (ed.). *Definite Descriptions: A Reader*. Cambridge, Mass.: MIT Press, 1998.

Overton, W. "Competence and Procedures: Constraints on the Development of Logical Reasoning." In Overton, *Reasoning, Necessity, and Logic: Developmental Perspectives*, 1–32.

———— (ed.). *Reasoning, Necessity, and Logic: Developmental Perspectives*. Hillsdale, N.J.: Lawrence Erlbaum, 1990.

Page, J. "Parsons on Mathematical Intuition." *Mind* 102 (1993): 223–232.

Papineau, D. *Philosophical Naturalism*. Oxford: Blackwell, 1993.

Parsons, C. "Finitism and Intuitive Knowledge." In M. Schirn (ed.), *The Philosophy of Mathematics Today*, 249–270. Oxford: Oxford University Press, 1998.

————. "Intuition in Constructive Mathematics." In Butterfield, *Language, Mind, and Logic*, 211–229.

————. "Mathematical Intuition." *Proceedings of the Aristotelian Society* 80 (1979–1980): 145–168.

Parsons, T. "What Is an Argument?" *Journal of Philosophy* 93 (1996): 164–185.

Peacocke, C. "Understanding Logical Constants: A Realist Account." *Proceedings of the British Academy* 73 (1987): 153–200.

Pears, D. *The False Prison: A Study of the Development of Wittgenstein's Philosophy*, 2 vols. Oxford: Oxford University Press, 1987.

Peirce, C. S. *Collected Papers of Charles Sanders Peirce*, 6 vols. Cambridge, Mass.: Belknap/Harvard University Press, 1961.

————. "Grounds for the Validity of the Laws of Logic." In Peirce, *Collected Papers of Charles Sanders Peirce*, vol. 5, §§318–357, 190–222.

Piaget, J. *Logic and Psychology*. New York: Basic Books, 1957.

Pinker, S. *How the Mind Works*. New York: W. W. Norton, 1997.

————. *The Language Instinct*. New York: Harper Perennial, 1994.

Pollard, P. "Human Reasoning: Some Possible Effects of Availability." *Cognition* 10 (1982): 65–96.

Potter, M. *Reason's Nearest Kin*. Oxford: Clarendon/Oxford University Press, 2000.

————. *Sets: An Introduction*. Oxford: Oxford University Press, 1990.

Priest, G. *In Contradiction*. Dordrecht: Martinus Nijhoff, 1987.

————. *An Introduction to Non-Classical Logic*. Cambridge: Cambridge University Press, 2001.

————. "The Logic of Paradox." *Journal of Philosophical Logic* 8 (1979): 219–241.

————. "What Is So Bad about Contradictions?" *Journal of Philosophy* 95 (1998): 410–426.

Prior, A. N. "Conjunction and Contonktion Revisited." *Analysis* 24 (1964): 191–195.

————. "The Runabout Inference-Ticket." *Analysis* 21 (1960): 38–39.

————. "What Is Logic?" In A. N. Prior, *Papers in Logic and Ethics*, 122–129. Amherst, Mass.: University of Amherst Press, 1976.

Putnam, H. "Analyticity and Apriority: Beyond Wittgenstein and Quine." In Putnam, *Realism and Reason: Philosophical Papers*, vol. 3, 115–138.

————. "The 'Innateness Hypothesis' and Explanatory Models in Linguistics." In Block, *Readings in Philosophy of Psychology*, vol. 2, 292–299.

————. *Mathematics, Matter, and Method: Philosophical Papers*, vol. 1, 2d. ed. Cambridge: Cambridge University Press, 1979.

————. *Mind, Language, and Reality: Philosophical Papers*, vol. 2. Cambridge: Cambridge University Press, 1975.

————. "Minds and Machines." In Putnam, *Mind, Language, and Reality: Philosophical Papers*, vol. 2, 362–385.

————. "The Nature of Mental States." In Putnam, *Mind, Language, and Reality: Philosophical Papers*, vol. 2, 429–440.

————. "On Properties." In Putnam, *Mathematics, Matter, and Method: Philosophical Papers*, vol. 1, 305–322.

————. "Philosophy of Logic." In Putnam, *Mathematics, Matter, and Method: Philosophical Papers*, vol. 1, 323–357.

————. *Realism and Reason: Philosophical Papers*, vol. 3. Cambridge: Cambridge University Press, 1983.

————. *Representation and Reality*. Cambridge, Mass.: MIT Press, 1988.

————. "Re-Thinking Mathematical Necessity." In H. Putnam, *Words and Life*, 245–263. Cambridge: Harvard University Press, 1994.

————. "There Is at Least One A Priori Truth." In Putnam, *Realism and Reason: Philosophical Papers*, vol. 3, 98–114.

Quine, W. V. O. "Carnap and Logical Truth." In Quine, *The Ways of Paradox*, 107–132.

————. "Epistemology Naturalized." In Quine, *Ontological Relativity and Other Essays*, 69–90.

————. *From a Logical Point of View*, 2d. ed. New York: Harper and Row, 1961.

———. *Methods of Logic*, 4th ed. Cambridge, Mass.: Harvard University Press, 1982.

———. "Ontological Relativity." In Quine, *Ontological Relativity and Other Essays*, 26–68.

———. *Ontological Relativity and Other Essays*. New York: Columbia, 1969.

———. *Philosophy of Logic*, 2d. ed. Cambridge, Mass.: Harvard University Press, 1986.

———. "Truth by Convention." In Quine, *The Ways of Paradox*, 77–106.

———. "Two Dogmas of Empiricism." In Quine, *From a Logical Point of View*, 20–46.

———. *The Ways of Paradox*, 2d. ed. Cambridge, Mass.: Harvard University Press, 1976.

———. *Word and Object*. Cambridge, Mass.: MIT Press, 1960.

Railton, P. "A Priori Rules: Wittgenstein on the Normativity of Logic." In Boghossian and Peacocke, *New Essays on the A Priori*, 170–196.

Rawls, J. "The Independence of Moral Theory." *Proceedings and Addresses of the American Philosophical Association* 48 (1974): 5–22.

———. "Kantian Constructivism in Moral Theory." *Journal of Philosophy* 77 (1980): 515–572.

———. "Themes in Kant's Moral Philosophy." In E. Förster (ed.), *Kant's Transcendental Deductions*, 81–113. Stanford: Stanford University Press, 1989.

———. *A Theory of Justice*. Cambridge, Mass.: Harvard University Press, 1971.

Raz, J. *Engaging Reason*. Oxford: Oxford University Press, 1999.

Recanati, F. *Direct Reference*. Oxford: Blackwell, 1993.

Rescher, N. *The Coherence Theory of Truth*. Oxford: Oxford University Press, 1973.

Resnik, M. "Logic: Normative or Descriptive? The Ethics of Belief or a Branch of Psychology?" *Philosophy of Science* 52 (1985): 221–238.

———. "Ought There to Be But One Logic?" In B. J. Copeland (ed.), *Logic and Reality: Essays on the Legacy of Arthur Prior*, 489–517. Oxford: Clarendon/Oxford University Press, 1996.

Richardson, A. *Carnap's Construction of the World*. Cambridge: Cambridge University Press, 1998.

Ricketts, T. "Frege, the *Tractatus*, and the Logocentric Predicament." *Philosophers' Annual* 8 (1985): 247–259.

Rips, L. J. "Cognitive Processes in Propositional Reasoning." *Psychological Review* 90 (1983): 38–71.

———. "Deduction." In R. Sternberg and E. Smith (eds.), *Psychology of Human Thought,* 116–152. Cambridge: Cambridge University Press, 1988.

———. *The Psychology of Proof: Deductive Reasoning in Human Thinking.* Cambridge, Mass.: MIT Press, 1994.

Rock, I. *The Logic of Perception.* Cambridge, Mass.: MIT Press, 1983.

Rockwell, W. T. *Neither Brain nor Ghost: A Nondualist Alternative to the Mind–Brain Identity Theory.* Cambridge, Mass.: MIT Press, 2005.

Russell, B. *Autobiography.* London: Unwin, 1975.

———. *Inquiry into Meaning and Truth.* London: George Allen and Unwin, 1940.

———. *Introduction to Mathematical Philosophy.* London: Routledge, 1993.

———. *Logic and Knowledge.* New York: G. P. Putnam's Sons, 1971.

———. "Mathematical Logic as Based on the Theory of Types." In Russell, *Logic and Knowledge,* 59–102.

———. "On Denoting." In Russell, *Logic and Knowledge,* 41–56.

———. *Principles of Mathematics,* 2d. ed. New York: W. W. Norton, 1996.

———. *The Problems of Philosophy.* Indianapolis: Hackett, 1995.

Sainsbury, M. *Logical Forms,* 2d. ed. Oxford: Blackwell, 2001.

Salmon, W. *Logic.* Englewood Cliffs, N.J.: Prentice Hall, 1963.

Schacter, D. "Perceptual Representation Systems and Implicit Memory: Towards a Resolution of the Multiple Memory Systems Debate." *Annals of the New York Academy of Science* 608 (1990): 543–571.

Schaffer, J. "Causes as Probability Raisers of Processes." *Journal of Philosophy* 98 (2001): 75–92.

Schilpp, P. (ed.). *The Philosophy of G. E. Moore.* New York: Tudor, 1952.

Searle, J. *Intentionality.* Cambridge: Cambridge University Press, 1983.

———. "Minds, Brains, and Programs." *Behavioral and Brain Sciences* 3 (1980): 417–424.

———. *Minds, Brains, and Science.* Cambridge, Mass.: Harvard University Press, 1984.

———. *Rationality in Action.* Cambridge, Mass.: MIT Press, 2001.

———. *The Rediscovery of the Mind.* Cambridge, Mass.: MIT Press, 1992.

Segall, M. H., D. T. Campbell, and M. J. Herskovits. *The Influence of Culture on Visual Perception*. Indianapolis: Bobbs-Merrill, 1966.

Sells, S. "The Atmosphere Effect: An Experimental Study of Reasoning." *Archives of Psychology* 29 (1936): 1–72.

Shalkowski, S. "Logic and Absolute Necessity." *Journal of Philosophy* 101 (2004): 55–82.

Shapiro, S. *Philosophy of Mathematics: Structure and Ontology*. New York: Oxford University Press, 1997.

———. "The Status of Logic." In Boghossian and Peacocke, *New Essays on the A Priori*, 333–366.

———. *Thinking about Mathematics*. Oxford: Oxford University Press, 2000.

Sheffer, H. M. "Review of *Principia Mathematica*, Volume I, second edition." *Isis* 8 (1926): 226–231.

Shepard, R. "The Mental Image." *American Psychologist* 33 (1978): 125–137.

Shepard, R., and S. Chipman. "Second Order Isomorphisms of Internal Representations: Shapes of States." *Cognitive Psychology* 1 (1970): 1–17.

Shepard, R., and L. Cooper. *Mental Images and Their Transformations*. Cambridge, Mass.: MIT Press, 1982.

Shepard, R., and J. Metzler. "Mental Rotation of Three-Dimensional Objects." *Science* 171 (1971): 701–703.

Sher, G. "Is Logic a Theory of the Obvious?" In Varzi, *The Nature of Logic*, 207–238.

Simon, H. *Reason in Human Affairs*. Stanford: Stanford University Press, 1983.

Smiley, T. "Relative Necessity." *Journal of Symbolic Logic* 28 (1963): 113–134.

———. "Sense without Denotation." *Analysis* 20 (1960): 125–135.

———. "A Tale of Two Tortoises." *Mind* 104 (1995): 725–736.

Smith, N., and I. M. Tsimpli. *The Mind of a Savant: Language Learning and Modularity*. Oxford: Blackwell, 1995.

Smullyan, R. *First-Order Logic*. Berlin: Springer Verlag, 1968.

Sober, E. "The Evolution of Rationality." *Synthese* 46 (1981): 95–120.

———. "Psychologism." *Journal for the Theory of Social Behavior* 8 (1978): 165–192.

Stalnaker, R. *Inquiry*. Cambridge, Mass.: MIT Press, 1984.

Stein, E. *Without Good Reason: The Rationality Debate in Philosophy and Cognitive Science*. Oxford: Clarendon/Oxford University Press, 1996.

Stern, R. (ed.). *Transcendental Arguments: Problems and Prospects*. Oxford: Oxford University Press, 1999.

————. *Transcendental Arguments and Skepticism*. Oxford: Oxford University Press, 2000.

Stevenson, C. "Roundabout the Runaround Inference Ticket." *Analysis* 21 (1961): 124–128.

Stich, S. "Could Man Be an Irrational Animal?" *Synthese* 64 (1985): 115–135.

————. *The Fragmentation of Reason*. Cambridge, Mass.: MIT Press, 1990.

Strawson, P. F. *The Bounds of Sense*. London: Methuen, 1966.

————. *Individuals*. London: Methuen, 1959.

————. *Introduction to Logical Theory*. London: Methuen, 1952.

Stroud, B. "The Charm of Naturalism." *Proceedings and Addresses of the American Philosophical Association* 70 (1996): 43–55.

————. Transcendental Arguments." *Journal of Philosophy* 65 (1968): 241–256.

Tarski, A. "The Concept of Truth in Formalised Languages." In Tarski, *Logic, Semantics, and Metamathematics*, 152–278.

————. "The Establishment of Scientific Semantics." In Tarski, *Logic, Semantics, and Metamathematics*, 401–408.

————. *Logic, Semantics, and Metamathematics*. Oxford: Clarendon/Oxford University Press, 1956.

————. "On the Concept of Logical Consequence." In Tarski, *Logic, Semantics, and Metamathematics*, 409–420.

————. "The Semantic Conception of Truth and the Foundations of Semantics." *Philosophy and Phenomenological Research* 4 (1943–44): 341–375.

————. "Truth and Proof." *Scientific American* 120 (1969): 63–77.

————. "What Are Logical Notions?" *History and Philosophy of Logic* 7 (1986): 143–154.

Tennant, N. *Anti-Realism and Logic*. Oxford: Oxford University Press, 1987.

Thomson, J. "What Achilles Should Have Said to the Tortoise." *Ratio* 3 (1960): 95–105.

Tichy, P. "Kripke on Necessity A Posteriori." *Philosophical Studies* 43 (1983): 225–241.

Toulmin, S. *The Uses of Argument*. Cambridge: Cambridge University Press, 1958.

Turing, A. "Computing Machinery and Intelligence." *Mind* 59 (1950): 433–460.

Tversky, A., and D. Kahneman. "Availability: A Heuristic for Judging Frequency and Probability." *Cognitive Psychology* 5 (1973): 207–232.

———. "Extensional vs. Intuitive Reasoning: The Conjunctive Fallacy in Probability Judgment." *Psychology Review* 90 (1983): 293–315.

van Fraassen, B. "Presuppositions, Supervaluations, and Free Logic." In K. Lambert (ed.), *The Logical Way of Doing Things*, 67–91. New Haven: Yale University Press, 1969.

———. "Singular Terms, Truth-Value Gaps, and Free Logic." *Journal of Philosophy* 63 (1966): 481–495.

Van Heijenoort, J. (ed.). *From Frege to Gödel*. Cambridge, Mass.: Harvard University Press, 1967.

———. "Logical Paradoxes." In Edwards, *Encyclopedia of Philosophy*, vol. 5, 45–51.

———. "Logic as Calculus and Logic as Language." *Synthese* 17 (1967): 324–330.

Varela, F., E. Thompson, and E. Rosch. *The Embodied Mind*. Cambridge, Mass.: MIT Press, 1991.

Varzi, A. C. (ed.). *The Nature of Logic*. Stanford: CSLI Publications, 1999.

Von Humboldt, W. *On Language*. Trans. P. Heath. Cambridge: Cambridge University Press, 1988.

Walton, D. "What Is Reasoning? What Is an Argument?" *Journal of Philosophy* 87 (1990): 399–419.

Ward, J. "Psychology." In *Encyclopedia Britannica*, 11th. ed., vol. 22, 547–604 (29 vols.). New York: Encyclopedia Britannica, 1911.

Warmbröd, K. "Logical Constants." *Mind* 108 (1999): 503–538.

Wason, P. "Reasoning." In B. M. Foss (ed.), *New Horizons in Psychology*, 135–151. Harmondsworth, Middlesex: Penguin, 1966.

———. "Reasoning about a Rule." *Quarterly Journal of Experimental Psychology* 20 (1968): 273–281.

Wason, P., and P. Johnson-Laird. *The Psychology of Reasoning*. Cambridge, Mass.: Harvard University Press, 1972.

Wedgwood, R. "The A Priori Rules of Rationality." *Philosophy and Phenomenological Research* 59 (1999): 113–131.

Weininger, O. *Sex and Character*. London: Heinemann, 1906.

Wetherick, N. "Human Rationality." In Manktelow and Over, *Rationality: Psychological and Philosophical Perspectives*, 83–109.

———. "Psychology and Syllogistic Reasoning." *Philosophical Psychology* 2 (1989): 111–124.

Whitehead, A. N., and B. Russell. *Principia Mathematica to *56*. Cambridge: Cambridge University Press, 1962.

Wilkins, M. C. "The Effect of Changed Material on Ability to Do Formal Syllogistic Reasoning." *Archives of Psychology* 16 (1928): 1–83.

Williams, B. *Ethics and the Limits of Philosophy*. London: Fontana Collins, 1985.

Wittgenstein, L. *Philosophical Investigations*. Trans. G. E. M. Anscombe. New York: Macmillan, 1953.

———. *Remarks on the Foundations of Mathematics*, 2d. ed. Trans. G. E. M. Anscombe. Cambridge, Mass.: MIT Press, 1983.

———. *Remarks on the Philosophy of Psychology*, 2 vols. Trans. G. E. M. Anscombe. Chicago: University of Chicago Press, 1980.

———. *Tractatus Logico-Philosophicus*. Trans. C. K. Ogden. London: Routledge and Kegan Paul, 1981.

Woodworth, R., and S. Sells. "An Atmosphere Effect in Formal Syllogistic Reasoning." *Journal of Experimental Psychology* 18 (1935): 451–460.

Wright, C. *Frege's Conception of Numbers as Objects*. Aberdeen: Aberdeen University Press, 1983.

———. "Inventing Logical Necessity." In Butterfield, *Language, Mind, and Logic*, 187–209.

———. *Wittgenstein on the Foundations of Mathematics*. Cambridge, Mass.: Harvard University Press, 1980.

Wynn, K. "Addition and Subtraction in Human Infants." *Nature* 358 (1992): 749–750.

Yablo, S. "Apriority and Existence." In Boghossian and Peacocke, *New Essays on the A Priori*, 197–228.

Yamada, J. *Laura: A Case for the Modularity of Language*. Cambridge, Mass.: MIT Press, 1990.

Index

Abductive inference, 98

Abstractness, platonic vs. nonplatonic, 192

Acknowledging the predicament, 55, 69–71

AI
  strong, 102–106, 108
  weak, 103, 105, 108

Analytic inferences, 225

Analyticity, 36–37

Analytic necessity, 197

Analytic philosophy, xx, 1, 3

Alice, 158–159, 185

Antinaturalism, 14–21

Antiperceptualism, 186, 189, 191–200

Antipsychologism, 14–21

Antirealism, 185, 188–189

Anti-supernaturalism, 9

Aphasias, 86, 106–107

A priori–a posteriori, xii, 272n20, 273n24

Apriority of protologic, 29, 44–45, 109, 201

Aristotle, xi, xiii–xv, 31–32, 124

Arnaud, P., xii, 204–205

Artificial intelligence. See AI

Art of Thinking (Arnaud and Nicole), xii

Aspects of the Theory of Syntax (Chomsky), 86

Autism, 112

Availability theory, 140

Avenarius, R., 4

Basic Laws of Arithmetic (Frege), 32

Begriffsschrift (Frege), xii, 32, 120, 207, 229

Belnap, N., 252n18

Benacerraf, P., xxi, 2, 22–25, 71, 155–156, 182–191, 193, 199–200

Benacerraf dilemma
  extended, 184
  original, 182–184

Block, N., 93

Bolzano, B., 5

Bonjour, L., 188

Boole, G., xi–xiii, 25, 31–32, 115–116, 131, 204–205, 216

Braine, M., 130

Brentano, F., 1, 4

Broca's area, 106

Byrne, R., 135

Carnap, R., 36–37, 54, 59–61, 65, 186, 210–211, 226–227

Carroll, L. (Charles Dodgson), 54–59, 61–62, 65, 69, 78, 158, 223

Carruthers, P., 111

Categorical normativity. See Normativity, categorical

Causation, 23, 189–190, 242n62

Central processes, 89, 100–102, 109

Chalmers, D., 12, 98, 277–278n78
Cheng, P., 141
Cherniak, C., 146, 270n88
Chinese brain argument, 106–107
Chinese nation argument, 93
Chinese room argument, 103–104
Chomsky, N., xi, xiii, xxi, 25–26, 30, 46–52, 78–79, 83–87, 90, 98, 109–111, 121, 128–130, 133, 172, 209, 250n57, 250n60
Church-Turing thesis, 102. *See also* Turing computability; Turing machines
Circular argument, 31
Classical logic, xii, xx, 29, 35–36, 40–46, 122, 148, 216, 222, 231, 233n10
Cogito, 224–225
Cognitive creativity, 86–88, 102
Cognitive faculties, 79–80
Cognitive generativity, 86, 102
Cognitive productivity. *See* Cognitive creativity
Cognitive relativism, 5, 8, 29, 238n18
Cognitive science, xx, 82, 231
Cognitivism, xii. *See also* Logical cognitivism
Cognitivist existential predicament, 136, 147–148, 200
Cognitivist solution to the logocentric predicament, 74
Cohen, L. J., 117, 128–129
Competence, 263n6
Connectionism, 260n38
Consequence. *See* Logical consequence
Constructivism, 79, 85–88, 107, 109
Conventionalism, 60–68, 185, 188–189, 210–213
Copi, I., 219
Cosmides, L., 142, 227
Cox, G. R., 125
Craik, K., 135
Creativity. *See* Productivity

Darwinian evolution, 97, 142–143
Davidson, D., 110, 173
Dedekind, R., 36
Dedicated cognitive capacity, 88, 101
Deduction, intuitive vs. rote, 199
Deferred ostension, 187, 190
De Morgan, A., 218
"Deny the doctrine, change the subject" argument, 42, 181
Descartes, R., 78, 98, 110, 134, 174, 176, 224–226, 229, 282n57
Deviant logic, xii, 40–41, 234n12, 246n34. *See also* Nonclassical logic
Deviant Logic (Haack), 40
Dialetheic logic, xii, 41, 223–224, 234n14
Diehard classicism (about logic), 41–43, 133, 222
Diehard nonclassicism (about logic), 41–43, 122, 133
Discourse on Method (Descartes), 224
Domain-specific cognitive capacities, 88–89, 101
Dretske, F., 187–188
Dualism, 262n56
Dummett, M., 3, 22, 54, 66–69, 253n21, 253n22, 253n24

Early modern conceptions of ourselves, 113–114
Einstein, A., 98
Elementary logic. *See* Classical logic
Embodied rationalism, 13, 283–284n72
Empirical psychology, 7, 52
Empiricism, 8, 10, 29, 84–85, 186, 188–189
Encapsulated cognitive capacities, 89, 101
Enquiry Concerning Human Understanding (Hume), 160
Epigenesis of pure reason, 86–87
E pluribus unum problem, xx, 29–52, 75, 77, 115, 122, 133, 138

Erdmann, B., 5
Ethics of logic, 201–231
Euler diagrams, 136
Evans, J., 126
Evnine, S., 254n33
Extensions of classical logic, xii, 40–41,
    148, 233n11

Fallacies, 217–220
Fast cognitive capacities, 88, 101
Fictionalism, 185, 188
Field, H., 185, 189
Fluent aphasia, 106–107
Fodor, J., xi, xiii, 26, 78–80, 88, 90–91,
    94, 110, 121
Folk psychology, 82
Foot, P., 210–212
Formalism, 244n7
Formal logic, 30–32
Form of life, 180
Foundations of Arithmetic (Frege), 7
Frankfurt, H., 281n36
Free logic, 41
Frege, G., xii, xix, 1–9, 14, 21, 24, 27,
    31–32, 36–37, 81, 90, 116, 174, 182,
    204–206, 208, 229, 238–239n29
Freud, S., 115

Garden path sentences, 220
Generativity, 249n52
Genetic logic, 120–122
Gentzen, G., 54, 63, 65, 131, 133, 206
Geuss, R., 228
Gödel, K., 36, 38–40, 134, 187, 235n23
Goodman, N., 66, 70, 161
Griggs, R. A., 125
Groundlessness of logic, 55, 69–72, 74–75

Haack, S., 1, 28, 54, 66, 68–70,
    246–247n35, 247n41, 253n23, 253n24
Hale, B., 39, 186
Harman, G., xxii, 202, 220–223
Hegel, G. W. F., xvii

Henle, M., 123, 130
Herbart, J., 5
Heuristics-and-biases theory, 139–146,
    226
Hilbert, D., 58, 60, 244n7
Holyoak, K., 141
Human interest postulates, 211
Hume, D., xvii, 84, 115, 160, 163, 164,
    167, 170, 186, 228
Husserl, E., xii, xix, 1–9, 14, 20, 24,
    27, 205, 206–209, 238n20

I-language, 90
I-logic, 51, 251n67
Imagination
    linguistic, 175
    logical, 35, 192–195, 199–200
Incompleteness theorems. *See* Gödel, K.
Indeterminacy of translation, 161
Indexicality, 145
Indirect perceptualism, 188, 190–191
Inductive logic, 30–31
Inference-tickets, 57, 63, 65, 207
Inference to the best explanation, 52, 98
Inferential role theory of logical con-
    stants, 63–65, 133–134, 156,
    168–170
Informal logic, 31
Informational encapsulation, 26,
    100–101, 132
Informational promiscuity, 100–102, 107
Innatism, 79, 83–85, 96–99, 107, 109,
    134, 138, 258–259n27
Inscrutability of reference, 161
Intensional logic, 31
Intentionalism. *See* Representationalism
Intentionality, 80–82, 256n9
Intuition, 171–182
    vs. deduction, 174–175
    logical (*see* Logical intuition)
    mathematical (*see* Mathematical
    intuition)
Intuitionist logic, 41

Intuition-of, 173
Intuition-that, 173
Investigation of the Laws of Thought
(Boole), xii, 32, 120
"Investigations into Logical Deduction"
(Gentzen), 63
Irrationalism, 124–130, 146–147, 150,
154, 226

Jackendoff, R., 86
James, W., 116–120, 122–123
Jäsche Logic (Kant), xii
Johnson-Laird, P., 116, 135, 193–194
Justification, 44
Justificational holism, 255n37
"The Justification of Deduction"
(Dummett), 66
"The Justification of Deduction"
(Haack), 68
Justification of deduction problem, 54,
66–69

Kahneman, D., 116, 140
Kant, I., xi–xiii, xvii, xx, xxi, 1, 5,
25–26, 31, 37, 45, 85–87, 110, 115,
131, 204–205, 208–214, 217, 230,
231, 236n32
Kantian constructivism in ethics, 212
Katz, J., 186, 189
Kim, J., 11
Kirk, R., 111
Kosslyn, S., 193–194
Kripke, S., 41, 98, 156, 158–168, 178
Kripkenstein, 161, 163–168, 170. *See
also* Rule following paradox
Kusch, M., 5

Language of thought (LOT), 26, 47, 78,
80, 90–92, 94–96, 107–110, 151, 213
L'art de penser (art of thinking), 204
Law of bivalence, 31, 41, 243–244n3
Law of excluded middle, 31, 41,
243–244n3

Law of noncontradiction, 31, 41–42,
243–244n3
Lawrence, D. H., 115
Laws of thought, 4, 25–26, 78, 110, 204
Leibniz, G. W., 5, 31
Lewis, C. I., 20, 33, 35, 40–41, 156
Liar paradox, 36, 38–39
Lingua mentis (mental language). *See*
Language of thought
Linguistic competence, 78, 121,
128–129, 250n55
Lipps, T., 5
Logical animals, 77–78, 201, 215
Logical bad faith, 166
Logical cognitivism, xix, 3, 24–29,
46–47, 52–53, 75, 77, 115, 156, 180,
202, 209, 213, 230
Logical communitarianism, 69–71
Logical competence, 51
Logical consequence, xv, 27, 133–34,
214–215, 231, 234n17
Logical expressivism, 70–71, 186, 188–189
Logical form, 51
Logical intuition, 156–157, 170–182,
168, 197–201
Logical knowledge, 155–200
Logical necessity, 158, 195–197
Logical nihilism, 227, 230
Logical nonfactualism. *See* Logical
expressivism
Logical possibility, 196, 225
Logical pragmatism, 70–71
Logical prudentialism, 69, 71
Logical psychologism, xii, xix–xx, 1–29,
75, 115–116, 206–208, 222
Logical structuralism, 156, 191–192,
199–200
Logical Syntax of Language (Carnap),
54, 59–61, 210
Logical truth, 61–63, 225
Logic faculty thesis, xiii, 3, 25–26, 28,
46–53, 73–75, 77, 107–108, 115,
154, 157, 168, 200–202, 230

Logicism, 36, 40
Logic-liberated-animals, xxii
Logic of thought, 51, 107–110,
    132–133, 135, 154, 213–214, 226
Logic-oriented conception of human
    rationality, xviii, 77–78, 113–115,
    156, 201–202
Logocentric predicament, 53–75, 115,
    157, 211
Lotze, H., 5
Lowe, E. J., 136

Mach, E., 5
Machine functionalism, 26, 103,
    243n74
Macnamara, J., 130, 266n37
Mad Hatter, 78
Manktelow, K.I., 126
Many-valued logic, 41
Mathematical intuition, 23
Mathematical logic, 3, 32–35
"Mathematical Truth" (Benacerraf), 182
Meditations on First Philosophy
    (Descartes), 224, 229
Mein Kampf, 229
Meinong, A., 41
Mentalese. See Language of thought
Mental language. See Language of
    thought
Mental Logic, 131–135, 138, 145
Mental logic theory, xxi, 130–136,
    145–146, 151, 266–267n45, 267n46
Mental models theory, 135–140, 145,
    267n51
Metalogic, 61
Metalogical principles of protologic,
    45–46
Metaphysical necessity, 196–197,
    234–235n21, 278n79
Mill, J. S., xii, 5, 13, 186, 204, 208
Millikan, R. G., 223
Milton, J., 227
Minimal rationality theory, 146–150

Modal downsizing, 8, 29
Modality, xv
Modes of presentation (MOPs), 81, 93
Modularity, 25, 79, 88–90, 100–102,
    107, 109, 131–132, 259n30
Monty Python's Flying Circus,
    115–116, 152
Mooney, C., 88
Mooney faces, 88
Moore, G. E., 2, 14–18, 20, 24, 27
"Morality as a System of Hypothetical
    Imperatives" (Foot), 210
Moral science conception of logic,
    205–206, 231
Mr. Spock, 215
Multiple embodiability, xviii, 25, 52

Nagel, T., 281n36
Nativism. See Innatism
Natorp, P., 5
Naturalistic fallacy, 2, 15–17
Natural logic, 51, 251n68
"The Nature of Judgement" (Moore), 15
Necker cube, 89, 219
Negation, 165–166, 180–182, 225
Negativity, 165–166, 180–182
Neo-Fregeanism, 39
Neo-Kantianism, 9, 13
Neo-Nietzscheanism, xii, 224,
    226–231
Nicole, J., xii, 204–205
Nietzsche, F., 226, 227–229
Noncausal perceptualism, 187, 190–191
Nonclassical logic, 40–46, 73,
    216, 231
Noncontact causal perceptualism, 187,
    189, 191
Nonepistemic perception, 187, 190
Nonformal logic, 30–31
Normativity
    categorical, xii, 202–205, 208–215, 231
    hypothetical, 203, 208
Nye, A., 201, 205–206, 226, 229

O'Brien, D. P., 130, 145
O'Neill, O., 212–213
One True Logic, 26, 41, 43, 222
Open question argument, 16–17
Overton, W., 130

Paraconsistent logic, xii, 41, 234n13
Paradise Lost (Milton), 227
Paradox of classes, 36–38, 237n15
Paratactic approach to logical intuition-that, 172–173
Parsons, C., 173–174, 188
Peano, G., 36, 38, 134
Peirce, C. S., 136, 225–226, 229
Permission schema, 142–143
Phenomenology of intuition, 177–178
Philo and Stoic logic, 31
Philosophical Investigations (Wittgenstein), xx, 159, 160–161
Philosophie der Arithmetik (Husserl), 4
Phrase-structure grammar, 249n51
Physicalism, 9–10, 239n32
Physical necessity, 196
Piaget, J., 116–118, 120–124, 130
Piagetian logic, 120–122, 124
Piagetian stages of cognitive development, 120–121
Plato, 227, 229
Platonism
  logical, xx, 2, 21–27, 29, 184
  mathematical, 22, 183–184
Plus–quus, 161–163. *See also* Rule-following paradox
Port Royal logicians, 204
Postmodern conceptions of ourselves, 113–114
Poverty-of-the-stimulus argument, 83–85, 97, 108, 257n18
Pragmatic reasoning schemas theory, 140–143
Presuppositional arguments, 72–73
  and transcendental arguments, 255n39

Principia Ethica (Moore), 2, 15
Principia Mathematica (Whitehead and Russell), 31, 35, 38, 54, 120
Principle of Tolerance, 211, 226
Principles of Psychology (James), 118
Prior, A. N., 54, 63–65, 69, 252n15, 252n16
Private language argument (PLA) (Wittgenstein), 160
Productivity, 249n52
Prolegomena to Pure Logic (Husserl), 5, 205–206
Property-identity criteria, 16–17, 241n50, 241n56
Prosopagnosia, 89
Protologic, xii, xviii, xxi–xxii, 30, 40, 43–46, 48, 50–53, 73–74, 77, 108–110, 112–113, 116, 133, 149–153, 201, 209, 213–215, 217, 222–223, 230–231
Protological competence theory, xxi, 116, 149–154
Psycholinguistics, xxi, 25–26, 30, 47–52
Psychological logicism, 116
Psychologism. *See* Logical psychologism
Psychology from an Empirical Standpoint (Brentano), 4
Psychology of reasoning, 26, 115–154, 200
Pure logic, 7, 116
Putnam, H., 70
Pylyshyn, Z., 194

Quine, W. V. O., xi, xiii, 9, 13, 25, 37, 41–44, 54, 59, 61–64, 69–71, 85, 161, 181–182, 186–188, 211, 239n30, 252n13

Rational anthropology, xx
Rationalism, 124–130, 135, 139, 146–147, 150, 154

Rationality, xv–xviii, 26–27, 77–78, 110–115, 117, 122–130, 153–154, 209, 235n24, 264n8, 270n88
Rawls, J., 212
Raz, J., 201
Realism, 156
Reasoning tests, 100, 116–117, 122–154, 270n91
Red Queen, 158–159, 185
Reduction
  explanatory, 6–7, 52, 157
  ontological, 7
Referential opacity, 81
Referential transparency, 81
Refined cognitivist model of the mind, 92–107. *See also* Standard cognitivist model of the mind
Reflective equilibrium, xvii, 235n25
"The Refutation of Idealism" (Moore), 15
Relevance logic, 41
Remarks on the Foundations of Mathematics (Wittgenstein), 165, 179
Remarks on the Philosophy of Psychology (Wittgenstein), 179
Representationalism, 79–82, 107, 109
Republic, 229
Rips, L., 125
Rule-based semantics, 159
Rule following paradox (RFP), 158–168
Rules for the Direction of the Mind (Descartes), 174, 224
Rules of etiquette, and moral rules, 210
"The Runabout Inference Ticket" (Prior), 63
Russell, B., 32, 36–38, 60, 216–217
Ryle, G., 57, 207

Satan, 227
Schmogic, 181
Science-forming faculty (SFF), 98
Science of logic, xiii–xv, 231

Scientific naturalism, xii, xx, 1–2, 9–14, 18–21, 27, 29, 52, 71, 99, 157, 207, 236n31
Scientism, 9
Searle, J., 102–106
Second-order logic, 40
Selection task, 125–128, 141–147, 152–153, 219
Self-evidence, 35
Sells, S. B., 123
Semantic holism, 67, 70
Semantic realism, 71
S4 modal logic, 20
Shapiro, S., 29, 44
Sheer logic, 25–26, 44, 242n69
Sheffer, H., 54–55, 69
Shepard, R., 155, 193
Signs, vs. symbols, 34–35, 92–94
Sigwart, C., 5
Smiley, T., 57–58
Sober, E., xii
Social contract schemas theory, 140, 142–146
Spencer, H., 5
Stalnaker, R., 111
Standard cognitivist model of the mind, xxi, 78–92, 146. *See also* Refined standard cognitivist model of the mind
Stein, E., 129
Stevenson, C., 252n18
Stich, S., 115, 129
Structuralism, 271n8
Superlogic, 29, 44
Supervenience, 7, 10–12, 19–21, 52, 63, 105, 239–240n36, 240n37
Survey of Symbolic Logic (Lewis), 33, 40
Syllogism, xi, xiii–xv, 31, 118, 120
Symbolic logic, 3, 32–35, 218
Symbols, vs. signs, 34–35, 92–94
Syntax languages, 61
System of Logic (Mill), xii, 9

Tarski, A., 21, 35–36, 39–40, 60, 71, 168, 182
Theory of mind module, 112
"Thoughts" (Frege), 21
Thrasymachus, 227
Three-valued logic, 41
Through the Looking-Glass (Carroll), 158, 228
Tonk, 64
Topic bias (of logic), 8, 29
Topic neutrality (of logic), 188
Tourette's syndrome, 106
Tractatus Logico-Philosophicus (Wittgenstein), xix, 34–35, 54, 59–60
Traité de logique (Piaget), 120
Transcendental psychology, xxi. *See also* Kant, I.
Treatise of Human Nature (Hume), 160
Truth tables, 136
Truth-value gluts, 41, 222
Turing computability, 87, 90–91, 102–103
Turing machines, 82, 103
Tversky, A., 117, 140
Two-dimensional semantics, 277–278n78

Unconstrained pluralism (about logic), 41–43
Universal grammar (UG), 44, 48–51, 78, 109–110, 121–122, 209, 249–250n53
Unrevisability, 44–45

Van Heijenoort, J., 3
Venn diagrams, 136
Visual illusions, 219–220
Von Humboldt, W., 86–87

Warmbröd, K., 248–249n48
Wason, P., 100, 116, 125, 127, 132, 147, 153, 219
Weininger, O., 201, 205–206, 217

Wernicke's aphasia. *See* Fluent aphasia
Wernicke's area, 106
Wetherick, N., 130
"What Numbers Could Not Be" (Benacerraf), 182
"What the Tortoise Said to Achilles" (Carroll), 55
Whitehead, A. N., 32
White Queen, 223, 228
White-queenism, 223–231
White queen psychology, 223–224
Wilkins, M. C., 123
Williams, B., 228
Will to truth, 226
Wittgenstein, L., xix, xxi, 34–37, 54, 59–60, 69–70, 155–170, 176–180, 157, 186, 216–217
Wittgenstein on Rules and Private Language (Kripke), 160
Woodworth, R. S., 123
Word and Object (Quine), 85
Words of Power (Nye), 228
Wright, C., 39, 186
Wundt, W., 5

Yablo, S., 185